AF685769

Biosecuring Animal Populations

Biosecuring Animal Populations

Guest Editors

Scott Dee
Gordon Spronk

MDPI

Basel • Beijing • Wuhan • Barcelona • Belgrade • Novi Sad • Cluj • Manchester

Guest Editors
Scott Dee
Pipestone Research
Pipestone
USA

Gordon Spronk
Pipestone Holdings
Pipestone
USA

Editorial Office
MDPI AG
Grosspeteranlage 5
4052 Basel, Switzerland

This is a reprint of the Special Issue, published open access by the journal *Animals* (ISSN 2076-2615), freely accessible at: https://www.mdpi.com/journal/animals/special_issues/939HM287RY.

For citation purposes, cite each article independently as indicated on the article page online and as indicated below:

Lastname, A.A.; Lastname, B.B. Article Title. *Journal Name* **Year**, *Volume Number*, Page Range.

ISBN 978-3-7258-5815-6 (Hbk)
ISBN 978-3-7258-5816-3 (PDF)
https://doi.org/10.3390/books978-3-7258-5816-3

© 2025 by the authors. Articles in this book are Open Access and distributed under the Creative Commons Attribution (CC BY) license. The book as a whole is distributed by MDPI under the terms and conditions of the Creative Commons Attribution-NonCommercial-NoDerivs (CC BY-NC-ND) license (https://creativecommons.org/licenses/by-nc-nd/4.0/).

Contents

About the Editors

Scott Dee

Scott Dee earned a DVM, MS, and PhD from the University of Minnesota, is board-certified in veterinary microbiology, is a past President of the American Association of Swine Veterinarians, past Chair of the American Veterinary Medical Association (AVMA) Council on Biologic and Therapeutic Agents, and currently serves on the AVMA House of Delegates where he has written policy promoting the use of science-based biosecurity for the benefit of the veterinary profession and all animal species. After 12 years in swine practice and another 12 years in academia, Scott joined Pipestone Veterinary Services as Director of Applied Research, where he served as Director of Discovery and Innovation for Pipestone for 12.5 years. Throughout his career, he has been awarded >12.5 M in research funding and published 192 peer reviewed papers, including the initial publications on the proof of concept of PEDV transmission in feed and the transboundary survival of ASFV in feed. He developed the initial knowledge and industry-wide protocols to understand and control Porcine Reproductive and Respiratory Syndrome Virus (PRRSV) infection at the farm level. The comprehensive application of these protocols has reduced annual PRRSV incidence below 10% across the 76 breeding herds in the Pipestone System, the 3rd largest pork production company in the US over a 3-year period. Scott has received the AASV Practitioner of the Year, the Howard Dunne Memorial, and Leman Science in Practice awards, a Warrior Chip from the FBI, and is a Master of the US Pork Industry. He has been awarded the Distinguished Service to the US Pork Industry Award from the National Pork Board and the Distinguished Service and Research award from the PRRS scientific community and recognized as the Distinguished Research Alumnus by the UMN College of Veterinary Medicine.

Gordon Spronk

Gordon Spronk is Chair and Chairman Emeritus of the Board of Pipestone Holdings, a diversified agribusiness headquartered in Pipestone, Minnesota, USA, where he also serves as Partner and Staff Veterinarian. With more than 45 years of leadership and entrepreneurial experience, he has been instrumental in growing Pipestone from a two-practitioner clinic into the largest swine veterinary team in the world, employing over 1500 people and comprising more than thirty-five related companies. His professional interests center on animal health, swine production management, and innovative business models for livestock producers. He has extensive international experience, having lectured, consulted, and managed swine farms across Asia, Europe, North and South America, and Australia. As a co-founder of the Pipestone System, Dr. Spronk has built a reputation as a progressive, forward-thinking leader in veterinary medicine and livestock production, combining strategic vision with hands-on expertise to improve animal health and support farmers globally.

animals

MDPI

Editorial

Biosecuring Animal Populations: The Future Is Now

Scott Dee * and Gordon Spronk

Pipestone Research, Pipestone, MN 56164, USA; gordon.spronk@pipestone.com
* Correspondence: scottdee7255@gmail.com

Dear Colleagues,

It was with great enthusiasm that Dr. Spronk and I agreed to lead this Special Issue on the topic of *Biosecuring Animal Populations*. This was, and still is, a very timely topic, based on the global challenges of domestic and transboundary pathogens of pigs, such as African swine fever virus and porcine reproductive and respiratory syndrome virus, as well as the threat of the One Health crisis secondary to highly pathogenic avian influenza virus across multiple animal species as well as humans ([1–4]). Looking back, this Special Issue has proven to be multi-disciplinary, involving over 100 authors across 25 countries from 3 continents, and transcends multiple species including pigs, dairy cattle, dairy goats, and poultry, and targets several significant viral pathogens of animals, including porcine reproductive and respiratory syndrome virus, porcine epidemic diarrhea virus, African swine fever virus, classical swine fever virus, Senecavirus A, bovine coronavirus, and hepatitis E virus.

The Special Issue contained 18 new articles that were both educational and diverse, including a Literature Review (Section 1, 1 article), New Research (Section 2, 6 articles), Bio-surveillance (Section 3, 5 articles), Human Behavior and Biosecurity (Section 4, 3 articles) and Industry Action (Section 5, 3 articles). Section 1 opened the Special Issue with a review of biosecurity principles and terminology, providing insight on the science-based practices currently in place to reduce the risk of porcine reproductive and respiratory syndrome virus entry into swine breeding herds in the US (Contribution 1). Section 2, New Research, brought forth 6 exciting new articles, providing new information on technological advancements to measure and enhance compliance of biosecurity protocols, identify "superspreader" herds, evaluated the standardization of biosecurity scoring systems, introducing protocols to improve the ability to decontaminate transport vehicles and feed, reviewing the practices of cleaning and disinfection in swine herds across 10 European countries, as well as describing new advancements in biocontainment to reduce the risk of African swine fever virus escape from controlled facilities (Contributions 3 and 4).

Following these articles, new information on the topic of bio-surveillance is discussed in Section 3, a highly relevant topic in agriculture today. Five articles highlight advancements in the use of pen-based sampling techniques collected from both the population and the environment, describe new approaches to active participatory regional surveillance, and provide updates on assessing the prevalence of novel pathogens, such as bovine coronavirus and hepatitis E virus (Contributions 10 and 11). Section 4 focuses on perhaps the most critical component to the success or failure of a biosecurity program—the role of human behavior in the implementation of known successful interventions at the farm level. Three articles are featured, describing the characteristics of farm personnel, along with a review of the awareness and attitudes of farmers and the role of veterinarians as it pertains

 https://doi.org/10.3390/ani15071048

to biosecurity. Clearly, due to its importance and complexity, the topic of human behavior and its impact on biosecurity could be a Special Issue on its own! Leadership is needed to cast a vision of successful implementation to change the intentional behavior of farm staff (Contribution 15).

Finally, Section 5 focuses on the field and highlights three novel biosecurity-based initiatives that are underway across the US and Denmark. This is clearly where the "rubber meets the road" as science is put into practice and actual results are measured over time. From Denmark comes a summary of industry actions to mitigate the introduction of African swine fever virus from neighboring European Union countries. The article describes how the Danish pig industry has used science to mitigate the significant risk factors of the spread of African swine fever virus—wild boar, transport, and feed imports (Contribution 16). This is followed by an article which describes experiences from the US Swine Health Improvement Plan (US SHIP), a pilot program based on the National Poultry Improvement Plan that focuses on the prevention of the incursion into the US of African swine fever virus and classical swine fever virus (Contribution 17). The final paper, again from the US, builds on a previous publication describing, for the first time since its emergence more than 35 years ago, the sustainable prevention of the introduction of porcine reproductive and respiratory syndrome virus into breeding herds from a large commercial pork production system following the implementation of science-based biosecurity protocols ([5] and Contribution 18). Truly this is a unique observation that brings hope and optimism to agricultural industries as it pertains to the control and prevention of critical livestock diseases.

In closing, this Special Issue was the perfect opportunity to address a pressing need in agriculture—the biosecuring of animal populations to prevent disease. It brings forth new information which answers several questions, but gaps in the knowledge base still exist at the level of the field, such as: What new technologies will be successfully implemented on-farm? How best to conduct surveillance of large populations? What is the best way to motivate and incentivize farm personnel to properly implement biosecurity plans? Will science-based biosecurity prove to be applicable and sustainable across the industry? Can the global industry successfully curtail the spread of African swine fever virus and what will be the role of US SHIP? Clearly, these are exciting times which require the input and expertise of all veterinary professionals.

Be hopeful,
Scott and Gordon

Acknowledgments: In closing, Dee and Spronk would like to sincerely thank all the authors for their high-quality contributions, as well as the *Animals* editorial team for all the help they have provided during this long, yet truly successful journey. We both hope that the readership finds this Special Issue to be truly special and that the information contained within will help to solve problems for your customers.

Conflicts of Interest: The authors declare no conflict of interest.

List of Contributions

Section 1: Literature Review

1. Otake, S.; Yoshida, M.; Dee, S. A Review of Swine Breeding Herd Biosecurity in the United States to Prevent Virus Entry Using Porcine Reproductive and Respiratory Syndrome Virus as a Model Pathogen. *Animals* **2024**, *14*, 2694. https://doi.org/10.3390/ani14182694.

Section 2: New Research

2. Soriano, O.; Batista, L.; Morales, J.; Quintana, E.; Piñeiro, C. A Dynamic System to Control the Entry of Non-Authorized Visitors and Detect Superspreader Farms in Strongly Interconnected Systems. *Animals* **2024**, *14*, 2932. https://doi.org/10.3390/ani14202932.
3. Makovska, I.; Chantziaras, I.; Caekebeke, N.; Dhaka, P.; Dewulf, J. Assessment of Cleaning and Disinfection Practices on Pig Farms across Ten European Countries. *Animals* **2024**, *14*, 593. https://doi.org/10.3390/ani14040593.
4. Pavone, S.; Iscaro, C.; Giammarioli, M.; Beato, M.S.; Righi, C.; Petrini, S.; Costarelli, S.; Feliziani, F. Biological Containment for African Swine Fever (ASF) Laboratories and Animal Facilities: The Italian Challenge in Bridging the Present Regulatory Gap and Enhancing Biosafety and Biosecurity Measures. *Animals* **2024**, *14*, 454. https://doi.org/10.3390/ani14030454.
5. Gebhardt, J.T.; Dee, S.A.; Little, E.; Scales, B.N.; Kern, D.R. Evaluation of Extended Storage of Swine Complete Feed for Inactivation of Viral Contamination and Effect on Nutritional, Microbiological, and Toxicological Profiles. *Animals* **2024**, *14*, 393. https://doi.org/10.3390/ani14030393.
6. Houston, G.E.; Jones, C.K.; Evans, C.; Otott, H.K.; Stark, C.R.; Bai, J.; Poulsen Porter, E.G.; de Almeida, M.N.; Zhang, J.; Gauger, P.C.; et al. Evaluation of Truck Cab Decontamination Procedures following Inoculation with Porcine Epidemic Diarrhea Virus and Porcine Reproductive and Respiratory Syndrome Virus. *Animals* **2024**, *14*, 280. https://doi.org/10.3390/ani14020280.
7. Dhaka, P.; Chantziaras, I.; Vijay, D.; Singh, M.; Bedi, J.S.; Caekebeke, N.; Dewulf, J. Situation Analysis and Recommendations for the Biosecurity Status of Dairy Farms in Punjab, India: A Cross-Sectional Survey. *Animals* **2023**, *13*, 3458. https://doi.org/10.3390/ani13223458.

Section 3: Bio-surveillance

8. Lupulović, D.; Gnjatović, M.; Prodanov-Radulović, J.; Ćujić, D.; Gajdov, V.; Samojlović, M.; Petrović, T. Seroepidemiological Survey of Hepatitis E Virus in Intensive Pig Farming in Vojvodina Province, Serbia. *Animals* **2025**, *15*, 151. https://doi.org/10.3390/ani15020151.
9. Berge, A.C.; Vertenten, G. Bovine Coronavirus Prevalence and Risk Factors in Calves on Dairy Farms in Europe. *Animals* **2024**, *14*, 2744. https://doi.org/10.3390/ani14182744.
10. Tarasiuk, G.; Remmenga, M.D.; O'Hara, K.C.; Talbert, M.K.; Rotolo, M.L.; Zaabel, P.; Zhang, D.; Giménez-Lirola, L.G.; Zimmerman, J.J. Pen-Based Swine Oral Fluid Samples Contain Both Environmental and Pig-Derived Targets. *Animals* **2024**, *14*, 766. https://doi.org/10.3390/ani14050766.
11. Trevisan, G.; Morris, P.; Silva, G.S.; Nakkirt, P.; Wang, C.; Main, R.; Zimmerman, J. Active Participatory Regional Surveillance for Notifiable Swine Pathogens. *Animals* **2024**, *14*, 233. https://doi.org/10.3390/ani14020233.
12. Panyasing, Y.; Gimenez-Lirola, L.; Thanawongnuwech, R.; Prakobsuk, P.; Kawilaphan, Y.; Kittawornrat, A.; Cheng, T.-Y.; Zimmerman, J. Performance of a Differentiation of Infected from Vaccinated Animals (DIVA) Classical Swine Fever Virus (CSFV) Serum and Oral Fluid Erns Antibody AlphaLISA Assay. *Animals* **2023**, *13*, 3802. https://doi.org/10.3390/ani13243802.

Section 4: Human Behavior and Biosecurity

13. Lianou, D.T.; Fthenakis, G.C. Associations of Socio-Demographic Characteristics of Dairy Goat Farmers in Greece with Biosecurity-Related Practices in the Farms. *Animals* **2024**, *14*, 2136. https://doi.org/10.3390/ani14142136.
14. Toson, M.; Dalla Pozza, M.; Ceschi, P. Farmers' Biosecurity Awareness in Small-Scale Alpine Dairy Farms and the Crucial Role of Veterinarians. *Animals* **2024**, *14*, 2032. https://doi.org/10.3390/ani14142032.
15. Herrmann, A.; Schulz, K.; Wischnewski, N.; Brüssau, J.; Zeiler, E.; Sauter-Louis, C. Risk Awareness and Attitude of German Farmers towards Biosecurity Measures. *Animals* **2024**, *14*, 1102. https://doi.org/10.3390/ani14071102.

Section 5: Industry Action

16. Dahl, J. Risk Assessments and Risk Mitigation to Prevent the Introduction of African Swine Fever into the Danish Pig Population. *Animals* **2024**, *14*, 2491. https://doi.org/10.3390/ani14172491.
17. Harlow, M.; Torremorell, M.; Rademacher, C.J.; Gebhardt, J.; Holck, T.; Linhares, L.C.M.; Main, R.G.; Trevisan, G. Biosecurity Insights from the United States Swine Health Improvement Plan: Analyzing Data to Enhance Industry Practices. *Animals* **2024**, *14*, 1134. https://doi.org/10.3390/ani14071134.
18. Dee, S.; Brands, L.; Edler, R.; Schelkopf, A.; Nerem, J.; Spronk, G.; Kikuti, M.; Corzo, C.A. Further Evidence That Science-Based Biosecurity Provides Sustainable Prevention of Porcine Reproductive and Respiratory Syndrome Virus Infection and Improved Productivity in Swine Breeding Herds. *Animals* **2024**, *14*, 2530. https://doi.org/10.3390/ani14172530.

References

1. Pfeiffer, D.U.; Otte, M.J.; Roland-Holst, D.; Zilberman, D. A one health perspective on HPAI H5N1 in the Greater Mekong sub-region. *Comp. Immunol. Microbiol. Infect. Dis.* **2013**, *36*, 309–319. [CrossRef] [PubMed]
2. Eisfeld, A.J.; Biswas, A.; Guan, L.; Gu, C.; Maemura, T.; Trifkovic, S.; Wang, T.; Babujee, L.; Dahn, R.; Halfmann, P.J.; et al. Pathogenicity and transmissibility of bovine H5N1 influenza virus. *Nature* **2024**, *633*, 426–432. [CrossRef] [PubMed]
3. Ruedas-Torres, I.; Rodríguez-Gómez, I.M.; Sánchez-Carvajal, J.M.; Larenas-Muñoz, F.; Pallarés, F.J.; Carrasco, L.; Gómez-Laguna, J. The jigsaw of PRRSV virulence. *J. Vet. Microbiol.* **2021**, *260*, 109168. [CrossRef] [PubMed]
4. Dixon, L.K.; Sun, H.; Roberts, H. African swine fever. *Antivir. Res.* **2019**, *165*, 34–41. [CrossRef] [PubMed]
5. Dee, S.; Brands, L.; Nerem, J.; Schelkopf, A.; Spronk, G.; Havas, K. Improvements in swine herd biosecurity reduce the incidence of porcine reproductive and respiratory syndrome virus in swine herds in the Midwestern United States. *J. Am. Vet. Med. Assoc.* **2024**, *262*, 520–525. [CrossRef] [PubMed]

Disclaimer/Publisher's Note: The statements, opinions and data contained in all publications are solely those of the individual author(s) and contributor(s) and not of MDPI and/or the editor(s). MDPI and/or the editor(s) disclaim responsibility for any injury to people or property resulting from any ideas, methods, instructions or products referred to in the content.

animals MDPI

Review

A Review of Swine Breeding Herd Biosecurity in the United States to Prevent Virus Entry Using Porcine Reproductive and Respiratory Syndrome Virus as a Model Pathogen

Satoshi Otake [1,*], Mio Yoshida [1] and Scott Dee [2]

[1] Swine Extension & Consulting, Inc., Shibata 957-0021, Niigata, Japan; mioyoshida@swext-consulting.co.jp
[2] Pipestone Research, Pipestone, MN 55482, USA; scottdee7255@gmail.com
* Correspondence: satoshiotake@swext-consulting.co.jp; Tel.: +81-(0)90-6136-6463

Simple Summary: Maintaining a sustainable supply of animal protein is the mission of the global swine production industry. The entry of infectious pathogens to swine populations can cause significant animal welfare issues, increase the use of antibiotics, challenge environmental stability, and interrupt/reduce the supply of pork; therefore, preventing pathogen entry is critical to achieve its mission using science-based biosecurity programs. Biosecurity is the application of science-based protocols to minimize the risk of pathogen entry. The objective of this review is to summarize basic biosecurity terms and concepts, review the transmission of porcine reproductive and respiratory syndrome virus (PRRSV) and the biosecurity protocols designed to mitigate these risk factors, and discuss how the swine industry is applying Next Generation Biosecurity to prevent PRRSV infection of the breeding herd.

Abstract: The prevention of disease introduction into swine herds requires the practice of science-based protocols of biosecurity that have been validated to reduce the risk of the entry of targeted pathogens. The fundamental pillars of biosecurity include bio-exclusion, biocontainment, and bio-management. Biosecurity protocols must be science-based, a way of life, continuously validated, cost-effective, and benchmarked over time. This paper will review these concepts, the direct and indirect routes of transmission of porcine reproductive and respiratory syndrome virus (PRRSV), and the interventions that have been designed and validated to prevent infection of the breeding herd. It will close with a review of Next Generation Biosecurity, describing how a science-based approach is being used to prevent PRRSV infection in breeding herds from a large commercial pork production system in the US.

Keywords: biosecurity; swine; breeding; herd; transmission; prevention; porcine; reproductive; respiratory; syndrome; virus; PRRSV

1. Introduction

Maintaining a sustainable supply of animal protein to consumers is the mission of the global swine industry. The introduction of infectious pathogens such as porcine reproductive and respiratory syndrome (PRRSV), porcine epidemic diarrhea virus (PEDV), and African swine fever virus (ASFV) to swine herds can disrupt the supply chain, resulting in reduced animal welfare, decreased production of pork, and significant economic loss to farmers. Therefore, preventing the introduction of infectious agents through the practice of biosecurity is critical to achieve this mission. Successful biosecurity requires a culture of confidence throughout the workforce, with all members of the farm team believing in a program of disease prevention, based on the understanding that it has been designed using science-based protocols validated to reduce the risk of targeted pathogen(s) entry.

Regarding difficult pathogens, throughout the global swine industry, no pathogen has challenged biosecurity and negatively impacted animal welfare to a greater degree than porcine reproductive and respiratory syndrome virus (PRRSV). This virus has been estimated to cost the US swine industry USD 660 million/year due to reduced performance, higher mortality, and increased costs of treatment and prevention [1]. In the absence of vaccines capable of inducing sterilizing immunity, the key to controlling PRRS is to prevent PRRSV entry to breeding herds, using biosecurity programs that utilize science-based protocols. Therefore, this review will use PRRSV as a model pathogen that has driven change in the biosecurity practices of the global swine industry.

To begin the discussion, this paper opens with a review of the concepts of biosecurity (Section 1) and then transitions into a focused review of the routes of transmission of PRRSV and the biosecurity protocols validated to reduce these risks at the level of the breeding herd (Section 2). It concludes with a review of how all this information was used to develop a new concept from the United States known as Next Generation Biosecurity (NGB) and the results following its application to a commercial system of pork production to specifically prevent virus entry to breeding herds (Section 3). The authors hope that this paper will provide a resource for farmers and veterinarians to refer to when reviewing biosecurity plans for PRRSV, as well as to instill hope that the prevention of PRRSV reinfection of breeding herds is possible, repeatable, and sustainable.

2. A Review of the Concepts of Biosecurity

The definition of biosecurity is "security from transmission or introduction of infectious diseases, parasites, and pests" [2]. This resource defines the three pillars of biosecurity as (1) bio-exclusion, (2) biocontainment, and (3) bio-management.

2.1. *Bio-Exclusion*

Bio-exclusion focuses on preventing the introduction of pathogens to animal populations. In the swine industry, it is routinely applied at the level of the breeding stock supplier, i.e., the artificial insemination center, the genetic nucleus sow farm, and the gilt multiplication unit. However, with the emergence of highly virulent variants of PRRSV, the philosophy of bio-exclusion has also become important at the level of the commercial sow farm, particularly if the herd is in a dense region of swine production.

2.2. *Biocontainment*

The concept of biocontainment has been well established in human clinical medicine and animal laboratory research [3]. As it refers to livestock, biocontainment focuses on internal biosecurity or the practice of biosecurity within the farm. Its goal is to minimize the transmission of pathogens that already exist in the herd population by attempting to prevent their spread within the population, as well as to other farms in the region. The concept of biocontainment is also important in wean-to-market populations, particularly during area regional disease control efforts in high swine dense regions.

2.3. *Bio-Management*

Bio-management focuses on the oversight and execution of the processes and protocols that are associated with a biosecurity program, particularly at the level of the farm personnel. For example, the education and training of the staff/team/organization, along with a regular program of auditing to document compliance, are critical components of a plan of bio-management [4,5].

3. What Are the Components of a Successful Biosecurity Program?

The success of the biosecurity program is based on the attitudes of the workforce at the farm and their ability to carry out recommended best management practices for the prevention of pathogen entry to their respective herds [6]. In their excellent review of the role of the farmer in the adoption of biosecurity based recommendations, Ritter et al. stated that "Approaches that appeal to farmers' internal motivators or that unconsciously elicit the desired behavior will increase the success of the intervention" [6]. After many years of experience in the swine industry, it is the opinion of the authors that there are five essential components of a successful biosecurity program, all of which are heavily dependent on the belief in, and the acceptance of, a program of disease prevention among all members of the farm workforce. Specifically, a biosecurity program must be:

(1) science-based,
(2) a way of life,
(3) continuously validated,
(4) cost-effective,
(5) benchmarked over time.

3.1. *Biosecurity Must Be Science-Based*

To prevent the risk of the transmission/introduction of pathogens between/within the units, it is first essential to understand the routes of transmission of the targeted pathogen(s). Once understood, intervention strategies can then be designed to minimize the risk(s). Therefore, all biosecurity protocols should be based on published evidence, as the scientific basis of the program is what helps build the confidence of herd owners and on-farm personnel, thereby driving acceptance and promoting compliance, even during periods of perceived program failure following an unexpected disease outbreak.

3.2. *Biosecurity Must Be a Way of Life*

The practice of biosecurity is challenging, as it requires extra time, continuous attention to detail, and consistent day-to-day application. For example, showering into a breeding herd to reduce the introduction of pathogens via fomites such as personnel clothing and footwear takes time. Filtering incoming air to prevent the airborne entry of pathogens requires a change in human behavior. The effort it takes to clean and disinfect transport vehicles after pig deliveries is physically demanding work. Therefore, a culture that believes that biosecurity is not just a job but is a way of life is an important characteristic of a successful farm. Personnel need to believe in the importance of biosecurity and practice it every day. For this to happen, people need to be educated in the science behind the interventions, be trained to consistently execute them properly, and live the belief that they are effective for preventing disease.

3.3. *Biosecurity Must Be Continuously Validated*

In the operational point of view, it is the daily execution of protocols by people (farm staff, team, organization, etc.) that determines whether a biosecurity program will succeed or fail. As mentioned, this often requires a change in culture on the farm regarding the importance of the program. Once established and implemented, there should be continuous education, the monitoring of compliance, and regular auditing to ensure proper execution. Audits should be unannounced and conducted by swine veterinarians and trained personnel originating from external, third-party sources. Following the audit, follow-up discussions between the farm veterinarian and the staff should take place to maximize learning and minimize future mistakes.

3.4. *Biosecurity Must Be Cost-Effective*

As biosecurity protocols frequently require capital investments, these costs should be calculated and, whenever possible, compared to the cost of the targeted disease(s). This helps all involved, including lenders, to understand the potential return on an investment if following a reduction in the incidence of the targeted disease(s).

3.5. *Biosecurity Must Be Benchmarked over Time*

Benchmarking the outcomes from a biosecurity program over time, be it at the population, herd, site, regional, or national level, is important to measure success and/or failure and to drive continuous improvement. One example of a useful benchmarking metric is the measurement of the percent incidence risk, i.e., the total number of new pathogen introductions divided by the number of herds observed. An excellent example of a system that uses this approach to measure incidence risk is the University of Minnesota Dr. Bob Morrison Swine Health Monitoring Project (MSHMP) [7]. This approach has been applied to multiple agents, including PRRSV, PEDV, and *Mycoplasma hyopneumoniae*. Also, the use of software-generated maps that display the location and pathogen status of neighboring herds in an area is particularly helpful for monitoring the progress of regional disease control programs.

4. Transmission of Porcine Reproductive and Respiratory Syndrome Virus and the Biosecurity Protocols Designed to Prevent Infection of the Breeding Herd

The disease of porcine reproductive and respiratory syndrome (PRRS) has been a driver of change throughout the global swine industry, especially regarding the practice of biosecurity at the level of breeding herds. PRRS is caused by PRRSV, a single-stranded enveloped RNA virus, classified in the genus *Arterivirus* and a member of family *Arteriviridae*, which was first identified in 1991 [8,9]. PRRSV undergoes genetic change that results in the regular emergence of new variants that often display enhanced pathogenicity and improved ability to spread [10,11]. Recent variants demonstrating these features include PRRSV 184, PRRSV 174, and PRRSV 144 lineage L1C [12,13]. For example, these variants have adapted efficient means to transmit within and between herds, including an improved ability to spread via the airborne route [14]. PRRSV is now known to be endemic in pig populations across five continents, and only African swine fever virus can rival its clinical and economic significance throughout the global swine industry [15]. The emergence of PRRSV, along with its continued adaptation and economic significance, has forced swine producers and veterinarians to focus intensively on preventing viral entry to breeding herds using science-based biosecurity. However, despite years of effort, a long-established paradigm has been that, while the elimination of PRRSV from infected herds is likely to be successful, the prevention of reinfection with a new variant is not. For the global swine industry to gain control of PRRS, this paradigm must change, and the first requirement to make this change is developing a comprehensive understanding of the routes of the transmission of PRRSV and the biosecurity protocols that have been proven to mitigate these risks and prevent infection. The paper will now review the published direct and indirect routes of PRRSV, reference the biosecurity protocols that have been validated to mitigate these risks, and describe industry actions that have been put into place at the level of the breeding herd.

4.1. Pigs

Pig-to-pig transmission is the principal means of transmission of many infectious pathogens of swine, including PRRSV. PRRSV produces a persistent infection and can survive in the tonsils and lymph nodes of infected pigs for extended periods, i.e., 157 days post-infection. In addition, shedding of the virus from carrier animals, resulting in the infection of naïve sentinels, can occur for up to 49–86 days [16,17]. Based on this information, it is essential that producers purchase breeding stock from a known PRRSV-negative source and quarantine/test these animals prior to their entry into the breeding herd.

4.2. Semen

PRRSV infection of breeding females can occur following insemination with PRRSV-positive semen [18]. The frequency of PRRSV shedding in semen is highest during the acute phases of the infection; however, intermittent shedding can occur days to months after infection. Swenson et al. detected viable virus in the semen from experimentally infected boars up to 43 days, and Christopher-Hennings et al. detected viral RNA by PCR in the semen from experimentally infected boars up to 92 days following inoculation [19,20]. Furthermore, Gradil et al. also demonstrated the transmission of PRRSV via extended semen from experimentally infected boars [21]. Therefore, as with breeding stock, only semen from PRRSV-negative boar populations should be used, and in the artificial insemination center, boars who provided semen should be blood tested on the day of collection, followed by PCR testing to document the absence of the virus, prior to shipment to breeding herds.

4.3. Needles

Needles are an efficient fomite for the transmission of swine pathogens, particularly during the viremic period of infection. In commercial swine farms, pigs receive numerous intramuscular injections of vaccines and antibiotics, and it is not a common practice to change needles between individual pigs, due to cost and labor constraints. It is well documented that needles can serve as mechanical vectors of PRRSV due to the prolonged viremia that occurs following infection. Under experimental conditions, Otake et al. demonstrated the transmission of PRRSV from viremic infected pigs to naïve pigs in the absence of a needle change between animals [22]. Furthermore, Madapong et al. reported that the use of a "needle-free" syringe prevented the hematogenous transfer of PRRSV between viremic and susceptible pigs, while sentinel pigs became infected following the injection of viremic pigs in the absence of a needle change [23]. With this in mind, needle-free devices should be considered for use in breeding herds and/or needles should be changed between individual sows, particularly those in the third trimester of gestation.

4.4. Fomites

Fomites such as coveralls, boots, or other inanimate objectives have also been considered as risk factors in the transmission of swine pathogens. Pirtle and Beran assessed the survivability of PRRSV in/on 16 fomites, including plastic, stainless steel, rubber, alfalfa, wood shavings, straw, corn, swine starter feed, denim cloth, phosphate-buffered saline, saline G, well water, city water, swine saliva, urine, and fecal slurry [24]. An attempt was done to isolate viable PRRSV from these samples after storage at 25 to 27° C. Viable PRRSV was isolated from phosphate-buffered saline through day three, saline G through day six, well water through day eight, and city water through day eleven following the inoculation. In contrast, viable PRRSV was not recovered from any other fomite samples beyond 30 min post-inoculation. Using a model involving a coordinated sequence of events routinely experienced in a "day in the life" of a pork producer, Dee et al. demonstrated that viable PRRSV could be mechanically transported from a truck wash environment onto the floormats of transport vehicles and the soles of personnel footwear and walked into a mock farm, resulting in the contamination of incoming farm supplies and the hands of personnel in 8 out of 10 replicates during cold weather and 2 out of 10 replicates during warm weather [25,26]. Other materials such as Styrofoam, cardboard, linoleum, concrete, and stainless steel have also been reported to harbor viable PRRSV for different amounts of time [25,27]. Dee et al. also indicated that the use of disposable boots, bleach boot baths, or bag-in-a-box shipping methods was highly efficacious in preventing the mechanical transmission of PRRSV [28]. Based on the risk of fomites such as incoming supplies, commercial farms have installed D&D (Disinfection and Downtime) rooms, a heated facility with a separate airspace from the farm facility proper. Incoming supplies are delivered to the D&D room from a commercial vendor to a designated employee of the farm, i.e., the manager, who receives the supplies, applies disinfectant to the exterior of the containers housing the supplies, and allows all materials to remain in the locked room for a designated period of time before entering the farm.

4.5. Personnel

Personnel movement within or between production units has long been considered as a risk factor for the transmission of swine pathogens; therefore, biosecurity protocols that control and regulate personnel entry to farms and movement between farms via downtime policies are frequently practiced. Under experimental conditions, Otake et al. documented that PRRSV could be transmitted to susceptible pigs by coveralls, boots, and hands of personnel that had been previously contaminated following contact with infected pigs. The study also demonstrated that changing coveralls and boots, along with washing hands or use of a shower in/shower out protocol between infected and susceptible groups prevented transmission of the virus [29]. Pitkin et al. also reported, under the conditions of the study, PRRSV RNA was readily detected on several different fomites and on the hands of personnel after contact with infected swine [30]. In addition, the transmission of PRRSV to sentinel pigs was observed in seven out of seven replicates in which biosecurity practices such as changing clothing and footwear or washing hands were not practiced. In contrast, when these basic practices were followed, transmission was prevented every time (0/7 replicates). Furthermore, Pitkin demonstrated that, following contact with groups of pigs that had been experimentally inoculated with PRRSV and *Mycoplasma hyopneumoniae*, the combination of a shower in/out protocol and a one-night downtime period prevented the spread of both pathogens to susceptible pigs [31]. Therefore, the use of personnel downtime prior to entry to breeding herds, in combination with a shower in/shower out protocol prior to entering the animal airspace, are common practices across modern swine production companies in the US. Another practice that has been devised based on this information is the "boot bench", a wooden bench that divides the farm anteroom into "clean" and "dirty" segments. The goal of the boot bench is to minimize the chance of the soles of personnel footwear to spread pathogens throughout the anteroom. Upon entry to the anteroom from the outside environment, personnel sit on the bench, remove their footwear, and place it on the floor of the "dirty" side of the bench, elevate their feet and rotate 180°, place their feet on the opposite ("clean") side of the bench, and walk to the shower facility.

4.6. Transport

Transport vehicles such as livestock trailers, feed trucks, and motorized vehicles used by farmworkers or veterinarians have been postulated to play a significant role in the transmission of swine pathogens. It is logical to assume that organisms can be carried on the frame of vehicles or in manure embedded in tire treads. The studies conducted by Dee et al. demonstrated that contaminated vehicles played a role in the transmission of PRRSV under controlled conditions depicting a "coordinated sequence of events" that swine farmers routinely practice under differing environmental conditions [25,26]. Both studies demonstrated how transport vehicles can act as fomites for the long-distance (50 km) spread of PRRSV. Dee et al. also demonstrated that pigs housed in model trailers intentionally contaminated with PRRSV could become infected [32]. This study involved housing experimentally PRRSV-infected seeder pigs in model trailers for two hours, followed by their removal and the introduction of naïve sentinel pigs in the absence of sanitation of the trailer model. Infection of the sentinels was demonstrated in three out of four replicates. The result concluded that transport vehicle sanitation is an important strategy to prevent the transmission of PRRSV to susceptible pigs. Galvis et al. developed and calibrated a mathematical model for the transmission of PRRSV, finding that 36% to 80% of PRRSV infections could be associated with the transport of animals between sites [33]. These collective observations provide evidence to support the importance of the risk of transport for spread of the virus. It is well accepted that transport vehicle sanitation is critical to mitigate risk. Therefore, in today's commercial industry, a program of transport sanitation, including the removal of organic material, followed by washing, the application of a validated product to disinfect the trailer, and allowing it to thoroughly dry, either through the short-term application of heated air (70 °C for 30 min) or the circulation of room temperature air (20 °C) for an overnight period, are common practices across commercial companies [34–36].

4.7. Avian Species

Birds have often been proposed as a risk factor for the transmission of swine pathogens between farms due to their ability to travel great distances. Direct transmission of pathogens from wild birds to pigs can occur through mechanical means, particularly in backyard farms where interactions between birds and pigs can occur. Zimmerman et al. reported that mallard ducks orally exposed to PRRSV shed the virus in feces. [37]. In addition, pigs intranasally exposed to PRRSV isolated from mallard feces became viremic, seroconverted, and transmitted the virus to sentinel swine. However, Trincado et al. attempted to reevaluate the potential of avian species in the transmission of PRRSV and concluded that mallard ducks were not likely to serve as reservoirs of PRRSV [38,39]. In this study, adult mallard ducks were exposed to PRRSV in the drinking water, as well as orally inoculated with PRRSV. However, the infection of PRRSV did not occur in the ducks, as PRRSV RNA was not detected in duck feces and naïve sentinel pigs exposed to the feces did not become infected. Additionally, PRRSV-infected pigs did not transmit the virus to mallard ducks following 21 days of direct contact under controlled conditions. Therefore, based on this contradictory information, the risk of PRRSV transmission by avian species is not currently considered of high importance.

4.8. Rodents

Based on published data, rodents do not appear to be vectors of PRRSV. Hooper et al. documented the inability to recover PRRSV from several tissues (sera, lung, thymus, and spleen) of mice and rats trapped on a swine farm with endemic PRRSV infection [40]. Furthermore, mice and rats experimentally inoculated with PRRSV did not become infected [40]. Wills et al. also examined the susceptibility of rats and mice for PRRSV and reported similar results [41]. However, despite these negative results, an on-farm rodent control program is very important for the control of other diseases, as well as to minimize the damage that rodents can cause to a swine facility.

4.9. Feral/Domestic Animals

In addition to rodents, feral and domestic animals have been hypothesized to be vectors of swine pathogens, especially as it pertains to wild pigs, which have been postulated to serve as disease reservoirs. In contrast to the well-documented risk of wild pigs for the spread of ASFV, serological surveys of wild pigs have demonstrated a low prevalence of PRRSV antibodies [42–44]. Regarding the risk of non-swine species as risk factors for PRRSV spread, Wills et al. inoculated dogs, cats, skunks, opossums, and racoons with PRRSV via the intranasal and intramuscular routes, and serum samples were collected over the course of 21 days post-inoculation, and tonsils and lymph nodes were collected upon euthanasia 21 days post-inoculation [41]. All samples were negative by PCR and virus isolation, except for two serum samples from two out of four opossums that were PCR-positive on days 3 and 14 post-inoculation. However, due to their weak Ct values, these values were concluded to be false positives; therefore, non-porcine mammalian species do not appear to serve as reservoirs of PRRSV.

4.10. Insects

It is well documented that various species of insects can serve as vectors of swine pathogens, including ASFV, Japanese B encephalitis virus, and vesicular stomatitis virus [45–47]. Regarding PRRSV, Otake et al. demonstrated PRRSV transmission to susceptible pigs following contact with houseflies (*Musca domestica*) allowed to feed on viremic pigs [48]. Otake et al. also showed the survivability of infectious PRRSV in the midgut of houseflies for up to 6 h [49]. Furthermore, Otake et al. demonstrated that 2 out of 10 individual houseflies, which had fed on an experimentally inoculated pig in the acute phase of viremia, could transmit PRRSV to naïve sentinels [50]. These studies collectively concluded that PRRSV can be mechanically transmitted from infected pigs to susceptible pigs by houseflies and PRRSV can survive within housefly populations for up to six hours post-feeding, supporting the hypothesis that house flies could contribute to the horizontal transmission of PRRSV among pigs within infected commercial farms. In addition, Schurrer et al. reported that flies could become contaminated with PRRSV from infected pigs and transport the virus out to 1.7 km from the population [51]. Furthermore, mosquitoes (*Aedes vexans*) have also been hypothesized to be mechanical vectors of PRRSV. Under experimental conditions, Otake et al. demonstrated PRRSV transmission to susceptible pigs via mosquitoes that had previously fed on viremic infected pigs [52]; however, it was determined that the virus was not capable of replicating in the salivary glands of the mosquitoes [53]. These findings indicate that mosquitoes may pose an ability to mechanically transmit PRRSV; however, they do not serve as biological vectors. Based on these collective observations, insect control programs on swine facilities involving the strategic use of insecticides and habitat management are important to minimize the PRRSV risk, maximize animal and personnel comfort, and reduce carcass damage at slaughter secondary to insect bites, particularly in tropical areas of the world where insects are prevalent year-round.

4.11. Swine Slurry and Lagoon Effluent

Despite limited research, it has been hypothesized that slurry or lagoon effluent could serve as risk factors for the transmission of PRRSV. Dee et al. reported that samples of lagoon effluent inoculated with PRRSV and stored at 4 °C and 20 °C were PCR-positive up to 12 days post-inoculation [54]. Exposure to PRRSV-contaminated swine effluent stored at 4 °C for up to eight days post-inoculation resulted in infection and seroconversion in pigs. At 20 °C storage, exposure to effluent 1–3 days post-inoculation resulted in PRRSV infection in pigs. These results indicated that (1) the viability of PRRSV in swine effluent is relatively short (1–8 days), (2) its survival in effluent appears to be dependent on temperature, and (3) pigs can become infected following contact with PRRSV-contaminated effluent. In conclusion, the risk of swine herds becoming infected with PRRSV following contact with contaminated lagoon effluent is a time- and temperature-sensitive risk. Recently, Linhares et al. assessed the infectivity of PRRSV in slurry at different temperatures [55]. In this study, PRRSV consistently displayed shorter half-lives in slurry compared to minimum essential media (MEM). Specifically, the PRRSV half-life in MEM and slurry was estimated at 112.6 and 120.5 h at 4 °C, 14.6 and 24.5 h at 20 °C, 1.6 and 1.7 h at 40 °C, 2.9 and 8.5 min at 60 °C, and 0.36–0.59 min at 80 °C, respectively. Therefore, based on these findings, along with the repeatable observation that PRRS outbreaks occur during pit pumping season, special care should be taken to avoid the mechanical risk of the virus spread in contaminated trucks and pumping equipment moving between sites.

4.12. Pig Meat

While the ability to transmit PRRSV in pig meat has been evaluated, limited data exist to support this hypothesis. PRRSV from meat samples of experimentally infected pigs was detected for up to 18 days stored at 4 °C and up to 60 days at −20 °C [56]. Furthermore, viable PRRSV was detectable at concentrations ranging from $10^{3.3}$ to $10^{4.3}$ $TCID_{50}/g$ of muscle tissue from experimentally PRRSV-infected pigs at day 11 post-inoculation, and susceptible pigs consuming these samples became infected [57]. Bloemraad et al. reported low PRRSV loads in muscle tissue collected from viremic pigs, and that viral titer remained stable following storage for up to 48 h at 4 °C [58]. Mager et al. isolated PRRSV from muscle samples collected from the experimentally inoculated pigs at 7 and 14 days post-inoculation; however, they did not detect the virus from muscle samples collected from 44 carcasses at slaughter [59]. In conclusion, the presence of the virus in pig meat from slaughter age animals appears to be very low. In contrast, viable PRRSV has been recovered from meat juice during the viremic period and from human hands after handling the meat, and the mechanical transmission of PRRSV from contaminated hands to susceptible pigs was documented [60]. Therefore, based on this collective information, the entry of pork products to commercial swine farms should be prohibited.

4.13. Aerosols

In the initial outbreaks of PRRSV in Europe, widespread transmission of the virus between herds in affected regions was initially attributed to airborne spread; however, experimental reproduction of the aerosol transmission of PRRSV was initially difficult to achieve and was documented only over short distances (1 m) [61,62]. At this time, the role of aerosols in the transmission of PRRSV has been well documented. With the emergence of new variants, i.e., PRRSV 184, Dee et al. demonstrated the infection of susceptible animals by aerosolized PRRSV over 150 m using a straight tube model at 27 km/h air speed [63]. Following this information, Otake et al. reported the airborne transport of a mixed infection of PRRSV 184 and *Mycoplasma hyopneumoniae* out to 9.2 km under controlled field conditions [64]. This work led to the investigation of whether the filtration of incoming air could reduce the risk of viral entry to herds. Dee et al. used a production region model to simulate a dense region of swine farming and demonstrated that air filtration systems could successfully prevent the airborne transmission of PRRSV and *Mycoplasma hyopneumoniae* [65]. Based on this information, the filtration of incoming air to breeding herds has been significantly advanced and is a critical component of biosecurity programs for breeding herds [66]. Under commercial field conditions, buildings are ventilated either by positive or negative pressure and utilize commercial filters with fractional efficiencies ranging primarily from MERV 14 to 16. The impact of this intervention has been significant, especially as it pertains to preventing PRRSV aerosol transmission between commercial breeding herds [67].

4.14. Feed

In 2014, Dee et al. first described the ability of feed to transmit PEDV to susceptible pigs [68]. Since that time, this hypothesis has been validated by several investigators across multiple viruses, such as Seneca virus A, PRRSV, Classical swine fever virus, Pseudorabies virus, and ASFV [69–71]. These studies evaluated the survival of important viral pathogens of livestock in animal feed ingredients imported daily into the United States under simulated transboundary conditions or during long-distance transport across the Continental US. In the transboundary model, PRRSV was recovered from soybean meal 37 days post-inoculation [71]. The experimental transmission of PRRSV through natural feeding behavior has been demonstrated as well [72,73]. In this study, tons of complete feed were contaminated with PRRSV, PEDV, and Seneca virus A using an "ice block challenge" model, which pigs were allowed to consume ad libitum. Finally, Dee et al. documented the transmission of PRRSV in domestic pigs under field conditions following the oral ingestion of feed material, followed by a proof-of-concept study under laboratory conditions [72]. This recent work has led to the regular use of validated feed additives having antiviral properties and being capable of neutralizing PRRSV in feed [72].

5. Next Generation Biosecurity: A Model for the Swine Industry

To advance the control of PRRS, a recent publication challenged US swine veterinarians to help farmers to eliminate PRRSV from breeding herds and prevent reinfection through improvements in biosecurity [74]. It had been historically demonstrated that PRRSV could be eliminated from infected breeding herds, resulting in the availability of PRRSV-naïve genetic sources, thereby mitigating viral entry via direct routes of transmission, i.e., infected breeding stock and contaminated semen [75–78]. However, the reinfection of naïve breeding herds with new variants of the virus despite the use of naïve replacement animals and semen was a frequent event (area spread), particularly when the herds were in regions of dense swine production [79,80]. To build confidence throughout the US industry, it was important to prove that the root cause(s) of area spread could be identified and that reinfection could be prevented/reduced. To solve this problem, a research plan was designed to identify potential indirect routes of PRRSV transmission, including mechanical (fomite-based) routes such as contaminated transport, personnel clothing and footwear and incoming supplies, PRRSV-positive aerosols, and contaminated feed, and to test and validate biosecurity protocols to mitigate these risks. Table 1 provides a timeline summarizing the history of area spread, the sequential milestones of the research plan, and the implementation of a comprehensive, science-based approach to biosecurity, known as Next Generation Biosecurity (NGB) [66].

Table 1. Timeline summarizing significant events and key research milestones that led to the development and validation of the Next Generation Biosecurity concept.

Years	Milestone
2000	PRRSV-naïve breeding stock and semen become commercially available.
2001	Area Spread of PRRSV is described.
2002–2004	Mechanical risks (contaminated transport, clothing/footwear, and incoming supplies) for the spread of PRRSV are identified.
2005–2007	Protocols of transport sanitation, personnel entry, and supply entry are validated.
2007–2010	Aerosol transmission and long-distance airborne transport of PRRSV are identified.
2009–2012	Air filtration for PRRSV prevention is validated.
2014–2023	The risk of contaminated feed for the transmission of viruses of veterinary significance is discovered.
2014–2022	Feed biosecurity protocols (additives, storage) are developed.
2021–2024	Next Generation Biosecurity concept developed and tested.

The concept of NGB was based on the application of validated mitigation strategies across "four pillars" of routes PRRSV entry to a breeding herd, including pillar 1: direct routes (infected pigs and contaminated semen); pillar 2: mechanical routes (transport, personnel, and supplies); pillar 3: the aerosol route; and pillar 4: the route of contaminated feed, depicted in Figure 1 [66]. NGB initially focused on the quarantining and testing incoming genetic stock originating from a documented PRRSV-naïve genetic source in combination with the use of semen from a PRRSV-naïve artificial insemination center. As new information became available from the research plan, NGB was then expanded to include protocols to manage mechanical/fomite-based risks of PRRSV, such as contaminated transport, personnel entry, boots and coveralls, and contaminated supplies. This was followed by the application of validated air filtration protocols to control airborne risk and, later, the use of additives capable of inactivating PRRSV in breeding herd diets. Through the application of this comprehensive, science-based approach, PRRSV incidence risk was documented at the incidence of new PRRSV variant infections across all participating herds at < 10% for 2 years (8.6% year 1 and 9.2% year 2) across a system of pig production that included 76 breeding herds and 381,404 sows, despite a historical incidence risk ranging as high as 40–55% per year.

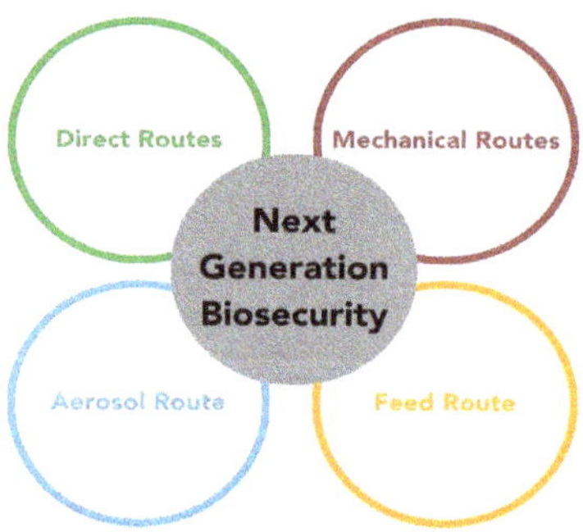

Figure 1. A graphic description of the Next Generation Biosecurity concept and its four pillars. Figure courtesy of Dr. Dee.

6. Conclusions

In closing, while no other pathogen has challenged the biosecurity protocols of swine breeding herds as aggressively as PRRSV, this review indicated that the literature regarding the routes of transmission is comprehensive and complete, and there is evidence that improvements in biosecurity can successfully lower the PRRSV incidence risk across large-scale commercial production for a period. In support of this conclusion, this topic of Next Generation Biosecurity will be revisited later in this Special Issue on Biosecuring Animal Populations, and follow-up data on the PRRSV incidence risk from the third year, as well as a cumulative summary of the incidence risk over all three years of the project will be provided. In addition, the impact of different levels of biosecurity on the PRRSV incidence risk and breeding herd productivity will be discussed. Clearly, progress is being made in preventing PRRSV infection of breeding herds, with the preliminary efforts also being put forward to biosecure wean-to-market herds. For the first time since its emergence, the paradigm is changing from "PRRSV is controlling the swine industry" to "the swine industry is controlling PRRSV", which is clearly a welcome change. We do acknowledge that an inherent limitation of this review is that it is "US-centric" and does not explore biosecurity approaches practiced in other countries. This is an excellent topic for a future publication, and perhaps reading this review will stimulate some activity from others.

Finally, it is the authors' opinions that these are exciting times when the concept of biosecurity has never been more relevant, not only regarding the domestic and transboundary diseases of pigs but to other species groups as well, such as the poultry and dairy industries, which are currently afflicted with highly pathogenic avian influenza virus. Therefore, as viruses will always be drivers of change in animal agriculture, it is time to band together, work as a team, and share our respective knowledge to protect the health and welfare of our global herds and flocks for the benefit of farmers and ranchers.

Author Contributions: Conceptualization, S.D. and S.O.; methodology, S.D., S.O., and M.Y.; validation, S.D., S.O., and M.Y; writing—original draft preparation, S.O., and M.Y.; writing—review and editing, S.O., S.D., and M.Y; visualization, S.O., S.D., and M.Y.; supervision, S.D.; project administration, S.O. All authors have read and agreed to the published version of the manuscript.

Funding: This research received no external funding.

Institutional Review Board Statement: Not applicable.

Informed Consent Statement: Not applicable.

Data Availability Statement: All data are included in the paper.

Acknowledgments: Otake would like to acknowledge Gordon Spronk for sharing his vision, encouragement, and great contribution to the field of swine biosecurity. Dee would also like to thank Spronk for his assistance in guest editing this Special Issue.

Conflicts of Interest: The authors declare no conflicts of interest. Authors Satoshi Otake and Mio Yoshida were employed by the company Swine Extension & Consulting, Inc. The remaining author declares that the research was conducted in the absence of any commercial or financial relationships that could be construed as a potential conflict of interest.

References

1. Holtkamp, D.; Kliebenstein, J.; Neumann, E.; Zimmerman, J.; Rotto, H.; Yoder, T.; Wang, C.; Yeske, P.; Mowrer, C.; Haley, C. Assessment of the economical impact of porcine reproductive and respiratory syndrome virus on United States pork producers. *J. Swine Health Prod.* **2013**, *21*, 72–84. [CrossRef] [PubMed]
2. Studdert, V.P.; Gay, C.C.; Hinchcliff, K.W. *Saunders Comprehensive Veterinary Dictionary*, 5th ed.; Elsevier: St. Luis, MO, USA, 2021; p. 130.
3. Cieslak, T.J., Kortepeter, M.G. A brief history of biocontainment. *Curr. Treat. Options Infect. Dis.* **2016**, *8*, 251–258. [CrossRef] [PubMed]
4. Holtkamp, D.J.; Lin, H.; Wang, C.; O'connor, A.M. Identifying questions in the American Association of Swine Veterinarian's PRRS risk assessment survey that are important for retrospective classifying swine herds according to whether they reported clinical PRRS outbreaks in the previous 3 years. *Prev. Veter Med.* **2012**, *106*, 42–52. [CrossRef] [PubMed]
5. Sasaki, Y.; Furutani, A.; Furuichi, T.; Hayakawa, Y.; Ishizeki, S.; Kano, R.; Koike, F.; Miyashita, M.; Mizukami, Y.; Watanabe, Y.; et al. Development of a biosecurity assessment tool and assessment of biosecurity levels by this tool on Japanese commercial swine farms. *Prev. Veter. Med.* **2020**, *175*, 104848. [CrossRef]
6. Ritter, C.; Jansen, J.; Roche, S.; Kelton, D.F.; Adams, C.L.; Orsel, K.; Erskine, R.J.; Benedictus, G.; Lam, T.J.; Barkema, H.W. Invited review: Determinants of farmers' adoption of management-based strategies for infectious disease prevention and control. *J. Dairy Sci.* **2017**, *100*, 3329–3347. [CrossRef]

7. Vilalta, C.; Arruda, A.G.; Tousignant, S.J.P.; Valdes-Donoso, P.; Muellner, P.; Muellner, U.; Alkhamis, M.A.; Morrison, R.B.; Perez, A.M. A Review of the quantitative tools used to assess the epidemiology of porcine reproductive and respiratory syndrome virus in U.S. swine farms using Dr. Morrison's Swine Health Monitoring Program Data. *Front. Veter. Sci.* **2017**, *4*, 94. [CrossRef]
8. Wensvoort, G.; Terpstra, C.; Pol, J.M.A.; Ter Laak, E.A.; Bloemraad, M.; De Kluyver, E.P.; Kragten, C.; Van Buiten, L.; den Besten, A.; Wagenaar, F.; et al. Mystery swine diseases in the Netherlands: The isolation of Lelystad virus. *Veter. Q.* **1991**, *13*, 121–130. [CrossRef]
9. Meulenberg, J.J. PRRSV, the virus. *Vet. Res.* **2000**, *31*, 11–21. [CrossRef] [PubMed]
10. Nelsen, C.J.; Murtaugh, M.P.; Faaberg, K.S. Porcine reproductive and respiratory virus comparison: Divergent evolution on two continents. *J. Virol.* **1999**, *73*, 270–280. [CrossRef]
11. Chang, C.-C.; Yoon, K.-J.; Zimmerman, J.J.; Harmon, K.M.; Dixon, P.M.; Dvorak, C.M.T.; Murtaugh, M.P. Evolution of porcine reproductive and respiratory syndrome virus during sequential passages in pigs. *J. Virol.* **2002**, *76*, 4750–4763. [CrossRef]
12. Rupasinghe, R.; Lee, K.; Liu, X.; Gauger, P.C.; Zhang, J.; Martínez-López, B. Molecular evolution of porcine reproductive and respiratory syndrome virus field strains from two swine production systems in the Midwestern United States from 2001 to 2020. *Microbiol. Spectr.* **2022**, *10*, e0263421. [CrossRef] [PubMed]
13. Trevisan, G.; Li, G.; Moura, C.A.A.; Coleman, K.; Thomas, P.; Zhang, J.; Gauger, P.; Zeller, M.; Linhares, D. Complete coding genome sequence of a novel porcine reproductive and respiratory syndrome virus 2 restriction fragment length polymorphism 1-4-4 lineage 1C variant identified in Iowa, USA. *Microbiol. Resour. Announc.* **2021**, *10*, e0044821. [CrossRef] [PubMed]
14. Cho, J.G.; Deen, J.; A Dee, S. Influence of isolate pathogenicity on the aerosol transmission of Porcine reproductive and respiratory syndrome virus. *Can. J. Vet. Res.* **2007**, *71*, 23–27. [PubMed]
15. Brar, M.S.; Murtaugh, M.P.; Shi, M.; Leung, F.C.-C. Evolutionary diversification of type 2 porcine reproductive and respiratory syndrome virus. *J. Gen. Virol.* **2015**, *96 Pt 7*, 1570–1580. [CrossRef]
16. Wills, R.; Zimmerman, J.; Yoon, K.-J.; Swenson, S.; McGinley, M.; Hill, H.; Platt, K.; Christopher-Hennings, J.; Nelson, E. Porcine reproductive respiratory syndrome virus: A persist infection. *Veter. Microbiol.* **1997**, *55*, 231–240. [CrossRef]
17. Bierk, M.D.; Dee, S.A.; Rossow, K.D.; Otake, S.; Collins, J.E.; Molitor, T.W. Transmission of porcine reproductive and syndrome virus from persistently infected sows to contact controls. *Can. J. Vet. Res.* **2001**, *65*, 261–266.
18. Christopher-Hennings, J.; Nelson, E.A.; Hines, R.J.; Nelson, J.K.; Swenson, S.L.; Zimmerman, J.J.; Chase, C.C.L.; Yaeger, M.J.; Benfield, D.A. Persistence of porcine reproductive and respiratory syndrome virus in serum and semen of adult boars. *J. Veter. Diagn. Investig.* **1995**, *7*, 456–464. [CrossRef]
19. Swenson, S.L.; Hill, H.T.; Zimmerman, J.J.; Evans, L.E.; Landgraf, J.G.; Wills, R.W.; Sanderson, T.P.; McGinley, M.J.; Brevik, A.K.; Ciszewski, D.K.; et al. Excretion of porcine reproductive and respiratory syndrome virus in semen after experimentally induced infection in boars. *J. Am. Veter. Med. Assoc.* **1994**, *204*, 1943–1948. [CrossRef]
20. Christopher-Hennings, J.; Nelson, E.A.; Hines, R.J.; Swenson, S.L.; Howard, T.H. Detection and duration of porcine reproductive and respiratory syndrome virus in semen by PCR. *J. Clin. Microbiol.* **1995**, *33*, 1730–1734. [CrossRef]
21. Gradil, C.; Dubuc, C.; Eaglesome, M.D. Porcine reproductive and respiratory syndrome virus: Seminal transmission. *Veter. Rec.* **1996**, *138*, 521–522. [CrossRef]
22. Otake, S.; Dee, S.A.; Rossow, K.D.; Joo, H.S.; Deen, J.; Molitor, T.W.; Pijoan, C. Transmission of porcine reproductive and respiratory syndrome virus by needles. *Veter. Rec.* **2002**, *150*, 114–115. [CrossRef]
23. Madapong, A.; Saeng-Chuto, K.; Tantituvanont, A.; Nilubol, D. Safety of PRRSV-2 MLV vaccines administrated via the intramuscular or intradermal route and evaluation of PRRSV transmission upon needle-free and needle delivery. *Sci. Rep.* **2021**, *11*, 23107. [CrossRef]
24. Pirtle, E.C.; Beran, G.W. Stability of porcine reproductive and respiratory syndrome virus in the presence of fomites commonly found on farms. *J. Am. Veter. Med. Assoc.* **1996**, *208*, 390–392. [CrossRef]
25. Dee, S.; Deen, J.; Rossow, K.; Wiese, C.; Otake, S.; Joo, H.S.; Pijoan, C. Mechanical transmission of porcine reproductive and respiratory syndrome virus throughout a coordinated sequence of events during cold weather. *Can. J. Vet. Res.* **2002**, *66*, 232–239. [PubMed]
26. Dee, S.; Deen, J.; Rossow, K.; Weise, C.; Eliason, R.; Otake, S.; Joo, H.S.; Pijoan, C. Mechanical transmission of porcine reproductive and respiratory syndrome virus throughout a coordinated sequence of events during warm weather. *Can. J. Vet. Res.* **2003**, *67*, 12–19. [PubMed]
27. Mesa, V.L.; Munoz, A.Q.; Sobhy, N.M.; Corzo, C.A.; Goyal, S.M. Survival of Porcine Reproductive and Respiratory Syndrome Virus (PRRSV) in the Environment. *Veter. Sci.* **2024**, *11*, 22. [CrossRef]
28. Dee, S.; Deen, J.; Pijoan, C. Evaluation of 4 intervention strategies to prevent the mechanical transmission of porcine reproductive and respiratory syndrome virus. *Can. J. Vet. Res.* **2004**, *68*, 19–26.
29. Otake, S.; Dee, S.A.; Rossow, K.D.; Deen, J. Transmission of porcine reproductive and respiratory syndrome virus by coveralls, boots, and personnel. *J. Swine Health Prod.* **2002**, *10*, 59–65. [CrossRef] [PubMed]
30. Pitkin, A.; Deen, J.; Dee, S. Further assessment of fomites and personnel as vehicles for the mechanical transport and transmission of porcine reproductive and respiratory syndrome virus. *Can. J. Vet. Res.* **2009**, *73*, 298–302.
31. Pitkin, A.; Otake, S.; Dee, S. A one-night downtime period prevents the spread of porcine reproductive and respiratory virus and *Mycoplasma hyopneumoniae* by personnel and fomites (boots and coveralls). *J. Swine Health Prod.* **2011**, *19*, 345–348. [CrossRef]

32. Dee, S.A.; Deen, J.; Otake, S.; Pijoan, C. An experimental model to evaluate the role of transport vehicles as a source of transmission of porcine reproductive and respiratory syndrome virus to susceptible pigs. *Can. J. Vet. Res.* **2004**, *68*, 128–133. [PubMed]
33. Galvis, J.A.; Corzo, C.A.; Machado, G. Modelling and assessing additional transmission routes for porcine reproductive and respiratory syndrome virus: Vehicle movements and feed ingredients. *Transbound. Emerg. Dis.* **2022**, *69*, e1549–e1560. [CrossRef] [PubMed]
34. Dee, S.; Deen, J.; Burns, D.; Douthit, G.; Pijoan, C. An evaluation of disinfectants for the sanitation of porcine reproductive and respiratory syndrome virus-contaminated transport vehicles at cold temperatures. *Can. J. Vet. Res.* **2005**, *69*, 64–70.
35. Dee, S.A.; Torremorell, M.; Thompson, B.; Deen, J.; Pijoan, C. An evaluation of thermos-assisted drying and decontamination for the elimination of PRRSV from contaminated livestock transport vehicles. *Can. J. Vet. Res.* **2005**, *69*, 58–63.
36. Holtkamp, D.J.; Myers, J.; Thomas, P.R.; A Karriker, L.; Ramirez, A.; Zhang, J.; Wang, C. Efficacy of an accelerated hydrogen peroxide disinfectant to inactivate porcine epidemic diarrhea virus in swine feces on metal surfaces. *Can. J. Vet. Res.* **2017**, *81*, 100–107.
37. Zimmerman, J.; Yoon, K.-J.; Pirtle, E.; Wills, R.; Sanderson, T.; McGinley, M. Studies of porcine reproductive and respiratory syndrome (PRRS) virus infection in avian species. *Veter. Microbiol.* **1997**, *55*, 329–336. [CrossRef]
38. Trincado, C.; Dee, S.A.; Jacobson, L.; Rossow, K.D.; Otake, S.; Halvorson, D.; Pijoan, C. Studies about porcine reproductive and respiratory syndrome virus (PRRS) transmission by aerosol and non-porcine vectors under field conditions. In Proceedings of the Annual Meeting of American Association of Swine Veterinarians, Orlando, FL, USA, September 2002; pp. 503–506.
39. Dee, S.A. Biosecurity: A critical review of today's practices. In Proceedings of the 2nd International Swine Disease Eradication Symposium, St. Paul, MN, USA, 14 September 2002; pp. 25–32.
40. Hooper, C.C.; Van Alstine, W.G.; Stevenson, G.W.; Kanitz, C.L. Mice and rats (laboratory and feral) are not a reservoir of PRRS virus. *J. Veter. Diagn. Investig.* **1994**, *6*, 13–15. [CrossRef]
41. Wills, R.W.; Osorio, F.A.; Doster, A.R. Susceptibility of selected non-swine species to infection with PRRS virus. In Proceedings of the31st Annual Meeting of American Association of Swine Practitioners, Indianapolis, IN, USA, 11–14 March 2020; pp. 411–413.
42. Reiner, G.; Fresen, C.; Bronnert, S.; Willems, H. Porcine reproductive and respiratory syndrome virus (PRRSV) infection in wild boars. *Veter. Microbiol.* **2009**, *136*, 250–258. [CrossRef] [PubMed]
43. Choi, E.J.; Lee, C.H.; Hyun, B.H.; Kim, J.J.; Lim, S.L.; Song, J.Y.; Shin, Y.K. A survey of porcine reproductive and respiratory syndrome virus among wild boar population in Korea. *J. Vet. Sci.* **2012**, *13*, 377–383. [CrossRef]
44. Kitamura, Y.; Saito, T.; Tanaka, E.; Takashima, Y. A serological survey of porcine reproductive and respiratory syndrome virus in wild boar in Gifu prefecture, Japan. *J. Veter. Med Sci.* **2022**, *84*, 1406–1409. [CrossRef]
45. Mellor, P.; Wilkinson, P. Experimental transmission of African swine fever virus by *Ornithodoros savignyi* (Audouin). *Res. Veter. Sci.* **1985**, *39*, 353–361. [CrossRef]
46. Clements, A. *The Biology of MOSQUITOES Volume 2: Sensory Reception and Behaviour*; CABI Publishing: New York, NY, USA, 1999; pp. 319–332.
47. Moon, R.D. Muscid flies, Muscidae. In *Medical and Veterinary Entomology*; Academic Press: New York, NY, USA, 2002; pp. 279–301.
48. Otake, S.; Dee, S.A.; Rossow, K.D.; Moon, R.D.; Trincado, C.; Pijoan, C. Transmission of porcine reproductive and respiratory syndrome virus by houseflies (*Musca domestica*). *Vet. Rec.* **2003**, *152*, 73–76. [CrossRef]
49. Otake, S.; Dee, S.A.; Moon, R.D.; Rossow, K.D.; Trincado, C.; Pijoan, C. Survival of porcine and reproductive and respiratory syndrome virus in houseflies (*Musca domestica*). *Can. J. Vet. Res.* **2003**, *67*, 198–203. [PubMed]
50. Otake, S.; Dee, S.A.; Moon, R.D.; Rossow, K.D.; Trincado, C.; Pijoan, C. Studies on the carriage and transmission of porcine reproductive and respiratory virus by individual houseflies (*Musca domestica*). *Veter. Rec.* **2004**, *154*, 80–85. [CrossRef] [PubMed]
51. Schurrer, J.A.; Dee, S.A.; Moon, R.D.; Rossow, K.D.; Mahlum, C.; Mondaca, E.; Otake, S.; Fano, E.; Collins, J.E.; Pijoan, C. Spatial dispersal of porcine reproductive and respiratory syndrome virus-contaminated flies after contact with experimentally infected pigs. *Am. J. Veter. Res.* **2004**, *65*, 1284–1292. [CrossRef]
52. Otake, S.; Dee, S.A.; Rossow, K.D.; Moon, R.D.; Pijoan, C. Mechanical transmission of porcine reproductive and respiratory syndrome virus by mosquitoes (*Aedes vexans*). *Can. J. Vet. Res.* **2002**, *66*, 191–195.
53. Otake, S.; Dee, S.; Moon, R.; Rossow, K.; Trincado, C.; Pijoan, C. Evaluation of mosquitoes (*Aedes vexans*) as biological vectors of porcine reproductive and respiratory syndrome virus. *Can. J. Vet. Res.* **2003**, *67*, 265–270.
54. Dee, S.A.; Martinez, B.C.; Clanton, C. Survival and infectivity of porcine reproductive and respiratory syndrome virus in swine lagoon effluent. *Veter. Rec.* **2005**, *156*, 56–57. [CrossRef] [PubMed]
55. Linhares, D.; Torremorell, M.; Joo, H.S.; Morrison, R. Infectivity of porcine reproductive and respiratory virus in pig manure at different temperature. *Vet. Microbiol.* **2012**, *160*, 23–28. [CrossRef]
56. Guarino, H.; Moura, J.; Cox, R.; Goyal, S.; Patnayak, D. Survival of porcine reproductive and respiratory syndrome virus in fresh pork. *Acta Veter. Hung.* **2014**, *62*, 257–263. [CrossRef]
57. Martin, R.G.; Steverink, P.J.G.M. Oral transmission of PRRS virus via the feeding of infected muscle to pigs. In Proceedings of the 17th International Pig Veterinarian Society Congress, Ames, IA, USA, 2–5 June 2002; p. 215.
58. Bloemraad, M.; De-Kluisver, E.P.; Peterson, A.; Burkhardt, G.B.; Wensvoot, G. Porcine reproductive and respiratory syndrome: Temperature and pH stability of Lelystad virus and its survival in tissue specimens from viremic pigs. *Vet. Microbiol.* **1994**, *42*, 361–371. [CrossRef] [PubMed]

59. Mager, R.; Robinson, Y.; Dubuc, C.; Larochelle, R. Evaluation of the persistence of porcine reproductive and respiratory syndrome virus in pig carcasses. *Vet. Rec.* **1995**, *137*, 559–561. [CrossRef] [PubMed]
60. Cano, J.P.; Murtaugh, M.P.; Dee, S.A. Evaluation of the survival of porcine reproductive and respiratory syndrome virus in non-processed pig meat. *Veter. Rec.* **2007**, *160*, 907–908. [CrossRef]
61. Alexander, T. A winter of windborne spread. *Br. Veter. J.* **1993**, *149*, 507–510. [CrossRef]
62. Torremorell, M.; Pijoan, C.; Janni, K.; Walker, R.; Joo, H.S. Airborne transmission of Actinobacillus pleuropneumoniae and porcine reproductive and respiratory syndrome virus in nursery pigs. *Am. J. Veter. Res.* **1997**, *58*, 828–832. [CrossRef]
63. Dee, S.A.; Deen, J.; Jacobson, L.; Rossow, K.D.; Mahlum, C.; Pijoan, C. Laboratory model to evaluate the role of aerosols in the transport of porcine reproductive and respiratory syndrome virus. *Veter. Rec.* **2005**, *156*, 501–504. [CrossRef]
64. Otake, S.; Dee, S.; Corzo, C.; Oliveira, S.; Deen, J. Long-distance airborne transport of infectious PRRSV and Mycoplasma hyopneumoniae from a swine population infected with multiple viral variants. *Veter. Microbiol.* **2010**, *145*, 198–208. [CrossRef]
65. Dee, S.; Otake, S.; Deen, J. Use of a production region model to assess the efficacy of various air filtration systems for preventing airborne transmission of porcine reproductive and respiratory syndrome virus and Mycoplasma hyopneumoniae: Results from a 2-year study. *Virus Res.* **2010**, *154*, 177–184. [CrossRef] [PubMed]
66. Dee, S.; Brands, L.; Nerem, J.; Schelkopf, A.; Spronk, G.; Kikuti, M.; Corzo, C.; Havas, K. Improvements in swine herd biosecurity reduce the incidence risk of porcine reproductive and respiratory syndrome virus in breeding herds in the Midwestern United States. *J. Am. Veter. Med. Assoc.* **2023**, *262*, 520–525. [CrossRef]
67. Dee, S.; Cano, J.P.; Spronk, G.; Reicks, D.; Ruen, P.; Pitkin, A.; Polson, D. Evaluation of the long-term effect of air filtration on the occurrence of new PRRSV infections in large breeding herds in swine-dense regions. *Viruses* **2012**, *4*, 654–662. [CrossRef]
68. Dee, S.A.; Clement, T.; Schelkopf, A.; Nerem, J.; Knudsen, D.; Christopher-Hennings, J.; Nelson, E. An evaluation of contaminated complete feed as a vehicle for porcine epidemic diarrhea virus infection of naïve pigs following consumption via natural feeding behavior: Proof of concept. *BMC Vet. Res. Vol.* **2014**, *10*, 176. [CrossRef] [PubMed]
69. Dee, S.; Shah, A.; Cochrane, R.; Clement, T.; Singrey, A.; Edler, R.; Spronk, G.; Niederwerder, M.; Nelson, E. Use of a demonstration project to evaluate viral survival in feed: Proof of concept. *Transbound. Emerg. Dis.* **2021**, *68*, 248–252. [CrossRef]
70. Jones, C.K.; Woodworth, J.; Dritz, S.S.; Paulk, C.B. Reviewing the risk of feed as a vehicle for swine pathogen transmission. *Veter. Med. Sci.* **2020**, *6*, 527–534. [CrossRef] [PubMed]
71. Dee, S.A.; Bauermann, F.V.; Niederwerder, M.C.; Singrey, A.; Clement, T.; Lima, M.D.; Long, C.; Patterson, G.; Sheahan, M.A.; Stoian, A.M.M.; et al. Survival of viral pathogens in animal feed ingredients under transboundary shipping models. *PLoS ONE* **2018**, *13*, e0194509. [CrossRef] [PubMed]
72. Dee, S.A.; Niederwerder, M.C.; Edler, R.; Hanson, D.; Singrey, A.; Cochrane, R.; Spronk, G.; Nelson, E. An evaluation of additives for mitigating the risk of virus-contaminated feed using an ice-block challenge model. *Transbound. Emerg. Dis.* **2020**, *68*, 833–845. [CrossRef]
73. Dee, S.; Clement, T.; Nelson, E. Transmission of porcine reproductive and respiratory syndrome virus in domestic pigs via oral ingestion of feed material. *J. Am. Veter. Med. Assoc.* **2023**, *262*, 1–4. [CrossRef]
74. Spronk, G.; Havas, K.; Patterson, G.; Dee, S.A. Will swine veterinarians lead by meeting the next generation needs of our industry? *J. Am. Vet. Med. Assoc.* **2022**, *261*, 424–429. [CrossRef]
75. Torremorell, M.; Baker, R. Eradication of PRRS virus by changing the pig flow and introduction of negative replacements into positive sow farms. *Proc. Allen D. Leman Swine Conf.* **2000**, *27*, 59–62.
76. Dee, S.A.; Molitor, T.W. Elimination of PRRS virus using a test and removal process. *Vet. Rec.* **1998**, *143*, 474–476. [CrossRef]
77. Dee, S.A.; Bierk, M.D.; Deen, J.; Molitor, T.W. An evaluation of test and removal for elimination of porcine reproductive and respiratory syndrome virus from 5 swine farms. *Can. J. Vet. Res.* **2001**, *65*, 22–27.
78. Corzo, C.A.; Mondaca, E.; Wayne, S.; Torremorell, M.; Dee, S.; Davies, P.; Morrison, R.B. Control and elimination of porcine reproductive and respiratory syndrome virus. *Virus Res.* **2010**, *154*, 185–192. [CrossRef] [PubMed]
79. Havas, K.A.; Brands, L.; Cochrane, R.; Spronk, G.D.; Nerem, J.; Dee, S.A. An assessment of enhanced biosecurity interventions and their impact on porcine reproductive and respiratory syndrome virus outbreaks within a managed group of farrow-to-wean farms, 2020–2021. *Front. Veter. Sci.* **2023**, *9*, 952383. [CrossRef] [PubMed]
80. Lambert, M.-È.; Poljak, Z.; Arsenault, J.; D'Allaire, S. Epidemiological investigations in regards to porcine reproductive and syndrome (PRRS) in Quebec, Canada. Part 1: Biosecurity practices and their geographical distribution in two areas of different swine density. *Prev. Veter. Med.* **2012**, *104*, 74–83. [CrossRef] [PubMed]

Disclaimer/Publisher's Note: The statements, opinions and data contained in all publications are solely those of the individual author(s) and contributor(s) and not of MDPI and/or the editor(s). MDPI and/or the editor(s) disclaim responsibility for any injury to people or property resulting from any ideas, methods, instructions or products referred to in the content.

animals MDPI

Article

A Dynamic System to Control the Entry of Non-Authorized Visitors and Detect Superspreader Farms in Strongly Interconnected Systems

Oscar Soriano [1], Laura Batista [2], Joaquin Morales [1], Eduardo Quintana [1] and Carlos Piñeiro [1,*]

1 Animal Data Analytics, SL, 40006 Segovia, Spain; oscar.soriano@ada-animaldata.com (O.S.); joaquin.morales@ada-animaldata.com (J.M.); eduardo.quintana@ada-animaldata.com (E.Q.)
2 Batista & Asociados, Sainte-Julie, QC J3E 0C6, Canada; laurabatistagarcia@gmail.com
* Correspondence: carlos.pineiro@ada-animaldata.com; Tel.: +34-629742031

Simple Summary: This paper addresses the need for effective biosecurity due to disease challenges, the emergence and re-emergence of new pathogens, the growing pressure of antibiotic restrictions, and the welfare and sustainability of livestock farming. It introduces the Biorisk® External platform, a cloud-based visitor control system that enhances data management and biosecurity compliance protocols. This system analyzes visitation patterns, revealing trends and categorizing visits by authorization and risk status. This network analysis also allows for the identification of 'superspreader' farms that represent a high epidemiological risk. The results advocate for integrating technology into biosecurity protocols to optimize standard operation procedures (SOPs) to improve animal health and mitigate economic losses, highlighting the importance of data-driven decision-making in modern livestock farming.

Abstract: This study explores the critical challenges the livestock sector faces, particularly those related to biosecurity, animal welfare, and antibiotic use restrictions. It highlights the need to implement advanced information and communication technologies to enhance operational sustainability and decision-making. We introduce the Biorisk® External platform, a cloud-based visit control system designed to optimize biosecurity management by accurately tracking visitor activity through QR codes and GPS geolocation. During a 6-month study period from July to December 2023, we analyzed visits to 142 different swine production sites and 30 vehicle movement patterns. The analysis revealed trends in visitation patterns and compliance with biosecurity SOPs. The software categorized visits as authorized (A), not authorized with access (NAWA), and not authorized without access (NAWOA), providing a framework to assess biosecurity risks. Additionally, network analysis identified interconnected farms, which were classified as 'superspreaders', highlighting their considerable risk of disease transmission. This study advocates for the integration of digital systems in livestock operations to improve biosecurity measures, facilitate real-time data input, and support informed decision-making. By enhancing biosecurity protocols through technology, the livestock industry can better safeguard animal health, increase operational efficiency, and reduce potential economic losses associated with disease outbreaks.

Keywords: biosecurity; information technology; livestock management

1. Introduction

The agricultural sector is currently facing significant challenges, with health considerations, such as increasing pressure due to antibiotic restrictions and the need to ensure the welfare and sustainability of livestock farming, being of paramount importance. Consequently, the implementation of intelligent biosecurity (IB) is non-negotiable since it is clear, as stated by Capper [1], that we will not move towards a better industry if we do not eliminate the diseases that are causing extreme losses to the industry. Intelligent biosecurity's goal is to improve animal health, welfare, and productivity through real-time traceability. Intelligent biosecurity gathers and screens information for signs of emerging issues to help anticipate problems through the analysis of collected data to minimize the threat of any future introduction of pathogens. In addition, IB utilizes technology and data-driven methods to monitor, analyze, and optimize different aspects of livestock production, and very importantly, it is an educational tool for all stakeholders in the swine industry, enhancing their understanding of effective biosecurity to change misguided conduct prototypes. The judicious use of information is crucial for making informed decisions that enhance the sustainability of businesses. In this sense, information and communication technologies play a pivotal role in companies' operations. A Deloitte report [2] described the potential impact of digitalization in meat chain production. In a comparable way, Gebauer et al. [3] emphasized the critical role of IT in boosting company revenues and productivity, while Drnevich and Croson [4] underscored the transformative role of technology in shaping business strategies, fostering novel value propositions, and enhancing the ability of companies to seize emerging opportunities.

Health status heavily influences the continuity of businesses, particularly concerning virus-related diseases [5]. These encompass devastating conditions like African swine fever (ASF) or foot-and-mouth disease (FMD), as well as epidemics such as porcine reproductive and respiratory syndrome (PRRS) or porcine epidemic diarrhea (PED) in swine and the highly pathogenic avian influenza or Newcastle disease in poultry. The impact of PRRS on business profitability has been extensively studied: Holtkamp et al. [6] estimated losses in the American swine industry associated with PRRS at USD 664 million annually, while in Europe, Nieuwenhuis et al. [7] identified economic losses of EUR 126 per sow due to the same disease. More recent work by Nathues et al. [8] reported losses of EUR 126.79 per sow per year and EUR 3.77 per fattening pig.

Moreover, there is growing pressure to restrict antibiotic use and improve animal welfare standards. In this context, biosecurity is a key component of this equation [9]. Historically, surveys have served as the primary method to drive improvements in biosecurity protocols [10–12]. Surveys conducted through standardized methods may fall short of addressing the human factor, which introduces a considerable degree of subjectivity. Conclusions drawn from surveys can be influenced by the surveyor's expertise and the respondents' subjectivity.

Ensuring compliance with biosecurity protocols presents challenges, including the reliability of information, the scope for data-driven training, the value of information, the necessary infrastructure, and operational ease, which surveys, farm audits, and surveillance cameras cannot entirely tackle. Sensor-based technology emerges as a promising solution for overcoming these challenges [13]. Leveraging information technologies in biosecurity enhances data quantity and quality, shortens response times—crucial for disease control and containment across farms—and reduces human biases, thereby boosting the reliability of conclusions. Consequently, this approach enhances the efficacy of action plans, lowers costs, and increases business profit margins.

External biosecurity focuses on shielding farms from the entry of diseases, a concern increasingly linked to transport, animals, feed, people, or material. The supervision of incoming visits to farms proves crucial in this regard, with the literature underscoring vehicles and individuals as primary sources of pathogen dissemination between farms [14–19].

This study establishes a framework for on-farm biosecurity analysis using a cloud-based platform tool. In addition to traditional numerical and descriptive statistical analyses, this study aims to determine patterns of visits to different facilities. This initial investigation highlights trends indicative of deficient farm visit management or non-compliance with biosecurity SOPs. Future research will include a network analysis that will regionally point out farms that require the most surveillance and changes in biosecurity SOPs.

2. Materials and Methods

Data collection is a key phase in any information system and has historically posed challenges in the livestock industry, representing one of the most important bottlenecks. While some areas like reproduction, health, and productive performance have generated effective metrics [20,21], the scenario differs significantly in biosecurity compliance. Ensuring adherence to biosecurity protocols represents a unique challenge for each farm or productive system. This issue is particularly challenging regarding visitor management, i.e., individuals and vehicles. Existing registration methods are paper-based, thus preventing non-compliant visits, tracking routes, or foreseeing potential risk issues difficult to accurately monitor any farm or system. With a combination of health, production, and swine industry consultants' expertise, we have developed an efficient data collection system:

1. Data collection system: Biorisk® External (Animal Data Analytics, SL; Segovia, Spain), a digital visitor control system, is a web application developed in .NET 5 framework by Animal Data Analytics SL in Segovia, Spain, that operates as Software as a Service (SaaS) in the cloud. Users can manually register visits via a cell phone using site-specific QR codes or leverage automatic registration utilizing GPS geolocation data from company or suppliers' vehicles. Biosecurity rules and downtime movements between sites are customized based on the farm PRRS status [22]. If the farm status changes, the system updates each visitor accordingly with the associated risks, classifying them into three categories: authorized (A), not authorized with access (NAWA), and not authorized without access (NAWOA).
2. Report generation: The application generates reports showing different metrics for the system's routine management, including alerts, summaries, distributions, outbreak traceability, and its evolution. It can also include a timeline of visits and their category. All this information can be easily exported to perform the network analysis.
3. Period: The system was implemented in two Spanish swine companies with a total of 142 sites and 30 vehicles (Table 1). Initially, a pilot project was conducted from September 2022 to July 2023, involving 30 sites and 4 vehicles. Thereafter, Biorisk® External was implemented in all sites of the two companies and visits to all sites were tracked up to December of 2023. Figure 1 shows a Biorisk® External map of the traced sites.

Figure 1. Biorisk® External map of the traced sites. (Figure 1 was obtained from Biorisk® External platform (2024) and reproduced with permission from Animal Data Analytics, SL).

Table 1. Site and vehicle categories.

Category	Number
Cleaning and disinfection centers	14
Feed mills	10
Sites I	24
Sites I and II	63
Sites III	24
Boar studs	2
Nucleus farms	3
Multipliers farms	2
Vehicles (no categorization)	30

2.1. Statistical Analysis

The methodology for each of the analyses performed is described below. For each of them, R statistical software version 4.1.2 was used.

2.1.1. Visit Evolution Timeline

The visits were studied qualitatively through time series plotting using the ggplot, plotly, and ggthemes libraries. In addition, the empirical 95% confidence interval was calculated using the 0.025 and 0.975 quantiles. This interval was validated using the Dvoretzky–Kiefer–Wolfowitz theorem (DKW inequality):

$$P\left(\sup_{x\in R}|F(x) - F_n(x)| > \varepsilon\right) \leq 2e^{2n\varepsilon^2}, \qquad \forall\varepsilon > 0$$

2.1.2. Comparative Study

The bar and sunburst plots were compared using the ggplot, ggthemes, and plotly libraries. The comparisons included the average number of visits per day of the week, and the type of visit (A, NAWA, and NAWOA). Also, the rate of visits per role (veterinarian, visitor, maintenance, others) and type of visit; the rate of visits per type of vehicle and type of visit; and the rate of visits per day of the week and type of vehicle were plotted. In turn, a quantitative study was carried out by performing a contrast of proportions. To perform this task, we used the R function prop.test which executes the Newcombe–Willson test.

2.1.3. Network Study

A graph is a pair $(V,\ E)$ where V is a set of generic elements, denoted by nodes, and $E \subset V \times V$ is a subset of pairs of elements of V, commonly denoted by a set of edges. An edge $(v_i,\ v_j) \in E$ symbolizes that the node v_i is related to the node v_j; the reciprocal relationship exists if and only if $(v_j,\ v_i) \in E$. Let $V' \subset V$; the induced subgraph is defined as $(V',\ E \cap (V' \times V'))$.

Given a graph $(V,\ E)$, it is said to be undirected if for every $(v_i,\ v_j) \in E$, $(v_j,\ v_i) \in E$. On the other hand, it is said to be weighted if there exists an application $\phi :\ E \to R$ such that $\phi(v_i,\ v_j) = \lambda_{i.j}$.

Given a graph (V,E), its adjacency matrix is defined as $A = (a_{i,j})_{1\leq i,j\leq \#V}$, where we have

$$a_{i,j} = \begin{cases} 1\ si\ (v_i,v_j) \in E \\ 0\ si\ (v_i,v_j) \notin E \end{cases}$$

if the graph is not weighted, or

$$a_{i,j} = \begin{cases} \phi(v_i,\ v_j) = \lambda_{i,j}\ si\ (v_i,v_j) \in E \\ 0\ si\ (v_i,v_j) \notin F \end{cases}$$

if the graph is weighted by an application ϕ.

Given a graph (V, E) with an adjacency matrix $A = (a_{i,j})$, the *degree* of a vertex $v_i \in V$ is defined as

$$\deg(v_i) = \sum_{j=0}^{\#V} a_{i,j}$$

Given a graph $(V,\ E)$ and $v_i,\ v_j \in\ V$, these are connected if $v_i = v_j$, $(v_i,\ v_j) \in\ E$ or if there is a path connecting v_i and v_j. The connection defines an equivalence relation in which the subgraphs induced by such equivalence classes are known as connected components associated with the graphs. A graph is a disjoint union of all its connected components.

Let (V, E) a graph, two vertices v_i, v_j are connected by a path if there exist vertices $v_{k_1}, \ldots v_{k_m} \in V$ such that $(v_i, v_{k_1}), (v_{k_1}, v_{k_2}), \ldots, \left(v_{k_{m-1}}, v_{k_m}\right), (v_{k_m}, v_j) \in E$. A graph is said to be strongly connected if there is a path connecting two pairs of vertices.

Being $\sigma(A)$ the spectrum (i.e., the set of eigenvalues) of the adjacency matrix, and supposing that it is non-negative irreducible, since the graph is strongly connected, then, by the Perron–Frobenius theorem, it exists a strictly dominant eigenvalue $\hat{\lambda} \in \sigma(A)$ and an eigenvector $x_{\hat{\lambda}}$ whose coordinates are all positive. The eigenvalue centrality associated with node v_i is the i-th coordinate of $x_{\hat{\lambda}}$. It must be considered that eigenvalue centrality is a qualitative measure of the importance of a vertex over a strongly connected graph, independently of the scale. Given an undirected graph $(V,\ E)$ with adjacency matrix A, its lagrangian matrix is defined as

$$L = A - (\deg(v_1),\ \ldots, \deg(v_m)) \cdot\ I$$

The subspace of eigenvectors associated with the null eigenvalue of the lagrangian matrix has a basis of vectors $\beta = < x_1,\ \ldots, x_n >$ satisfying the following properties:

- The vectors x_i have only ones or zeros for their coordinates;
- If the vector x_i has a 1 in the j-th coordinate, then the rest of the vectors in the basis have zeros in that coordinate.

Each of these vectors is in correspondence with one of the connected components, which are defined as the subgraphs generated by the vertex set associated with the non-zero coordinates of the vectors (i.e., if $x_1 = (1,\ 1,\ 1,\ 0,\ 0)$ is one of the above vectors, then the graph generated by the vertex set $\{v_1,\ v_2,\ v_3\}$ would be one of the connected components).

It should be noted that, by definition, the connected components of a graph are themselves strongly connected graphs; therefore, it makes sense to speak of eigenvalue centrality in each of these. In the context of this study, set V consists of the sites monitored by Biorisk® External (Animal Data Analytics, SL; Segovia, Spain). Two sites $A,\ B \in\ V$, are considered epidemiologically related if a vehicle has gone from A to B. This fact will be expressed as $(A,\ B) \in\ E \subset\ V \times V$.

Given two epidemiologically connected sites, we define the weighting on $\phi(A, B)$ as the number of times a vehicle has gone from A to B plus the number of times that the vehicle has gone from B to A. The adjacency matrix associated with this problem is symmetric; therefore, the graph to be studied is undirected. Once the adjacency matrix has been composed, the lagrangian matrix is calculated to obtain the connected components of the network. In the physical sense, each connected component is understood as a zone of interaction between sites.

Finally, the eigenvalue centrality is calculated for each of the connected components. A superspreader within a connected component will be the farm or farms with the highest eigenvalue centrality, which translates as those farms that epidemiologically interact the most with the others.

3. Results

- Visit analysis

The number of visits was calculated by detecting the vehicle plate number, the visited site, and visit date. The dispersion of daily visits was studied with a 95% empirical confidence interval, which resulted in 1 to 60 visits in one day. In addition, this interval was contrasted using the DKW inequality, which found that the probability of deviation more than one visit from the theoretical confidence interval is lower or equal $2 \cdot \exp(-2 \cdot 7812 \cdot 1) << 0.0001$.

Based on these results, visits were plotted on a daily time series. Figure 2 shows the days with unusual personnel flow, and its trend was calculated by local regression. This analysis also shows, first, an increased visit pattern starting in July of 2023, second, unusually high activity compared to the pre-established level from the end of September to the beginning of November, and third, the sigmoid curve with stabilization at the end of the study period. After the pilot project, the system was implemented in all the company sites as shown in the stabilization curve to the end of December 2023 (Figure 2).

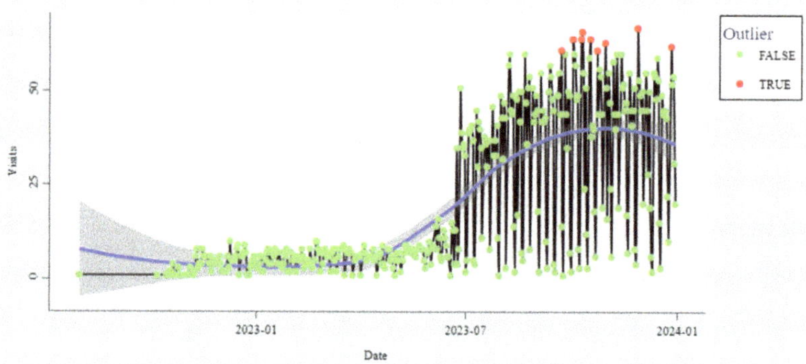

Figure 2. Visit timeline (number).

Visits were also analyzed by date and weekday (Table 2 and Figure 3).

Table 2. Days with unusually high visit activity.

Date	Number of Visits
22 September 2023	61
02 October 2023	64
09 October 2023	64
10 October 2023	66
17 October 2023	64
23 October 2023	61
30 October 2023	63
27 November 2023	67
26 December 2023	62

The Newcombe–Willson test results show a lower number of visits on Saturdays and Sundays ($p << 0.05$), in agreement with regular farm standards; Mondays have the highest visit percentage l ($p << 0.05$), and Fridays have significantly more visits than Tuesdays, Wednesdays, and Thursdays ($p = 0.031$). No significant differences were detected between Tuesday, Wednesday, and Thursday ($p = 0.549$).

Figure 4 shows a sunburst chart of visit distribution by category: authorized (A), not authorized with access (NAWA), and not authorized without access (NAWOA). The results of the Newcombe–Willson test show that category A visits, which mostly consist of veterinarian visits, live animal introduction, and feed delivery ($p << 0.05$), are the principal category ($p << 0.05$), as can be seen in Figure 5.

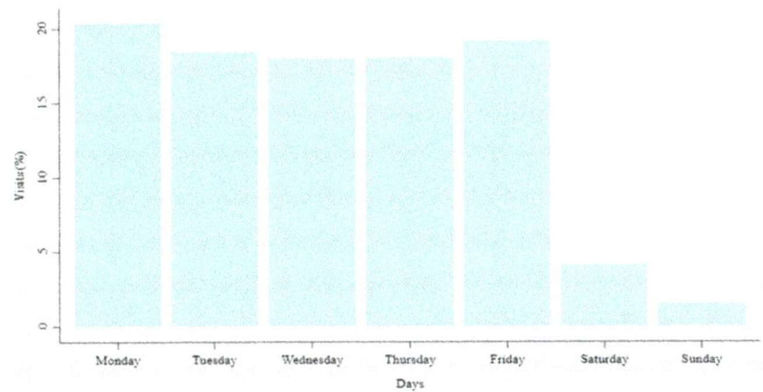

Figure 3. Percentage distribution per weekday.

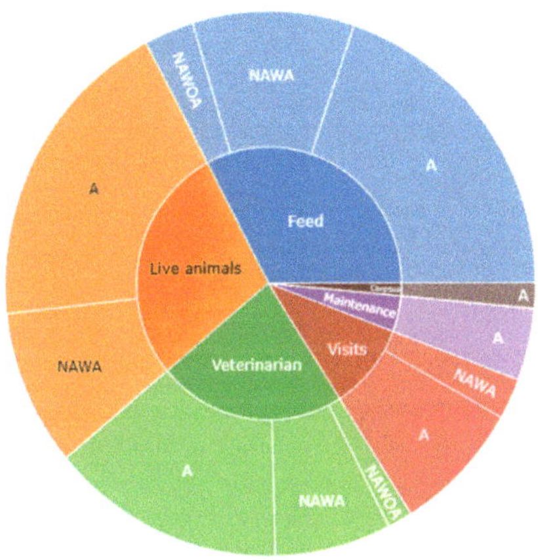

Figure 4. Visit sunburst chart by categories.

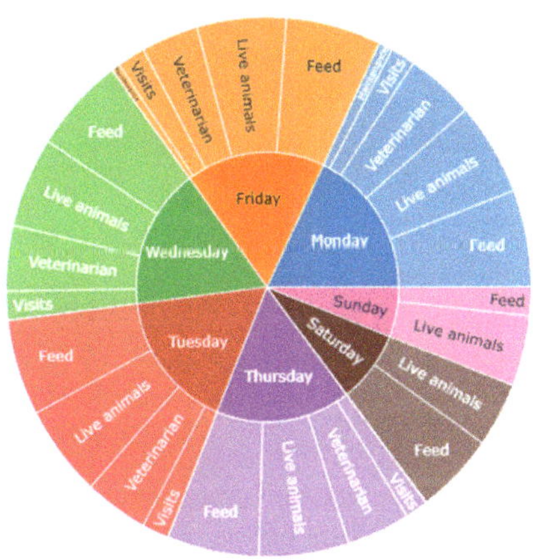

Figure 5. Weekday and category sunburst chart.

The implementation of Biorisk® External (Animal Data Analytics, SL; Segovia, Spain) resulted in a significant improvement in data collection and its characterization. As shown in Figure 5, the digitalization of visitor logs facilitated timely and accurate tracking of biosecurity protocols and their compliance, therefore reducing the risk of pathogen introduction into a farm.

- Route Analysis

In this study, a route was defined as a pair of time-ordered visits between two different sites performed by the same single vehicle on the same day. Figure 6 shows that the company's sites were distributed in eight connected groups. Since each component is strongly connected, by construction, we calculated the eigenvector centrality for each site. As mentioned before, eigenvector centrality indicates the influence of a site within its group, the greater it is, the more likely that it will be affected by a new epidemiological relationship. As shown in Figure 7, a farm was considered a superspreader if its centrality was above the third quantile of its group, therefore representing a higher risk of disease dissemination.

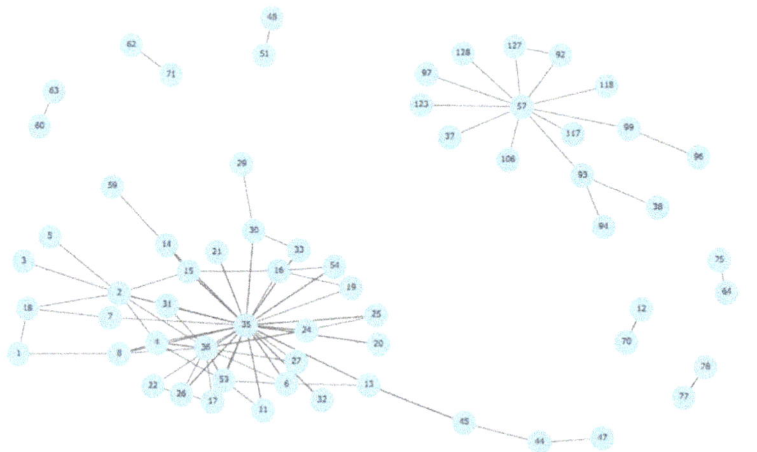

Figure 6. Graph route epidemiological relationships between sites (Each sphere represents a site and its identification number).

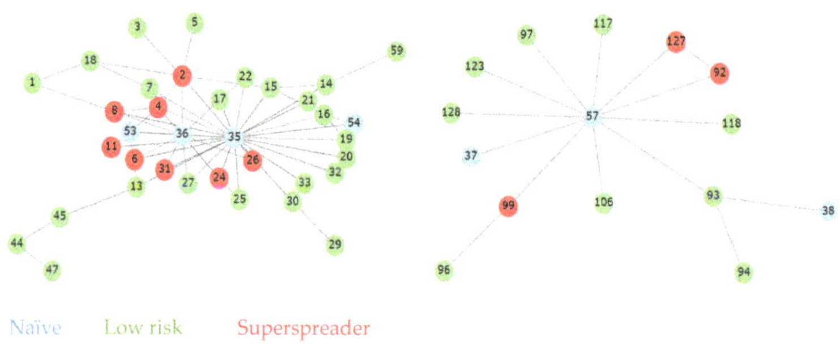

Figure 7. Qualitative representation of massive components (Each sphere represents a site and its identification number).

4. Discussion

As the livestock production sector faces health, sustainability, and regulatory compliance pressures, integrating advanced data management systems emerges as an imperative strategy to promote animal welfare and efficiency. Traditional paper-based systems have long been identified as inadequate due to human error and inefficiency in capturing real-time data. Consequently, the findings of this study highlight the increasing role of incorporating information and communication technologies to reinforce biosecurity protocols within the sector. The use of Biorisk® External (Animal Data Analytics, SL; Segovia, Spain) improves the accuracy and the quality of people and vehicle movement data management. The digitalization of visitor logs not only facilitates data collection, but it also allows for an immediate response to biosecurity threats. The transition to a cloud-based framework enables stakeholders to make objective data-driven assessments and decisions which are necessary in an era characterized by rapid technological advancement.

Categorizing visits provides a framework to evaluate compliance and determine areas which require investigation and intervention. Furthermore, the analysis of visitor patterns reveals crucial insights into operational workflows and SOP compliance in livestock-producing companies. Our results support the studies of [9–11], who concluded that human behavior plays a crucial role in the effectiveness of biosecurity measures. Therefore, the continuous training and education of personnel becomes essential in minimizing the risks associated with authorized and non-authorized visits. Additionally, the network analysis of epidemiological connections among farms developed in our study introduces a novel approach to understanding the dynamics of disease transmission pathways. Identifying farms as 'superspreaders' provides actionable intelligence for prioritizing resources and interventions. Previous studies have shown the importance of minimizing contact between farms to prevent the spread of diseases [14,15]. This innovative epidemiological mapping of relationships allows the establishment of biosecurity buffer zones and comprehensive vehicle movement protocols to understand and avoid potential lateral pathogen transmission.

New mathematical models based on graphs and dynamic systems have made great contributions to epidemiological studies; however, some generated incorrect conclusions. By transforming the adjacence matrix into a generalized symmetric adjacence matrix and calculating the eigenvalue centrality for each connected component in the system, we avoid incorrect conclusions, thus accurately identifying the 'superspreader' farms for every group of epidemiological-related sites.

Integrating quantitative data and qualitative observations goes beyond theoretical information, thus allowing farm managers to implement accurate biosecurity interventions, respond proactively to emerging threats, and engage in strategic planning to mitigate risks. The capacity to generate timely reports and alerts within the Biorisk® External system (Animal Data Analytics, SL; Segovia, Spain) enhances operational alertness, allowing for real-time decision-making in a permanently changing environment. In addition, this system allows us to understand the high impact of continuous education on biosecurity decisions related to livestock disease dynamics [23].

The next steps in our project will include other areas of farm management, such as slurry and mortality management, as the present management practices pose significant pathogen transmission risks. Therefore, the traceability of vehicles involved in these tasks could also be highly relevant. Additionally, it is important to assess how the biosecurity information and protocols provided by each system affect compliance with effective biosecurity regulations. As Bucini et al. [23] noted, individual risk perception influences adherence to these regulations. Thus, risk characterization based on objective data can influence behavioral changes that enhance overall farm biosecurity.

In summary, this research highlights the transformative potential of integrating advanced data management systems in the livestock sector. By enhancing biosecurity protocols through informed decision-making, the industry can better protect animal health, ensure sustainability, and mitigate the significant economic losses associated with disease outbreaks. The findings of this study advocate for a change in our biosecurity mindset, emphasizing the critical combination of technology, data analytics, and proactive operational strategies in overcoming the multiple challenges posed by modern livestock farming.

5. Conclusions

This study highlighted the necessity of transitioning conventional paper-based recording to digital frameworks to ensure reliable data capture. The successful implementation of Biorisk® External (Animal Data Analytics, SL; Segovia, Spain) optimizes the combination of advanced information and communication technologies that can significantly improve biosecurity protocols. This system can be used in any livestock sector, thus shielding operations from the risk of pathogen introduction. Also, farm managers will be able to make informed decisions to optimize resource allocation and operational practices by utilizing accurate and real-time data according to visitor patterns in their system. This effective data management will enhance their ability to quickly respond to potential biosecurity threats, reducing economic losses and protecting livestock welfare. Also, the results highlight the critical need for continuous farm personnel education and training regarding biosecurity measures. Mitigating human-related risks associated with unauthorized visits is essential, suggesting that educational programs should be integrated into farm management practices to foster a culture of compliance and awareness. New knowledge on how to identify 'superspreader' farms underlines the importance of recognizing interconnectedness among farms. Knowledge that enables stakeholders to prioritize monitoring and intervention strategies for specific facilities enhances the overall resilience of livestock operations against infectious diseases. Finally, our results advocate for developing strategies that support adopting innovative technologies in agriculture.

Future research should continue to explore the positive long-term impact of implementing these technological tools on both animal health improvement and economic performance, fostering a deeper understanding of the interaction between technology, compliance, and sustainability of biosecurity in the livestock industry.

Author Contributions: Conceptualization, O.S. and C.P.; methodology, L.B. and J.M.; software, E.Q. and C.P.; validation, O.S., L.B. and J.M.; formal analysis, E.Q.; investigation, O.S., J.M. and C.P.; data curation, E.Q.; writing—original draft preparation, C.P.; writing—review and editing, O.S., L.B., J.M. and E.Q.; visualization, J.M.; supervision, L.B. and C.P.; project administration, O.S.; funding acquisition, C.P. All authors have read and agreed to the published version of the manuscript.

Funding: This research was funded by the Ministry of Industry, Commerce and Tourism of the Government of Spain (ORDER ICT/1117/2021 modified by ORDER ICT/474/2022), and by the European Union under the Horizon Europe grant 101083923 (BIOSECURE). The views and opinions expressed are, however, those of the authors only and do not necessarily reflect those of the European Union or the European Research Executive Agency (REA). Neither the European Union nor the granting authority can be held responsible for them.

Institutional Review Board Statement: Ethical review and approval were waived for the present study as the project did not include procedures with animals according to EC/2010/63.

Informed Consent Statement: Not applicable.

Data Availability Statement: The data presented in this study are available on request from the corresponding author. The data are not publicly available due to confidentiality agreements with the pig companies involved in the study.

Conflicts of Interest: The authors O.S., J.M., E.Q. and C.P. are employees of Animal Data Analytics that participated in the study. The authors declare no conflicts of interest. Animal Data Analytics partially financed the project. The funders had no role in the design of the study; in the collection, analyses, or interpretation of data; in the writing of the manuscript, or in the decision to publish the resuts.

References

1. Capper, J.L. The impact of controlling diseases of significant global importance on greenhouse gas emissions from livestock production. *One Health Outlook* **2023**, *5*, 17. [CrossRef] [PubMed]
2. Gebauer, H.; Arzt, A.; Kohtamäki, M.; Lamprecht, C.; Parida, V.; Witell, L.; Wortmann, F. How to convert digital offerings into revenue enhancement–Conceptualizing business model dynamics through explorative case studies. *Ind. Mark. Manag.* **2020**, *19*, 429–441. [CrossRef]

3. Wehberg, G.; Vaessen, W.; Nijland, F.; Berger, T. Smart Livestock Farming: Potential of Digitalization for Global Meat Supply (No. 11). Discussion Paper; Deloitte. 2017. Available online: https://www.google.com/url?sa=t&rct=j&q=&esrc=s&source=web&cd=&ved=2ahUKEwiJ8LDnvZOBAxXItKQKHSGaA68QFnoECBcQAQ&url=https://www2.deloitte.com/content/dam/Deloitte/de/Documents/operations/Smart-livestock-farming_Deloitte.pdf&usg=AOvVaw1OOpLRI7FC6BQQGzZU0ZCG&opi=89978449 (accessed on 28 August 2024).
4. Drnevich, P.L.; Croson, D.C. Information technology and business-level strategy: Toward an integrated theoretical perspective. *MIS Q.* **2013**, *37*, 483–509. [CrossRef]
5. Pan, C.; McCracken, C.; Gocsik, E. *Rabobank Global Pork Quarterly Q4 2023 Report*; RaboResearch Food & Agribusiness: Amsterdam, The Netherlands, 2023.
6. Holtkamp, D.; Kliebenstein, J.B.; Yoder, K.J.; Polson, D.D.; Mardassi, H.; Torremorell, M.; Thacker, E.L. Assessment of the economic impact of porcine reproductive and respiratory syndrome virus on United States swine production. *J. Swine Health Prod.* **2013**, *21*, 3–12.
7. Nieuwenhuis, N.; Duinhof, T.F.; van Nes, A. Economic analysis of outbreaks of porcine reproductive and respiratory syndrome virus in nine sow herds. *Vet. Rec.* **2012**, *170*, 225. [CrossRef]
8. Nathues, H.; Alarcon, P.; Rushton, J.; Jolie, R.; Fiebig, K.; Jimenez, M.; Geurts, V.; Nathues, C. Cost of porcine reproductive and respiratory syndrome virus at individual farm level—An economic disease model. *Prev. Vet. Med.* **2017**, *142*, 16–29. [CrossRef] [PubMed]
9. Alarcón, L.V.; Allepuz, A.; Mateu, E. Biosecurity in pig farms: A review. *Porcine Health Manag.* **2021**, *7*, 5. [CrossRef] [PubMed]
10. Pinto, R.M.; Oliveira, S.C.; Pinto, R.S. Biosecurity: A review of concepts and application to swine production. *Livest. Res. Rural Dev.* **2003**, *15*, 1–9.
11. Gelaude, A.; Corbeil, S.; Dubois, C. A survey of biosecurity practices on swine farms in Quebec. *Can. J. Vet. Res.* **2014**, *78*, 184–190.
12. Correge, C.; LeJeune, J.T.; Torremorell, M. Biosecurity practices, perceptions, and barriers to implementation on swine farms in the United States. *Animals* **2019**, *9*, 17.
13. Vaillancourt, J.P. Improving biosecurity through Bob Morrison's legacy. In Proceedings of the Allen D. Leman Swine Conference, Saint Paul, MN, USA, 16–19 September 2023; p. 68.
14. Elbers, A.R.; Stegeman, J.A.; de Jong, M.C. Factors associated with the introduction of classical swine fever virus into pig herds in the central area of the 1997/98 epidemic in the Netherlands. *Vet. Rec.* **2001**, *149*, 377–382. [CrossRef] [PubMed]
15. Dee, S.A.; Clark, L.; Paterson, J.L.; Martin, S.W.; Dohoo, I.R. Risk factors for the introduction of *S. tiphymurium* into swine herds in Prince Edward Island. *Can. J. Vet. Res.* **2004**, *68*, 117–122.
16. Mannion, P.D.; Reid, S.W.; Gould, L. The potential for the spread of *Salmonella* spp. in the pig meat production chain. *Int. J. Food Microbiol.* **2008**, *128*, 140–149.
17. Pitkin, J.B.; Duffield, S.S.; Tormo, M.A. Biosecurity on U.S. swine farms: Factors associated with the control of porcine reproductive and respiratory syndrome virus. *J. Swine Health Prod.* **2009**, *17*, 125–131.
18. Lowe, J.; Hicks, S.; Stott, A. The potential for the transmission of Salmonellaa via pig transport vehicles. *J. Appl. Microbiol.* **2017**, *122*, 107–116.
19. Havas, K.A.; Brands, L.; Cochrane, R.; Spronk, G.D.; Nerem, J.; Dee, S.A. An assessment of enhanced biosecurity interventions and their impact on porcine reproductive and respiratory syndrome virus outbreaks within a managed group of farrow-to-wean farms, 2020–2021. *Front. Vet. Sci.* **2023**, *9*, 952383. [CrossRef] [PubMed]
20. Sasaki, Y.; Koketsu, Y. Reproductive profile and lifetime efficiency of female pigs by culling reason in high-performing commercial breeding herds. *J. Swine Health Prod.* **2011**, *19*, 284–291. [CrossRef] [PubMed]
21. Wang, C.; Wu, Y.; Shu, D.; Wei, H.; Zhou, Y.; Peng, J. An analysis of culling patterns during the breeding cycle and lifetime production from the aspect of culling reasons for gilts and sows in Southwest China. *Animals* **2019**, *9*, 160. [CrossRef] [PubMed]
22. Holtkamp, D.; Torremorell, M.; Corzo, C.A.; Linhares, D.C.L.; Almeida, M.N.; Yeske, P.; Polson, D.D.; Becton, L.; Snelson, H.; Donovan, T.; et al. Proposed modifications to porcine reproductive and respiratory syndrome virus herd classification. *J. Swine Health Prod.* **2021**, *29*, 261–270. [CrossRef] [PubMed]
23. Bucini, G.; Clark, E.M.; Merrill, S.C.; Langle-Chimal, O.; Zia, A.; Koliba, C.; Cheney, N.; Wiltshire, S.; Trinity, L.; Smith, J.M. Connecting livestock disease dynamics to human learning and biosecurity decisions. *Front. Vet. Sci.* **2023**, *9*, 1067364. [CrossRef]

Disclaimer/Publisher's Note: The statements, opinions and data contained in all publications are solely those of the individual author(s) and contributor(s) and not of MDPI and/or the editor(s). MDPI and/or the editor(s) disclaim responsibility for any injury to people or property resulting from any ideas, methods, instructions or products referred to in the content.

animals

Article

Assessment of Cleaning and Disinfection Practices on Pig Farms across Ten European Countries

Iryna Makovska [1,*], Ilias Chantziaras [1], Nele Caekebeke [2], Pankaj Dhaka [1,3] and Jeroen Dewulf [1]

1 Veterinary Epidemiology Unit, Faculty of Veterinary Medicine, Ghent University, Salisburylaan 133, 9820 Merelbeke, Belgium; ilias.chantziaras@ugent.be (I.C.); pankaj.dhaka2@gmail.com (P.D.); jeroen.dewulf@ugent.be (J.D.)

2 Biocheck.Gent BV, 8720 Dentergem, Belgium; nele.caekebeke@biocheckgent.com

3 Centre for One Health, College of Veterinary Science, Guru Angad Dev Veterinary and Animal Sciences University, Ludhiana 141004, India

* Correspondence: iryna.makovska@ugent.be; Tel.: +32-467846308

Simple Summary: Biosecurity measures play a pivotal role in minimizing the risk of introducing and spreading infectious agents. Within external and especially internal biosecurity, cleaning and disinfection (C&D) procedures play an important role. The current study aimed to assess the implementation of C&D procedures on pig farms in Europe during 2019–2022, with a focus on identifying areas that warrant improvement. Increased C&D implementation is observed in countries like Belgium, Finland, Italy, and Spain, signaling a high level of farmers' awareness. The study identifies stronger implementation of C&D measures in the framework of external biosecurity but notes gaps in the application of C&D measures for material introduction practices (22% of farms). In the frame of internal biosecurity, more gaps were highlighted in the presence and use of hand hygiene stations (19% of farms), boots disinfection equipment (40% of farms), cleaning between compartments, and evaluation of the efficacy of C&D measures. Notably, only 1% of farms evaluate hygiene after C&D procedures, revealing an important area for improvement. In conclusion, the analysis highlights C&D implementation in European pig farming, emphasizing progress areas and advocating for targeted awareness campaigns and training initiatives to improve biosecurity practices.

Abstract: This study delves into the assessment of cleaning and disinfection (C&D) measures within the context of European pig farming, employing the Biocheck.UGent™ tool as an effective instrument for evaluation. A comprehensive set of relevant parameters was examined to enable meaningful comparisons across farms from 10 European countries during four years (2019–2022). Findings indicate a notable increase in C&D measure implementation in select countries (Belgium, Finland, Italy, and Spain), reflecting heightened awareness and responsibility among farmers. Additionally, the overall score for the C&D subcategory highlights variation across countries, with Italy (75), Poland (74), and Belgium (72) displaying the highest scores, while Ireland (56), Slovenia (55), and Serbia (50) reported the lowest scores. However, the considerable variation in the number of participating farms necessitates cautious comparisons. The study identifies well-implemented C&D measures in the frame of external biosecurity but underscores gaps in the application of C&D measures for the material introduction practices across farms (22% of farms), which are attributed to awareness gaps and resource limitations. In the areas of internal biosecurity, strong points include C&D procedures after each production cycle (79%), implementing different stages in the C&D process (65%) and sufficient sanitary break (82%), while gaps are evident in the presence and using of hand hygiene stations (19% of farms) and boots disinfection equipment (40% of farms) between compartments/units. Notably, the study reveals a lack of evaluation of hygiene after C&D procedures (only 1% of farms), signaling critical knowledge gaps among farmers regarding proper assessment tools and methods. In conclusion, this comprehensive analysis sheds light on the implementation status of C&D measures in European pig farming, offering insights into both areas of progress and those requiring improvement. The findings emphasize the need for targeted awareness campaigns and training initiatives to bolster biosecurity practices within the industry.

Keywords: biosecurity; cleaning; disinfection; Biocheck.Ugent; pigs; stable hygiene

1. Introduction

Both in the framework of reducing antimicrobial usage (AMU) and preventing the spread of epidemic and endemic diseases, prioritizing preventive veterinary management strategies is crucial for the sustainability of animal health [1–3]. A primary approach to reducing infections and the need for treatment is establishing and maintaining biosecurity on animal farms [4–7]. Past studies indicate that improving biosecurity on livestock farms is particularly crucial in combating antimicrobial resistance as it can help to reduce the AMU as well as curb the dissemination and persistence of resistant microbes within farms [8–11].

Farm biosecurity comprises a set of measures aimed at minimizing the risk of introducing pathogenic agents (external biosecurity) and preventing the spread of these pathogenic agents within a herd (internal biosecurity) [2,3,12]. Among the various parameters of external and internal biosecurity, the role and impact of cleaning and disinfection (C&D) measures have been highlighted in various studies [13–17]. The C&D measures involve a series of step-by-step procedures such as dry and wet cleaning, rinsing, disinfecting, and evaluating the effectiveness of the process, along with additional measures that must be applied for premises, vehicles, equipment, personnel, visitors, protective clothing, manure, etc. [17–19]. Therefore, proper adherence to C&D procedures is crucial in reducing the transmission of infectious agents within and between animal farms [14,17,20,21].

In Europe, biosecurity measures on pig farms vary across countries and regions, as there are no standardized procedures or regulations [2,14,22]. Each country typically has its own national guidelines or codes of practice for biosecurity, and the implementation of these measures depends on factors like geographical location, farm size, production type, techno-managerial capabilities, and the epidemiological situation [16,17,23,24]. Moreover, there is a scarcity of data regarding the level of implementation of biosecurity measures on conventional indoor pig farms in various European countries [6,10,23,25].

To address these gaps, a specialized free tool with public access has been developed, enabling quantitative assessment of both external and internal biosecurity measures at the farm. This tool, Biocheck.UGent™, is widely used in Europe. Many field-based epidemiological studies have been conducted using this system to evaluate the biosecurity measures across farms with different animal species [4,8,26–28]. However, the available published data concerning C&D on pig farms in European countries remain limited. With this background, the study was conducted to assess the implementation of C&D measures and identify the current gaps in the C&D procedures on pig farms in European countries using Biocheck.UGent™ data.

2. Materials and Methods

2.1. Scoring System

Biocheck.UGent™ is a risk-based biosecurity scoring system developed by the Faculty of Veterinary Medicine at Ghent University, Belgium. This standardized tool is based on scientific research, which allows the quantitative assessment of biosecurity measures implementation on farms and thereby facilitates comparative analysis between different farms or regions or countries [26,27].

In brief, the Biocheck.UGent™ scoring system for pig's farms consists of 109 questions, divided into six subcategories for external biosecurity and six subcategories for internal biosecurity. External biosecurity encompasses subcategories such as the purchase of breeding pigs, piglets, and semen; the transportation of animals, removal of carcasses and manure; feed, water and equipment supply; visitors and farmworkers; vermin and bird control; and location of the farm. Internal biosecurity includes subcategories such as disease management; farrowing and suckling period; nursery unit; finishing unit; measures between compartments, working lines and use of equipment; as well as cleaning and disinfection. The comprehensive information on each (sub)category can be assessed at https://biocheckgent.com/en, accessed on 31 December 2022. A score is given to each question, with 0 indicating the absence of a biosecurity measure and a score between 0.5 and 1.0, indicating the presence of measures and ranking based on their importance in disease prevention. The assigned score reflects the significance of the question [29], allowing for a comprehensive evaluation of biosecurity measures through weighted scores for each subcategory. Scores for each question are calculated using a weighting factor based on the importance of the biosecurity measure. The survey for pig indoor farms can be accessed at https://biocheckgent.com/en/questionnaires/pigs, accessed on 31 December 2022, and is available in multiple languages, including English, Dutch, French, Polish, German, Italian, Norwegian, Spanish, Vietnamese, Hungarian, and Chinese.

2.2. Data Collection

Our analysis utilized the dataset sourced from the Biocheck.UGent™ database. The required permission to access the data was obtained in December 2022. The dataset encompassed responses from the voluntary online completion of the Biocheck.UGent™ questionnaire. For this study, data included between 1 January 2019 and 31 December 2022 were utilized. A precondition for incorporation into the trend analysis involves achieving a specified minimum number of observations, originating from 40 farms in total and a minimum of 20 farms on an annual basis for each country (Figure 1). In this, it has to be noted that sometimes the same farm can come back in the next year, so not all observations are necessarily different farms.

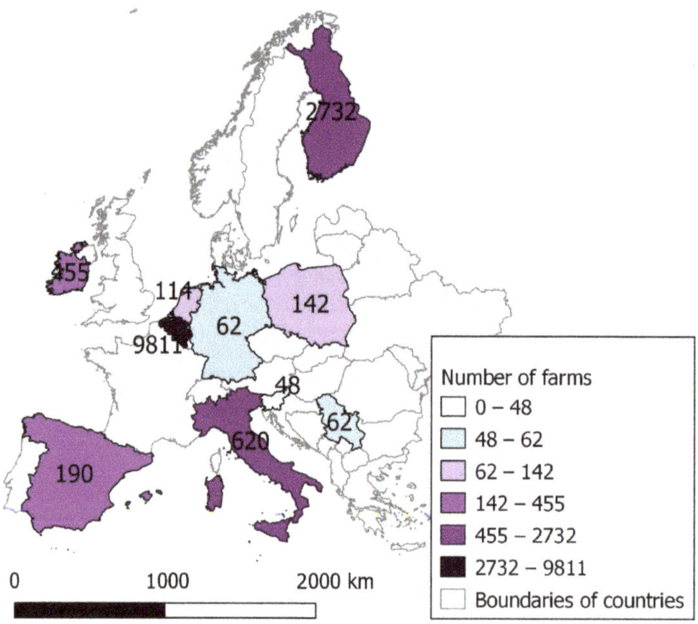

Figure 1. Sampling regions for 2019–2022.

For the present study, we selected questions relevant to C&D measures for external and internal pig farm biosecurity. In total, we analyzed twelve parameters: seven specifically for the "cleaning and disinfection" subcategory (Table 1), three from "measures between compartments, working lines, and use of equipment", one from "feed, water, and equipment supply", and one from "visitors and farmworkers". The analyses of the C&D measures implemented at the external biosecurity primarily revolve around assessing four key aspects: the presence and proper utilization of hygiene locks, the availability and usage of disinfection baths or boot washers, the regularity of fluid changes in the disinfection baths, and the implementation of measures for the introduction of materials. The assessment of C&D measures for maintaining internal biosecurity included measures between compartments/units, measures in corridors and loading areas, measures conducted after each production cycle, adherence to all stages of the C&D procedure, sufficient sanitary breaks, and the evaluation of the efficacy of all implemented measures.

Table 1. Assessment of parameters for C&D measures * on pig farms.

External Biosecurity Measures	Internal Biosecurity Measures
1. Is there a hygiene lock available and is it always used by visitors when they enter the stables?	5. Are hands washed and/or disinfected between different compartments/units?
2. Are specific measures taken for the introduction of material (e.g., cleaning and disinfection, quarantine period at a specific location)?	6. Are either disinfection baths and/or boot washers used between different compartments/units or are boots changed between compartments?
3. Are disinfection baths/boot washers present at the entrance of the farm and are they used?	7. Is there a protocol for the cleaning and disinfection of equipment (such as brooms, spades) after their use and is this protocol abided by?
4. Is the fluid of the disinfection baths immediately changed when visually contaminated?	8. Are the stables/compartments cleaned and disinfected after each production cycle?
	9. Are the corridors and the loading area cleaned and disinfected after pigs are moved?
	10. Is the sanitary break long enough, i.e., is there sufficient time for drying and adjusting the temperature before pigs enter the stable?
	11. Are the different stages in the cleaning and disinfection process respected and is there sufficient time (according to the used product specifications) provided for each stage?
	12. Is the efficacy of cleaning and disinfection checked with for example a hygienogram?

* questions 3, 4, 8, 9, 10, 11 and 12 are included in the cleaning and disinfection subcategory. The answers to these questions are used to calculate the C&D score.

The score for the C&D subcategory is determined by assessing various internal and external parameters (questions 3, 4, 8, 9, 10, 11, and 12) (Table 1). The risk-based biosecurity quantification system generates scores that correspond to the extent of biosecurity measures implemented, with scores ranging from 0 (indicating a complete absence of any biosecurity measures) to 100 (representing the full implementation of all available biosecurity measures).

Data analysis: The analysis of the frequency of the responses for each targeted parameter was carried out. The options "No" or "Never" were considered for non-implementation, "Yes" or "Always" for complete implementation, and "Sometimes" for partial implementation. In the Biocheck.UGent™ scoring system, the option "Sometimes" is allocated the same score as "Never". The collected data were transformed into a Microsoft® Excel 2019 (Microsoft Corp., Santa Rosa, CA, USA) for statistical analysis.

3. Results

By the end of December 2022, a total of 37,283 submissions from conventional indoor pig farms were recorded. A total of ten European countries viz., Belgium, Finland, Ireland, Italy, Spain, the Netherlands, Germany, Poland, Slovenia, and Serbia were included in the analysis, meeting the inclusion criteria. Among these, 14,236 observations relevant to the present study were selected, excluding 'fictional' data (data generated through using the Biocheck.UGent™ survey as an exercise not necessarily representing a real herd). More than half of the responses were received in 2021 (n = 6403). Among the surveyed countries, Belgium, Finland, Italy, and Ireland contributed the highest number of observations, indicating a major participation from these countries. On the other hand, fewer surveys were provided by Slovenia, Serbia, and Germany (Table 2). It is important to mention that in Belgium, Finland, Ireland, and Italy, the use of the Biocheck.UGent™ score system for the evaluation of conventional indoor pig farms is included in a national biosecurity evaluation procedure, requiring annual mandatory completion.

Table 2. Number of responses per country from 2019 to 2022.

No.	Countries	2019	2020	2021	2022	Total
1	Belgium	93	45	4971	4702	9811
2	Finland	398	674	871	789	2732
3	Italy	85	64	120	351	620
4	Ireland	116	151	182	6 *	455
5	Spain	34	45	87	24	190
6	Poland	1 *	21	108	12 *	142
7	The Netherlands	22	34	27	31	114
8	Germany	17 *	9 *	9 *	27	62
9	Serbia	51	0 *	6 *	5 *	62
10	Slovenia	0 *	0 *	41	7 *	48
	Total	817	1043	6422	5954	14,236

*—not used for evaluation of the C&D subcategory score, as only years with at least 20 observations were included.

3.1. Analysis of Responses for C&D

The study revealed notable variations in the mean scores for the C&D subcategory across different countries and responding years (Figure 2). Italy (75), Poland (74), and Belgium (72) exhibited the highest overall scores, while Serbia (50), Slovenia (55), and Ireland (56) reported the lowest scores during 2019–2022.

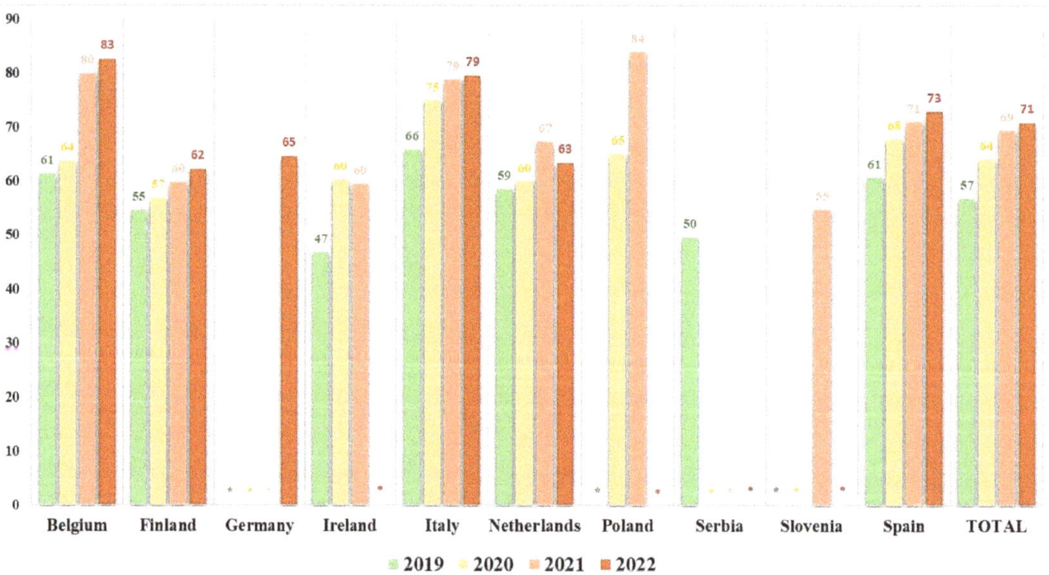

Figure 2. The mean score for the C&D subcategory by year per country. *—not used for evaluation of the C&D subcategory score as only years with at least 20 observations were included.

Over the four-year period, a general improvement in C&D measures can be seen, especially in Belgium, Finland, Italy, and Spain. The Netherlands demonstrated an increase in the level of C&D measures from 2019 to 2021 but experienced a decrease in scores from 2021 to 2022. Other countries, such as Ireland and Poland, also demonstrated an increase in C&D scores over several years (Figure 3).

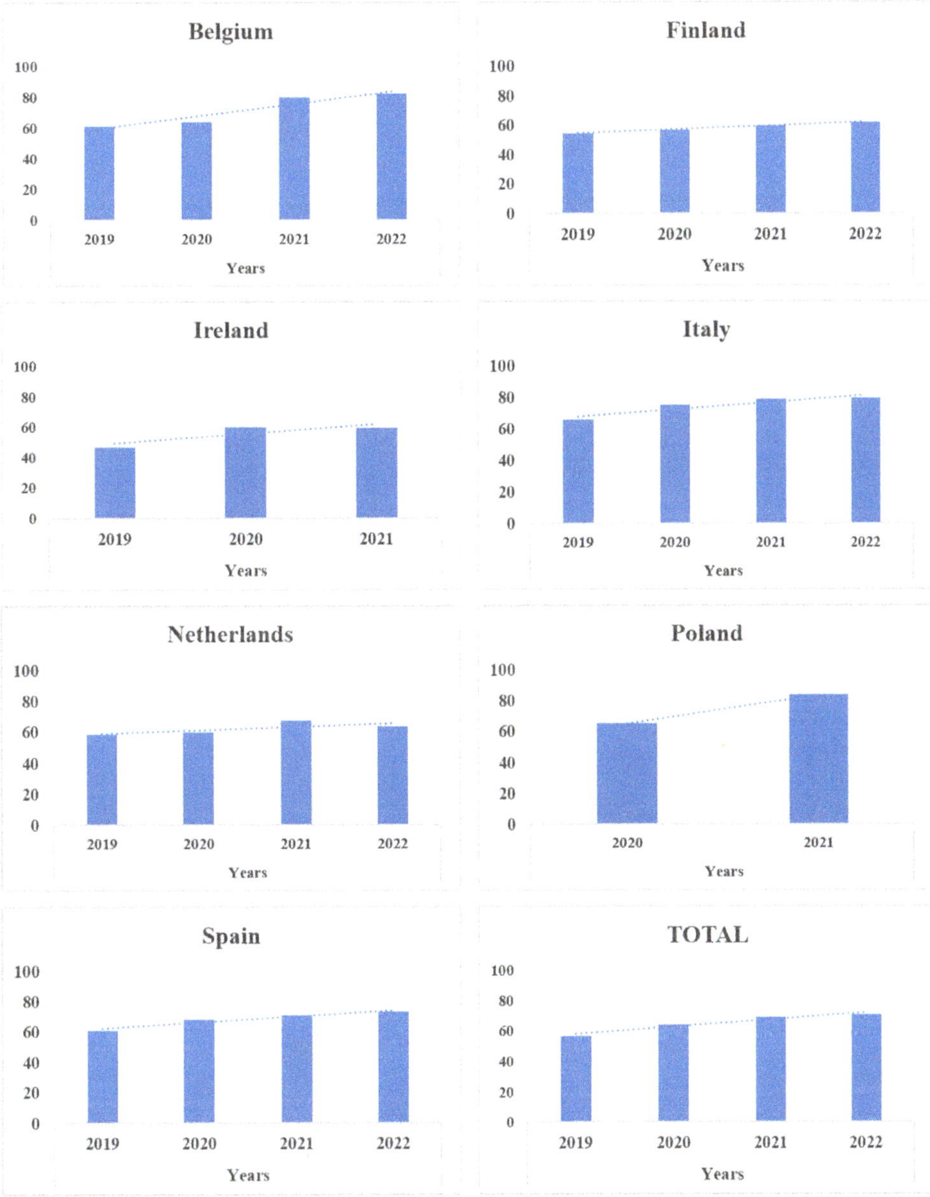

Figure 3. The trend of the mean score for the C&D subcategory over the four-year period per country (evaluation of the trend of the mean scores for the C&D subcategory in a 4-year period specifically focused only on countries with a minimum of 2 years of data (with at least 20 observations included)).

3.2. C&D Measures for External Biosecurity

The analyses of the C&D measures implemented at the external biosecurity are provided in Table 3.

Table 3. Percentage of farms with C&D measures in the frame of external biosecurity (2019–2022) (The color gradient in the illustration represents the percentage of farms implementing C&D measures. The intensity of green indicates a high level of implementation, yellow signifies a medium level, and red indicates a low level).

	Country	1. Presence of Hygiene Lock	2. Presence of Disinfection Baths/Boot Washers Present at the Entrance of the Farm	3. Appropriate Change of Fluid in Disinfection Baths *	4. C&D Measures Taken for the Introduction of Materials
1	Belgium	85%	88%	89%	21%
2	Finland	88%	79%	60%	19%
3	Germany	94%	42%	69%	19%
4	Ireland	56%	49%	92%	13%
5	Italy	77%	30%	98%	11%
6	The Netherlands	94%	61%	67%	19%
7	Poland	87%	75%	82%	65%
8	Serbia	55%	55%	26%	16%
9	Slovenia	27%	19%	78%	21%
10	Spain	79%	52%	79%	19%
	Average	74%	55%	74%	22%

*—only applicable to farms that reported the presence of disinfection baths.

The findings regarding the 'presence of hygiene locks and consistent usage of them' by visitors upon entering the stables have shown a good level of application, with 74% (n = 11,866) of respondents confirming adherence to this measure. The data from Table 3 also suggest that the implementation of these measures has been successful in most countries, especially in the Netherlands (94%, n = 107) and Germany (94%, n = 58), contrary to Slovenia, where only 27% (n = 13) of respondents reported the presence of a hygiene lock and its utilization by visitors.

The results of the 'presence and proper usage of disinfection baths or boot washers' at the entrance of farms indicate that in 55% (n = 11,511) of cases, the farms had disinfection baths or boot washers installed and they were being appropriately utilized. The observations indicate that these measures were well implemented in Belgium (88%, n = 8614), Finland (79%, n = 2146), and Poland (75%, n = 107). Conversely, the implementation of these measures was found to be inadequate in Slovenia (19%, n = 9) and Italy (30%, n = 183).

The outcomes pertaining to the 'regularity of fluid changes in disinfection baths' revealed a well-implemented practice in all countries. Overall, 74% of farms (n = 9629) reported promptly changing the disinfectant upon visual contamination. Interestingly, Italian (98%, n = 178) and Irish (92%, n = 208) farms demonstrated the highest level of implementation in this regard.

The observations regarding the implementation of measures for the 'introduction of materials' revealed that compliance with the application of specific measures (e.g., cleaning and disinfection, quarantine period at a specific location) was noted in 22% (n = 2846) of farms. The Polish farms observed a higher level of implementation of these measures (65%, n = 93) as compared to other countries.

3.3. C&D Measures for Internal Biosecurity

The assessment of C&D measures for maintaining internal biosecurity is presented in Table 4.

Table 4. Percentage of farms with C&D measures in the frame of internal biosecurity (2019–2022) (The color gradient in the illustration represents the percentage of farms implementing C&D measures. The intensity of green indicates a high level of implementation, yellow signifies a medium level, and red indicates a low level).

	Country	1. Presence of Protocol for the C&D of Equipment	2. Presence of Hand Washing Stations and/or Hand Disinfection Equipment between Compartments/Units	3. Presence of Disinfection Baths/Boot Washers between Compartments/Units	4. C&D Measures in Corridors and Loading Areas	5. Conducted C&D after Each Production Cycle	6. Provided Different Stages in the C&D Process	7. Long Enough Sanitary Break	8. Checking the Efficacy of C&D
1	Belgium	57%	24%	38%	78%	74%	89%	96%	2%
2	Finland	34%	12%	27%	43%	49%	58%	90%	0%
3	Germany	58%	10%	32%	52%	92%	53%	74%	0%
4	Ireland	46%	20%	35%	51%	45%	61%	87%	1%
5	Italy	56%	31%	38%	87%	96%	95%	96%	2%
6	The Netherlands	47%	12%	54%	61%	82%	61%	69%	1%
7	Poland	49%	49%	68%	64%	96%	87%	92%	2%
8	Serbia	77%	5%	42%	69%	97%	23%	63%	2%
9	Slovenia	25%	21%	17%	50%	65%	56%	63%	0%
10	Spain	56%	11%	46%	47%	97%	67%	89%	2%
	Average	51%	19%	40%	60%	79%	65%	82%	1%

It was observed that the 'protocol for C&D of equipment after use' was present in almost half of the farms (51%, $n = 7420$), with higher implementation levels among Serbian farms (77%, $n = 48$). However, inadequate implementation of the practices was identified regarding 'hand and boot cleaning and disinfection between compartments/units' in most countries. Only 40% of farms ($n = 5124$) reported the presence of disinfection baths and/or boot washers or the practice of changing boots, and even fewer farms (19%, $n = 3090$) had hand washing stations and/or hand disinfection equipment between compartments/units. Notably, Polish farms showed the best implementation for these practices (68%, $n = 96$ and 49%, $n = 69$, respectively).

The assessment also emphasized the importance of 'implementing C&D measures in corridors and loading areas following pig movement'. Over half of farms (60%, $n = 9974$) confirmed the implementation of these measures, with Italian (87%) and Belgian (78%) farms demonstrating adequate performance.

In terms of implementing 'C&D procedures after each production cycle', overall, 79% of farms ($n = 9920$) reported compliance, with Spanish (97%, $n = 184$), Serbian (97%, $n = 60$), Polish (96%, $n = 136$), Italian (96%, $n = 594$), and German (92%, $n = 57$) farms showing the high rates of implementation.

More than half of farms (65%, $n = 11{,}516$) reported following proper C&D procedures, primarily observed on Italian, Belgian, and Polish farms. These procedures demonstrated maintaining hygiene standards, incorporating 'different stages of C&D process' (such as dry cleaning, soaking, high-pressure cleaning, the first drying of the stable, disinfection of the stable, the second drying of the stable, and testing of the efficiency of the procedure) and allocating sufficient time for each stage. The majority of farms (82%, $n = 13{,}382$) reported having enough time for drying and adjusting the temperature before pigs entered the stables. However, the effectiveness of all C&D practices was only validated by 1% of the farms ($n = 218$).

4. Discussion

The Biocheck.UGent™ tool surveys are effective instruments for evaluating the execution of C&D measures. The tool offers a comprehensive set of relevant parameters, enabling meaningful comparisons across farms, regions, and countries. However, we have to take into account the possible presence of biases in the data used. Firstly, the submissions could have originated from the best-performing farms, potentially leading to an overestimation of the overall biosecurity measures, especially in those countries where the Biocheck.UGent™ scoring system was used on a voluntary basis (Germany, the

Netherlands, Serbia, Slovenia, Spain, and Poland). In these countries, it is also possible that the submissions might have come from specific regions within a country, which may not accurately represent the biosecurity practices of all farms nationwide. The latter risk is not so much present in those countries (Belgium, Finland, Ireland, and Italy) where Biocheck.UGent™ biosecurity evaluations have become part of a mandatory national auditing campaign. In these countries, the data provide a more or less complete overview of the actual situation. Yet, on the other hand, here we have the risk that some respondents provide answers that are better than the real situation and rather reflect what they believe they are to respond instead of what is really applied. Moreover, it is crucial to acknowledge that the responses to the survey may vary based on the respondent's profile, such as being a veterinarian, practitioner, farmer, etc. In our specific context, in countries with mandatory national auditing campaigns, the survey for pig farms is typically completed by veterinarians. However, in countries with voluntary participation, the surveying of pig farms is generally completed by farmers in collaboration with the vets/advisors, which mitigates potential bias. In addition, socio-economic considerations may contribute to the varying levels of C&D measure implementation and correspondingly influence survey responses. Nonetheless, our dataset presents substantial advantages primarily attributed to the validated methodology, ensuring a high degree of accuracy and reliability. In addition, our analysis is based on data from a huge database, affording us the capacity to check and determine the level of the implementation of C&D measures within European countries at the vanguard of pig industry production and research [4,27,30]. The overall findings highlight both areas of progress and areas that require improvement, offering opportunities to strengthen external and internal biosecurity measures and reduce the risk of disease transmission within the pig farming industry.

In general, the analysis of the C&D subcategory scores revealed an upward level of implementation measures in the majority of the included countries. This may be the result of an increasing level of awareness and responsibility on farms. It may also demonstrate a biosecurity audit, wherein farmers and veterinarians try to improve the situation by working on the biosecurity gaps [8,10,31].

The overall score for the C&D subcategory ranged from 47 to 84, with the highest scores observed in Italy, Poland, and Belgium. Our results are in agreement with the results provided by Chantziaras et al. (2020) on the biosecurity levels of pig fattening farms from four EU countries (Belgium, the United Kingdom, Finland, and Poland), where the score for the C&D subcategory ranged from 40 to 87, and the highest scores for C&D measures were observed in Belgian and Polish farms [32].

Our study also revealed that there were more deficiencies in implementing C&D measures in the framework of internal biosecurity as compared to external biosecurity. This observation may be explained by the fact that farmers find it easier to enforce rules on others (e.g., visitors, vehicles), which primarily relate to external biosecurity, as opposed to changing their own habits, which are more closely linked to internal biosecurity [4,20].

The study findings regarding the C&D measures for maintaining external biosecurity indicated well-implemented measures with regard to the presence of hygiene locks and their consistent usage by visitors, the presence of footbaths, and regular fluid changes in disinfection baths in most countries, except for Serbia and Slovenia. The measures for the introduction of materials (such as cleaning and disinfection) were poorly applied across farms in all countries except Poland (65%).

The assessment of C&D measures related to internal biosecurity revealed strong points as well as some gaps. Strong points were observed with regard to the implementation of the different stages in the C&D process, conducting C&D after each production cycle, C&D measures in corridors and loading areas, and long enough sanitary breaks. However, important gaps were observed regarding the presence of hand washing stations and/or hand disinfection equipment between compartments/units (19% in total), particularly in Serbia (5%), Germany (10%), Spain (11%), Finland (12%), and the Netherlands (12%); only Polish farms (49%) exhibited better implementation. In addition, fewer but also noteworthy gaps were identified regarding the presence of disinfection baths or boot washers between compartments or units, with a total of only 40% of the farms applying these measures. This tendency suggests a considerable need for improvement in maintaining proper biosecurity measures between compartments.

Finally, an important shortcoming was observed in the assessment of the effectiveness of C&D measures, with only 1% of farms reporting that they conducted evaluations of hygiene after C&D procedures (e.g., with hygienogram). This may be linked to the limited information available to farmers regarding proper tools or methods for evaluating hygiene status in animal houses. As reported by Alarcón et al. (2021), pig farmers often adopt the assessment of the effectiveness of hygiene procedures with low perseverance and consistency [20]. Practical challenges, including resource availability, time constraints, and complex data collection methods, may hinder comprehensive assessments [33]. Yet, the effect of the limited evaluation of the quality of the C&D procedures is that farmers and farm workers often have no clue of whether they are performing the procedures sufficiently well. Designing farm-specific assessment approaches can lead to more accurate evaluations by considering factors like farm size, layout, disease risks, and biosecurity measures [20,21,34,35].

In general, motivating farmers to adopt a standard hygiene protocol and alter their daily routines poses a well-known challenge [23,36,37]. As reported by Garforth et al. (2013) many farmers believe they are already taking all reasonable measures to minimize disease risk, considering other practices irrelevant [38]. Additionally, limited resources, such as insufficient supplies of cleaning agents, disinfectants, personal protective equipment, or appropriate infrastructure for quarantine areas, can make it challenging to adhere to the recommended protocols [23,36,39]. As reported by Alarcón et al. (2021) farm personnel play a critical role in maintaining internal biosecurity, with responsibilities for implementing biosecurity rules while also having the potential to inadvertently spread pathogens within the farm [20]. Cross-contamination risks may not be fully understood by farmers and employees, leading to inadequate hand and boot cleaning before moving to different parts of the farm, especially in small herds [2,40–42]. In addition, time restrictions may cause them to speed up or completely omit phases in the hygiene procedure [43]. Thereby, inconsistent practices, lack of time, and inadequate training contribute to these issues, increasing disease transmission risk [2,20,23]. Recent studies have found that non-compliance with biosecurity measures is often related to inadequate training of farm personnel and poor communication with advisors [36,44–46]. Additionally, factors like farm infrastructure, condition variability, insufficient supervision, monitoring, and documentation also contribute to the gaps [34,47,48]. A study conducted by Chen et al. (2023) highlighted that biosecurity cognition plays a substantial mediating role in the adoption of C&D methods by pig farmers who have undergone specialized training [49]. Therefore, addressing all these factors through improved knowledge dissemination and effective evaluation techniques can enhance biosecurity practices on pig farms [17,36,45].

5. Conclusions

The assessment of C&D measures on conventional indoor pig farms reveals both areas of progress and scope for improvement. While some farms demonstrate commendable commitment to hygiene standards, there is a need for enhanced practices in areas such as hand and boot cleaning between compartments/units, verification of effectiveness, and the implementation of C&D measures across all compartments and units. Sharing best practices and promoting knowledge exchange among countries can contribute to elevating biosecurity standards industry-wide. By addressing these key findings and implementing targeted interventions, farmers can strengthen biosecurity measures, safeguard the animal health, and mitigate the risk of disease transmission.

Author Contributions: Conceptualization, I.M., I.C., P.D. and J.D.; Methodology, I.M., I.C., N.C., P.D. and J.D.; Software, I.M., I.C., N.C., P.D. and J.D.; Validation, I.C., N.C. and J.D.; Formal analysis, I.M., I.C., N.C., P.D. and J.D.; Investigation, I.M., I.C., N.C., P.D. and J.D.; Resources, I.M., N.C., P.D. and J.D.; Data curation, I.M., I.C., N.C., P.D. and J.D.; Writing—original draft, I.M., I.C., P.D. and J.D.; Writing—review & editing, I.M., I.C., N.C., P.D. and J.D.; Visualization, I.M., I.C., P.D. and J.D.; Supervision, I.C., P.D. and J.D.; Project administration, N.C. and J.D.; Funding acquisition, J.D. All authors have read and agreed to the published version of the manuscript.

Funding: This research received no external funding.

Institutional Review Board Statement: The study was survey-based, and there was no involvement of any invasive procedures or experimental protocols on the study subjects.

Informed Consent Statement: The study was completely voluntary and proper consent of the participants was obtained before enrollment in the study.

Data Availability Statement: The data presented in this study are available upon request from the corresponding author. The data are not publicly available due to privacy and confidentiality agreements with the participants.

Acknowledgments: The authors thank Ghent University (Belgium) for providing the necessary facilities for undertaking the research. We would also like to thank the farmers involved in this study for their collaboration.

Conflicts of Interest: The authors declare no conflicts of interest. Author Nele Caekebeke is an employee of BiocheckGent BV (spin-off company from Ghent University) and Jeroen Dewulf is shareholder of BiocheckGent BV. The remaining authors declare that the research was conducted in the absence of any commercial or financial relationships that could be construed as a potential conflict of interest.

References

1. Davies, R.; Wales, A. Antimicrobial Resistance on Farms: A Review Including Biosecurity and the Potential Role of Disinfectants in Resistance Selection. *Compr. Rev. Food Sci. Food Saf.* **2019**, *18*, 753–774. [CrossRef]
2. Dewulf, J.; Immerseel, F.V. *Biosecurity in Animal Production and Veterinary Medicine*; Acco: Leuven, Belgium, 2018.
3. Renault, V.; Humblet, M.-F.; Saegerman, C. Biosecurity Concept: Origins, Evolution and Perspectives. *Animals* **2021**, *12*, 63. [CrossRef]
4. Filippitzi, M.E.; Brinch Kruse, A.; Postma, M.; Sarrazin, S.; Maes, D.; Alban, L.; Nielsen, L.R.; Dewulf, J. Review of Transmission Routes of 24 Infectious Diseases Preventable by Biosecurity Measures and Comparison of the Implementation of These Measures in Pig Herds in Six European Countries. *Transbound Emerg. Dis.* **2018**, *65*, 381–398. [CrossRef]
5. Huber, N.; Andraud, M.; Sassu, E.L.; Prigge, C.; Zoche-Golob, V.; Käsbohrer, A.; D'Angelantonio, D.; Viltrop, A.; Żmudzki, J.; Jones, H.; et al. What Is a Biosecurity Measure? A Definition Proposal for Animal Production and Linked Processing Operations. *One Health* **2022**, *15*, 100433. [CrossRef]
6. Sahlström, L.; Virtanen, T.; Kyyrö, J.; Lyytikäinen, T. Biosecurity on Finnish Cattle, Pig and Sheep Farms—Results from a Questionnaire. *Prev. Vet. Med.* **2014**, *117*, 59–67. [CrossRef]
7. Maes, D.; Piñeiro, C.; Haesebrouck, F.; Boyen, F.; Rubio, P.; Manzanilla, E.G. Control and Prevention of Bacterial Diseases in Swine. In *Advancements and Technologies in Pig and Poultry Bacterial Disease Control*; Elsevier: Amsterdam, The Netherlands, 2021; pp. 171–198. [CrossRef]
8. Caekebeke, N.; Jonquiere, F.J.; Ringenier, M.; Tobias, T.J.; Postma, M.; Van Den Hoogen, A.; Houben, M.A.M.; Velkers, F.C.; Sleeckx, N.; Stegeman, J.A.; et al. Comparing Farm Biosecurity and Antimicrobial Use in High-Antimicrobial-Consuming Broiler and Pig Farms in the Belgian–Dutch Border Region. *Front. Vet. Sci.* **2020**, *7*, 558455. [CrossRef] [PubMed]
9. Dhaka, P.; Chantziaras, I.; Vijay, D.; Bedi, J.S.; Makovska, I.; Biebaut, E.; Dewulf, J. Can Improved Farm Biosecurity Reduce the Need for Antimicrobials in Food Animals? A Scoping Review. *Antibiotics* **2023**, *12*, 893. [CrossRef] [PubMed]
10. Yun, J.; Muurinen, J.; Nykäsenoja, S.; Seppä-Lassila, L.; Sali, V.; Suomi, J.; Tuominen, P.; Joutsen, S.; Hämäläinen, M.; Olkkola, S.; et al. Antimicrobial Use, Biosecurity, Herd Characteristics, and Antimicrobial Resistance in Indicator Escherichia Coli in Ten Finnish Pig Farms. *Prev. Vet. Med.* **2021**, *193*, 105408. [CrossRef] [PubMed]
11. Kruse, A.B.; Nielsen, L.R.; Alban, L. Herd Typologies Based on Multivariate Analysis of Biosecurity, Productivity, Antimicrobial and Vaccine Use Data from Danish Sow Herds. *Prev. Vet. Med.* **2020**, *181*, 104487. [CrossRef] [PubMed]
12. Saegerman, C.; Parisi, G.; Niemi, J.; Humblet, M.-F.; Ron-Román, J.; Souley Kouato, B.; Allepuz, A.; Porphyre, V.; Rodrigues Da Costa, M.; Renault, V. Evaluation Survey on Agreement with Existing Definitions of Biosecurity with a Focus on Livestock. *Animals* **2023**, *13*, 1518. [CrossRef] [PubMed]
13. Alarcon, P.; Marco-Jimenez, F.; Horigan, V.; Ortiz-Pelaez, A.; Rajanayagam, B.; Dryden, A.; Simmons, H.; Konold, T.; Marco, C.; Charnley, J.; et al. A Review of Cleaning and Disinfection Guidelines and Recommendations Following an Outbreak of Classical Scrapie. *Prev. Vet. Med.* **2021**, *193*, 105388. [CrossRef]

14. De Lorenzi, G.; Borella, L.; Alborali, G.L.; Prodanov-Radulović, J.; Štukelj, M.; Bellini, S. African Swine Fever: A Review of Cleaning and Disinfection Procedures in Commercial Pig Holdings. *Res. Vet. Sci.* **2020**, *132*, 262–267. [CrossRef]
15. Huneau-Salaün, A.; Scoizec, A.; Thomas, R.; Martenot, C.; Schmitz, A.; Pierre, I.; Allée, C.; Busson, R.; Massin, P.; Briand, F.-X.; et al. Avian Influenza Outbreaks: Evaluating the Efficacy of Cleaning and Disinfection of Vehicles and Transport Crates. *Poult. Sci.* **2022**, *101*, 101569. [CrossRef]
16. Mannion, C.; Leonard, F.C.; Lynch, P.B.; Egan, J. Efficacy of Cleaning and Disinfection on Pig Farms in Ireland. *Vet. Rec.* **2007**, *161*, 371–375. [CrossRef]
17. Martelli, F.; Lambert, M.; Butt, P.; Cheney, T.; Tatone, F.A.; Callaby, R. Evaluation of an Enhanced Cleaning and Disinfection Protocol in Salmonella Contaminated Pig Holdings in the United Kingdom. *PLoS ONE* **2017**, *12*, e0178897. [CrossRef]
18. Dorado-García, A.; Graveland, H.; Bos, M.E.H.; Verstappen, K.M.; Van Cleef, B.A.G.L.; Kluytmans, J.A.J.W.; Wagenaar, J.A.; Heederik, D.J.J. Effects of Reducing Antimicrobial Use and Applying a Cleaning and Disinfection Program in Veal Calf Farming: Experiences from an Intervention Study to Control Livestock-Associated MRSA. *PLoS ONE* **2015**, *10*, e0135826. [CrossRef]
19. Štukelj, M.; Prodanov-Radulović, J.; Bellini, S. Cleaning and Disinfection in the Domestic Pig Sector. In *Understanding and Combatting African Swine Fever*; Iacolina, L., Penrith, M.-L., Bellini, S., Chenais, E., Jori, F., Montoya, M., Ståhl, K., Gavier-Widén, D., Eds.; Wageningen Academic Publishers: Wageningen, The Netherlands, 2021; pp. 283–304. [CrossRef]
20. Alarcón, L.V.; Allepuz, A.; Mateu, E. Biosecurity in Pig Farms: A Review. *Porc. Health Manag.* **2021**, *7*, 5. [CrossRef]
21. Galipó, E.; Zoche-Golob, V.; Sassu, E.L.; Prigge, C.; Sjölund, M.; Tobias, T.; Rzeżutka, A.; Smith, R.P.; Burow, E. Prioritization of Pig Farm Biosecurity for Control of Salmonella and Hepatitis E Virus Infections: Results of a European Expert Opinion Elicitation. *Porc. Health Manag.* **2023**, *9*, 8. [CrossRef] [PubMed]
22. Boklund, A.; Alban, L.; Mortensen, S.; Houe, H. Biosecurity in 116 Danish Fattening Swineherds: Descriptive Results and Factor Analysis. *Prev. Vet. Med.* **2004**, *66*, 49–62. [CrossRef] [PubMed]
23. Casal, J.; De Manuel, A.; Mateu, E.; Martín, M. Biosecurity Measures on Swine Farms in Spain: Perceptions by Farmers and Their Relationship to Current on-Farm Measures. *Prev. Vet. Med.* **2007**, *82*, 138–150. [CrossRef] [PubMed]
24. Horrillo, A.; Obregón, P.; Escribano, M.; Gaspar, P. A Biosecurity Survey on Iberian Pig Farms in Spain: Farmers' Attitudes towards the Implementation of Measures and Investment. *Res. Vet. Sci.* **2022**, *145*, 82–90. [CrossRef]
25. Viltrop, A.; Reimus, K.; Niine, T.; Mõtus, K. Biosecurity Levels and Farm Characteristics of African Swine Fever Outbreak and Unaffected Farms in Estonia—What Can Be Learned from Them? *Animals* **2022**, *12*, 68. [CrossRef]
26. Alloui, N.; Sellaoui, S.S.; Ayachi, A.; Bennoune, O. Evaluation of Biosecurity Practices in a Laying Hens Farm Using Biocheck. *UGent. Multi* **2021**, *3*, e2021014. [CrossRef]
27. Rodrigues Da Costa, M.; Gasa, J.; Calderón Díaz, J.A.; Postma, M.; Dewulf, J.; McCutcheon, G.; Manzanilla, E.G. Using the Biocheck.UGent™ Scoring Tool in Irish Farrow-to-Finish Pig Farms: Assessing Biosecurity and Its Relation to Productive Performance. *Porc. Health Manag.* **2019**, *5*, 4. [CrossRef]
28. Calderon Diaz, J.A.; O'Neill, L.; Da Costa, M.R.; Manzanilla, E.G. Biosecurity Practices Associated with Antimicrobial Usage in Farrow-to-Finish Pig Farms. *J. Anim. Sci.* **2020**, *98* (Suppl. 3), 5–6. [CrossRef]
29. Gelaude, P.; Schlepers, M.; Verlinden, M.; Laanen, M.; Dewulf, J. Biocheck.UGent: A Quantitative Tool to Measure Biosecurity at Broiler Farms and the Relationship with Technical Performances and Antimicrobial Use. *Poult. Sci.* **2014**, *93*, 2740–2751. [CrossRef] [PubMed]
30. Postma, M. Evaluation of the Relationship between the Biosecurity Status, Production Parameters, Herd Characteristics and Antimicrobial Usage in Farrow-to-Finish Pig Production in Four EU Countries. *Porc. Health Manag.* **2016**, *2*, 9. [CrossRef] [PubMed]
31. Caekebeke, N.; Ringenier, M.; Jonquiere, F.; Tobias, T.; Postma, M.; Van Den Hoogen, A.; Houben, M.; Velkers, F.; Sleeckx, N.; Stegeman, A.; et al. Coaching Belgian and Dutch Broiler Farmers Aimed at Antimicrobial Stewardship and Disease Prevention. *Antibiotics* **2021**, *10*, 590. [CrossRef] [PubMed]
32. Chantziaras, I.; Dewulf, J.; Van Limbergen, T.; Stadejek, T.; Niemi, J.; Kyriazakis, I.; Maes, D. Biosecurity Levels of Pig Fattening Farms from Four EU Countries and Links with the Farm Characteristics. *Livest. Sci.* **2020**, *237*, 104037. [CrossRef]
33. Niemi, J.K.; Sahlström, L.; Kyyrö, J.; Lyytikäinen, T.; Sinisalo, A. Farm Characteristics and Perceptions Regarding Costs Contribute to the Adoption of Biosecurity in Finnish Pig and Cattle Farms. *Rev. Agric. Food Env. Stud.* **2016**, *97*, 215–223. [CrossRef]
34. O'Hara, K.C.; Beltrán-Alcrudo, D.; Hovari, M.; Tabakovski, B.; Martínez-López, B. Descriptive and Multivariate Analysis of the Pig Sector in North Macedonia and Its Implications for African Swine Fever Transmission. *Front. Vet. Sci.* **2021**, *8*, 733157. [CrossRef] [PubMed]
35. Prodanov-Radulović, J.; Petrović, J.; Grubač, S.; Hristov, S.; Bojkovski, J. Relevant Biosecurity Measures to Prevent The Spread of African Swine Fever In The Domestic Pig Production Sector In Serbia. In Proceedings of the 14th International Symposium Modern Trends in Livestock Production, Belgrade, Serbia, 4–6 October 2023.
36. Scollo, A.; Perrucci, A.; Stella, M.C.; Ferrari, P.; Robino, P.; Nebbia, P. Biosecurity and Hygiene Procedures in Pig Farms: Effects of a Tailor-Made Approach as Monitored by Environmental Samples. *Animals* **2023**, *13*, 1262. [CrossRef] [PubMed]
37. Klein, L.; Hessling-Zeinen, S.; Adler, F.; Gerdes, U.; Blome, S.; Beilage, E.G.; Campe, A. Exploring Pig Farmers' Decision-Making Concerning Biosecurity Measures against African Swine Fever. *Prev. Vet. Med.* **2023**, *217*, 105949. [CrossRef] [PubMed]
38. Garforth, C.J.; Bailey, A.P.; Tranter, R.B. Farmers' Attitudes to Disease Risk Management in England: A Comparative Analysis of Sheep and Pig Farmers. *Prev. Vet. Med.* **2013**, *110*, 456–466. [CrossRef] [PubMed]

39. Delsart, M.; Pol, F.; Dufour, B.; Rose, N.; Fablet, C. Pig Farming in Alternative Systems: Strengths and Challenges in Terms of Animal Welfare, Biosecurity, Animal Health and Pork Safety. *Agriculture* **2020**, *10*, 261. [CrossRef]
40. Simon-Grifé, M.; Martín-Valls, G.E.; Vilar, M.J.; García-Bocanegra, I.; Martín, M.; Mateu, E.; Casal, J. Biosecurity Practices in Spanish Pig Herds: Perceptions of Farmers and Veterinarians of the Most Important Biosecurity Measures. *Prev. Vet. Med.* **2013**, *110*, 223–231. [CrossRef] [PubMed]
41. Nöremark, M.; Sternberg Lewerin, S.; Ernholm, L.; Frössling, J. Swedish Farmers' Opinions about Biosecurity and Their Intention to Make Professionals Use Clean Protective Clothing When Entering the Stable. *Front. Vet. Sci.* **2016**, *3*, 46. [CrossRef]
42. Correia-Gomes, C.; Henry, M.K.; Auty, H.K.; Gunn, G.J. Exploring the Role of Small-Scale Livestock Keepers for National Biosecurity—The Pig Case. *Prev. Vet. Med.* **2017**, *145*, 7–15. [CrossRef]
43. Bernaerdt, E.; Díaz, I.; Piñeiro, C.; Collell, M.; Dewulf, J.; Maes, D. Optimizing Internal Biosecurity on Pig Farms by Assessing Movements of Farm Staff. *Porc. Health Manag.* **2023**, *9*, 11. [CrossRef]
44. Heinemann, C.; Meyer, I.; Bögel, F.T.; Schmid, S.M.; Hayer, J.J.; Steinhoff-Wagner, J. Individual Training for Farmers Based on Results from Protein and ATP Rapid Tests and Microbiological Conventional Cultural Methods Improves Hygiene in Pig Fattening Pens. *J. Anim. Sci.* **2019**, *98*, skz389. [CrossRef]
45. Militzer, N.; McLaws, M.; Rozstalnyy, A.; Li, Y.; Dhingra, M.; Auplish, A.; Mintiens, K.; Sabirovic, M.; Von Dobschuetz, S.; Heilmann, M. Characterising Biosecurity Initiatives Globally to Support the Development of a Progressive Management Pathway for Terrestrial Animals: A Scoping Review. *Animals* **2023**, *13*, 2672. [CrossRef] [PubMed]
46. Kuster, K.; Cousin, M.-E.; Jemmi, T.; Schüpbach-Regula, G.; Magouras, I. Expert Opinion on the Perceived Effectiveness and Importance of On-Farm Biosecurity Measures for Cattle and Swine Farms in Switzerland. *PLoS ONE* **2015**, *10*, e0144533. [CrossRef] [PubMed]
47. Plut, J.; Knific, T.; Golinar Oven, I.; Vengušt, G.; Štukelj, M. Evaluation of Biosecurity Measures in Pig Holdings in Slovenia as a Risk Assessment for the Introduction of African Swine Fever Virus. *Pathogens* **2023**, *12*, 434. [CrossRef] [PubMed]
48. Bucherer, M.; Holzhausen, J.; Conraths, F.; Probst, C. Infrastructure of Animal Farms: Key Constructional Elements in Terms of Biosecurity Based on Experience from Germany. *Berl. Und Münchener Tierärztliche Wochenschr.* **2021**, *134*, 1–12. [CrossRef]
49. Chen, Y.; Xia, R.; Ding, J.; Meng, Z.; Liu, Y.; Wang, H. How Does Epidemic Prevention Training for Pig Breeding Affect Cleaning and Disinfection Procedures Adoption? Evidence from Chinese Pig Farms. *Vet. Sci.* **2023**, *10*, 516. [CrossRef]

Disclaimer/Publisher's Note: The statements, opinions and data contained in all publications are solely those of the individual author(s) and contributor(s) and not of MDPI and/or the editor(s). MDPI and/or the editor(s) disclaim responsibility for any injury to people or property resulting from any ideas, methods, instructions or products referred to in the content.

animals

MDPI

Article

Biological Containment for African Swine Fever (ASF) Laboratories and Animal Facilities: The Italian Challenge in Bridging the Present Regulatory Gap and Enhancing Biosafety and Biosecurity Measures

Silvia Pavone [1,*], Carmen Iscaro [1], Monica Giammarioli [1], Maria Serena Beato [1], Cecilia Righi [1], Stefano Petrini [1], Silva Costarelli [2,†] and Francesco Feliziani [1,†]

[1] National Reference Laboratory for Pestivirus and Asfivirus, Istituto Zooprofilattico Sperimentale dell'Umbria e delle Marche "Togo Rosati" (IZSUM), Via G. Salvemini 1, 06126 Perugia, Italy; c.iscaro@izsum.it (C.I.); m.giammarioli@izsum.it (M.G.); ms.beato@izsum.it (M.S.B.); c.righi@izsum.it (C.R.); s.petrini@izsum.it (S.P.); f.feliziani@izsum.it (F.F.)

[2] Animal Health Diagnostic Laboratory, Istituto Zooprofilattico Sperimentale dell'Umbria e delle Marche "Togo Rosati" (IZSUM), Via G. Salvemini 1, 06126 Perugia, Italy; s.costarelli@izsum.it

* Correspondence: s.pavone@izsum.it; Tel.: +39-074343202

† These authors contributed equally to this work.

Simple Summary: The African Swine Fever Virus (ASFV) is a DNA virus belonging to the *Asfivirus* genus and Asfarviridae family. In recent years, ASF has grown to be a significant concern for the pig industry worldwide. The ever-growing handling of ASFV pathological material in laboratories, for either diagnostic or research activities, requires particular caution to avoid accidental virus release from laboratories with consequent detrimental economic and environmental effects. This paper aims to describe the Italian National Reference Laboratory (NRL) efforts to develop a biological containment tool for ASF laboratories and animal facilities. The easy-to-apply recommendations focusing on the risk assessment system and internal audits used for enhancing Italian biosafety and biosecurity measures are reported.

Abstract: The African Swine Fever Virus (ASFV) is a DNA virus of the Asfarviridae family, *Asfivirus* genus. It is responsible for massive losses in pig populations and drastic direct and indirect economic impacts. The ever-growing handling of ASFV pathological material in laboratories, necessary for either diagnostic or research activities, requires particular attention to avoid accidental virus release from laboratories and its detrimental economic and environmental effects. Recently, the Commission Delegated Regulation (EU) 2020/689 of 17 December 2019 repealed the Commission Decision of 26 May 2003 reporting an ASF diagnostic manual (2003/422/EC) with the minimum and supplementary requirements for ASF laboratories. This decision generated a regulatory gap that has not been addressed yet. This paper aims to describe the Italian National Reference Laboratory (NRL) efforts to develop an effective and reliable biological containment tool for ASF laboratories and animal facilities. The tool consists of comprehensive and harmonized structural and procedural requirements for ASF laboratories and animal facilities that have been developed based on both current and repealed legislation, further entailing a risk assessment and internal audit as indispensable tools to design, adjust, and improve biological containment measures.

Keywords: African Swine Fever; animal facility; biological containment measures; biorisk management; biosafety; biosecurity; internal audit; laboratory; risk assessment

1. Introduction

Veterinary laboratories and animal facilities routinely handle biological material that may contain infectious agents and toxins that may pose risks to animal and/or public health as well as causing adverse economic effects [1]. Their occurrence is framed as a biorisk. Biorisk management (BRM) is a systematic way of addressing and controlling risks encountered in laboratories, a process that is necessary to protect laboratory workers, animals, and the human population from inadvertent exposures and infections whilst also protecting the environment from accidental or intentional release and subsequent spread of biological agents and their toxins from laboratories [1]. BRM includes a range of documented processes and physical infrastructure which are developed to ensure that both biosafety and biosecurity risks are adequately controlled. According to the World Health Organization (WHO), biosafety is defined as the safe working practices associated with the handling of biological materials, particularly infectious agents. It addresses containment principles, technologies and practices that are implemented to prevent the unintentional exposure to pathogens and toxins, or their accidental release [2]. Biosafety involves administrative controls, standard operating procedures (SOPs), personal protective equipment (PPE), and engineering control [3]. On the other hand, biosecurity focuses on personal and institutional security measurements to protect bioagents from unauthorized access, misuse, thievery, and intentional release [4].

The biosafety and biosecurity of laboratories and animal facilities are based on risk assessment and containment. Risk assessment is the process that allows for the identification of the hazard and its source, establishes the probability of intentional or accidental spreading, and identifies the probability of exposure to contagion (animals or humans) or contamination (environment) [5]. Moreover, it allows for the severity of the epidemiological and economic consequences to be established, as well as for the practices that require implementation to prevent laboratory-associated infections (LAI) to be identified. The primary risk criteria used to define the four ascending levels of containment, referred to as Biosafety Levels 1 to 4 (BSL-1, BSL-2, BSL-3, BSL-4), are: 1. infectivity, 2. severity of disease, 3. transmissibility, and 4. nature of the work being conducted. Another important risk factor for the agents that cause moderate to severe disease is the agent's origin, i.e., whether it is indigenous or exotic. Each level of containment describes the microbiological practices, safety equipment, and facility safeguards for the corresponding level of risk associated with the handling of an agent. The facility safeguards associated with BSL-1 through BSL-4 help protect the non-laboratory occupants of the facility, the public health, and the environment [6]. On the other hand, the fundamentals of containment include microbiological practices, safety equipment, and facility safeguards (PPE) that protect laboratory workers, the environment, and the people from exposure to infectious microorganisms that are handled and stored in the laboratory and/or animal facilities [7]. Worldwide, the accepted classification systems adopted by most international organizations and countries for the different laboratory protection levels are BSL-1, BSL-2, BSL-3, and BSL-4 for facilities designed to handle suitable pathogens in vitro; similarly, ABSL-1, ABSL-2, ABSL-3, and ABSL-4 (animal biosafety laboratory) are used for facilities designed to handle suitable pathogens for in vivo experiments [8].

The African Swine Fever virus (ASFV) is a DNA virus of the Asfarviridae family, *Asfivirus* genus [9]. ASFV is the sole member of its family and the only known DNA arbovirus [10]. The disease does not pose a risk to human health, but is characterized by high morbidity and lethality rates in swines [9]. ASF occurs through transmission cycles involving domestic pigs, wild boars, wild African suids, and soft ticks (*Ornithodorus* spp.) [11]. The ASFV can spread by direct contact between sick and healthy animals as well as indirectly by feeding on garbage containing infected meat (ASFV can remain infectious for 3–6 months in uncooked pork products), mechanic (premises, vehicles, implements, clothes, etc.), or biological vectors (ticks) [12–14]. Responsible for massive losses in pig populations and drastic economic consequences, ASF has become a major concern for the swine industry in recent years [15]. ASF is currently affecting several regions around the world and, with no effective vaccine, the disease is not only affecting animal health and welfare, but also, in certain areas, biodiversity and farmers' livelihoods [16].

Based on a risk assessment for ASFV, the Centers for Disease Control and Prevention National Institutes of Health (CDC) considers ASFV an agent that needs biosafety level 3 (BSL-3) in laboratories and biosafety level 3 (ABSL-3) in animal facilities [6] where swines are used in vaccine efficacy and pathogenicity studies, as well as in the production of reference material.

Recently, the Commission Delegated Regulation (EU) 2020/689 of 17 December 2019 [17] repealed the Commission Decision of 26 May 2003 reporting an African Swine Fever diagnostic manual (2003/422/EC) [18] and the minimum/supplementary safety requirements for ASFV BSL-3 laboratories and animal facilities. This decision created a regulatory gap that has not been addressed yet. Currently, Directive 2000/54/EC of the European Parliament and of the Council of 18 September 2000 on the protection of workers from risks related to exposure to biological agents at work [19] is the only legally recognized measure with respect to containment levels. However, these provisions are specific to the protection of laboratory workers rather than public, animal, and environmental health.

In the current epidemiological scenario, with ongoing ASF outbreaks in several countries across Asia, the Caribbean, Europe, and the Pacific [20–22], it is necessary to promote research on this disease and develop biosecurity measures that are clear, applicable, and cost-effective. Since vaccine development remains a major gap in ASF control and eradication, in vivo and in vitro experiments are essential to increase the knowledge of this disease [23]. On the other hand, the handling of pathogens that pose a high risk to animal health requires strict and specific guidelines recognized by the scientific community.

This paper aims to describe the Italian National Reference Laboratory (NRL) efforts to develop an effective and reliable biological containment tool for ASF laboratories and animal facilities. The tool consists of comprehensive and harmonized structural and managerial requirements for ASF laboratories and animal facilities, further entailing a risk assessment and internal audit system which represents an indispensable tool to design, adjust, and improve biological containment measures. The easy-to-apply structural and managerial requirements were developed considering both current and repealed legislation, the broad biosafety manuals and guidelines published by national and international agencies, and the long-time experience of the Italian NRL with regard to ASF.

2. Materials and Methods

The biological containment tool adopted by the Italian NRL for ASF was developed based on the following three main points: 1. risk assessment, 2. the easy-to-apply requirements for ASF laboratories and animal facilities based on both current and repealed legislation, the standard international guidelines as a reference framework, and the long-time experience of the Italian NRL, and 3. internal audit program.

2.1. The Italian Risk Assessment Guidelines for Accidental Release of ASFV Infected Material from the BSL-3 Containment Area

The developed risk assessment was based on the flowchart described in the WOAH Terrestrial Manual, Chapter 1.1.4. [24], suitably adapted to the scope. The present risk assessment used a qualitative system. For each source of an identified hazard, a judgment was expressed considering the likelihood of the risk-related event occurring and the severity of its consequences. For a better definition of the risk levels, a severity scale with five risk levels ("low", "acceptable", "medium", "remarkable", and "severe") was defined. Low and acceptable levels were ascribed to the activities or areas identified as sources of hazards for which it is not necessary to implement targeted actions aimed at reducing the risk. Conversely, the risk attributed to the activities or areas identified as sources of hazards for which it is essential to implement targeted actions to reduce the risk was defined as medium, remarkable, and severe.

The risk level was based on the combination of the two variables, as reported in Figure 1: (1) the assessment of the likelihood of the risk-related event occurring and (2) the severity of the consequences engendered by the occurrence of the hazard. The first variable was defined by three increasing levels of probability (improbable, possible, and probable), the second variable was defined on a scale of three increasing levels (mild, medium, and marked).

		Severity of consequences		
		Mild	**Medium**	**Marked**
Risk-related event occurring	Improbable	Low	Acceptable	Medium
	Possible	Acceptable	Medium	Remarkable
	Probable	Medium	Remarkable	Severe

Risk level

Figure 1. Diagram used to obtain the risk level.

The risk assessment must be completed prior to any laboratory activities taking place, as well as each time that new procedures and/or risks are identified or changes to the facility, equipment, or agents occur. The biosafety officer conducts the assessment annually, or whenever deemed appropriate, and must identify risk factors and potential sources of contamination that are linked to the identified hazard(s). This activity was carried out jointly with the laboratory and animal facility staff. When implementing the requirements for ASF laboratories and animal facilities, a specific risk assessment was carried out—following the scheme in Figure 1—for each source of hazards identified. The risk assessment, and, therefore, the identification of any preventive measures, was implemented to control and reduce the risk to an acceptable level. When risk mitigation was not possible, the biosafety officer informed the facility director about the identified hazard, as the director is able to order a suspension of work activities related to the process associated with an unacceptable risk factor. Here, we report the list of the eleven identified hazards for which a risk assessment was performed over time in the BSL-3 containment area of the Italian NRL for ASF.

1. Management, storage, and handling of samples contaminated or potentially contaminated with ASFV related to diagnostic and research activities.
2. Infectious or potentially infectious waste (solid and liquid) produced within the BSL-3 containment area.
3. Personnel involved in institutional activities or having access to the BSL-3 containment area.
4. Material (documentation, equipment, tools, etc.) leaving the BSL-3 structure.
5. Contamination resulting from experimental infections conducted in animal housing facilities.
6. Intrusion by unauthorized personnel and possible theft of infected material.
7. Spillage of contaminated material from the sewage system that leads the wastewater to the sterilization system.
8. Entry of personnel involved in research activities using experimental animals housed in the ABSL-3.
9. Interruption of services essential to the containment measures (electricity, water);
10. Non-epidemic emergencies.
11. Epidemic emergencies.

2.2. Requirements for ASF Laboratories and Animal Facility

The Directive 2000/54/EC of the European Parliament and of the Council of 18 September 2000 on the minimum health and safety requirements regarding the exposure of workers to the risks arising from physical/chemical/biological agents [19], as well as the repealed Commission Decision of 26 May 2003 [18] describing the ASF diagnostic manual (2003/422/EC) and the minimum safety requirements for ASF laboratories, were the regulations considered to develop the new ASF biosafety and biosecurity tool.

To provide updated, harmonized, and more comprehensive specific guidelines, the provisions in both legislations were reported in the present recommendations, avoiding redundancy to seamlessly integrate both guidelines and other indications extrapolated from the WHO Laboratory Biosafety Manual [25] and Biosafety in Microbiological and Biomedical Laboratories [6] that were deemed relevant. The scheme followed integrated the sections considered in the ASF diagnostic manual (2000/54/EC) ("General environment", Laboratory clothing", "Control of personnel", "Equipment" for the diagnostic laboratories, and "General environment", Laboratory clothing", "Control of personnel", "Equipment" and "Animals" for the experimental animal rooms) with newly added entries ("Specific disinfection procedures" and "Handling procedures of infected material" for the diagnostic laboratories, and "Specific disinfection procedures" for the experimental animal rooms) necessary to better clarify every aspect of the biosafety for ASF laboratories and animal facilities.

The present requirements did not include general standard practices and procedures, basic training, and basic PPE necessary to ensure the biosafety of laboratories which was not specific to ASF. Therefore, some indications such as the ban on eating, drinking, smoking, contact lenses handling, cosmetics application, and storage of food for human consumption were not reported, unlike other biosafety manuals.

2.3. *Internal Audit Activity of the NRL for ASF*

The internal audit activity was also developed to ensure the proper functioning of the biological containment for ASF laboratories and animal facilities to avoid, in turn, accidental or intentional release of ASFV infected material outside the BSL-3 containment area. The checklists were designed based on the Veterinary Public Health Department of the Istituto Superiore di Sanità (ISS) biosecurity audit system and suitably adapted to our aim. Specific checklists for the general environment, policies, responsibilities, and SOP, personnel, and ASFV repositories were used. Here, we report the issues considered in the Italian NRL internal audit activity in relation to all the above mentioned aspects.

- Biosafety management system requirements, including risk management, analysis of accidents and injuries, monitoring and control of activities and emergency management, and change in management.
- Infrastructure management that includes general issues (such as availability of testing documents on site and commissioning of the system, suitability of laboratory surfaces and general environment, etc.), availability of procedures on commissioning, disposal, and release of materials, access to the working area, and management of premises and equipment.
- Storage and handling of ASFV, including traceability of the biological agents, SOP, inactivation of ASFV, waste management, clothing and PPE, transport, and shipment of ASFV.
- Security management to minimize the risk of unlawful removal of ASFV from the facility.

3. Results

Herein we report: 1. the easy-to-apply requirements for ASF laboratories and animal facilities developed by the Italian NRL, 2. the results of the risk assessment applied to our ASF containment area, and 3. the results of the internal audit program performed over time in our laboratories.

3.1. *Application of Risk Assessment Guidelines for Accidental Release of Infected Material by ASFV Outside the BSL-3 Containment Area*

Here we report the risk assessment for ASFV performed at the start of the activity in the Italian NRL BSL-3 containment area, having identified eleven hazards:

1. Management, storage, and handling of samples contaminated or potentially contaminated with ASFV related to institutional and research activities. All these activities represent a high risk source of environmental contamination outside the BSL-3 facility if the containment measures are not properly applied.
2. Infectious or potentially infectious waste (solid and liquid) produced within the BSL-3 containment area.
3. Personnel involved in diagnostic and research activities or having access to the BSL-3 containment area. The personnel operating in the BSL-3 containment area or any person entering even occasionally the facility is to be considered contaminated or potentially contaminated and, therefore, a possible carrier of the virus outside the facility, with the potential for infection of susceptible animals.
4. Material (documentation, equipment, tools, etc.) leaving the BSL-3 infrastructure. All material leaving the BSL-3 containment area is to be considered potentially contaminated with ASFV and, therefore, a potential source of high risk environmental contamination.
5. Contamination resulting from in vivo experimental studies conducted in the animal facility. The in vivo experimental studies with ASFV in the animal facilities represent a significant risk of environmental contamination, as the animals may become high virus shedders.
6. Intrusion by unauthorized personnel and possible theft of infected material. The theft and intentional release in the environment of ASFV or contaminated material for bioterrorism purposes cannot be excluded.
7. Spillage of contaminated material from the sewage system leading the wastewater to the sterilization system.
8. Entry of personnel involved in research activities using experimental animals housed in the ABSL-3. The research activity requires entry to and exit from the animal facility of personnel who is to be considered potentially contaminated with ASFV and, therefore, a potential source of environmental contamination.
9. Interruption of services essential to the containment measures (electricity and water).
10. Non-epidemic emergencies. The most common non-epidemic emergencies are accidental spillage of material, fire, and earthquakes. In particular, the Italian NRL for ASF is located in Perugia, a town where severe earthquakes may occur.
11. Epidemic emergencies. During an epidemic emergency, increasing analytical activities, flows, handling of samples, etc. may compromise the maintenance of biosafety measures in the BSL-3 premises.

The risk level associated with each hazard, except for hazard 7, was considered remarkable based on the developed diagram (Figure 1), as the possibility of environmental contamination results from the intersection of two events defined as "possible" (probability of the event) and "marked" (severity of the consequences). To reduce the risk to an acceptable level, specific procedures and measures were implemented.

More specifically, the following SOPs were identified and outlined:

(a) Procedure reporting the provisions for biosafety management, storage, and handling of biological material infected or potentially infected with ASFV. This procedure was linked to a further practice reporting the provisions for dangerous goods transport by road, i.e., ADR (European Agreement concerning the International Carriage of Dangerous Goods by Road), rail, i.e., RID (Agreements Concerning the International Carriage of Dangerous Goods by Rail), air, i.e., ICAO (International Civil Aviation Organization), and sea, i.e., IMDG (International Maritime Dangerous Goodsk).
(b) Specific waste management procedure reporting the requirements for management, decontamination, and disposal of waste infected or potentially infected with ASFV.
(c) Procedure regulating access to/exit from the BSL-3 containment area.
(d) SOPs on personnel behavior within the containment area.
(e) SOPs on cleaning and disinfection of environments, surfaces, equipment, and tools.
(f) SOPs on cleaning and disinfection of the animal facility.

(g) SOPs on waste decontamination (solid and liquid effluents).
(h) SOPs on the proper use and maintenance of the incinerator for carcasses disposal; SOPs on management of non-epidemic emergencies.
(i) SOPs on the management of epidemic emergencies. A Crisis Unit designated to strategic choices and a special Emergency Team consisting of technical personnel adequately trained were identified.

Moreover, the following preventive measures were applied to:

- Prevent unauthorized access to the BSL-3 area. A video surveillance system was installed to constantly monitor the perimeter outside the BSL-3 facility.
- Ensure the supply of electrical and water services. Therefore, an electric generator was installed in the BSL-3 facility. When an interruption to the electricity occurs, the electric generator guarantees prolonged autonomy. Similarly, an adequate water supply was set up to guarantee occurrence of essential operations in support of the disinfection procedures.

Regarding point 7, the risk level was considered acceptable because the risk associated with occurrence of environmental contamination results from the intersection of two events defined as "improbable" (probability of the event) and "medium" (severity of the consequences). Indeed, although the spillage of contaminated material from the sewage system cannot be excluded, the system is underground and, therefore, a surface soil contamination is unlikely. Moreover, the BSL-3 containment area is located far from groundwater; therefore, any spillage of contaminated fluid effluents is likely to remain confined within the small underground area, thus posing no significant risk to suids. To further reduce the risk, periodic checks (or checks prompted by suspicion) of the tightness of the sewage system were envisaged.

3.2. *Requirements for BSL-3 Containment Area for ASFV*

The BSL-3 containment area can include an area for the laboratory and one for the animal facility, where the animals are housed during in vivo experimental trials. Within the facility where ASFV is handled, minimum biosecurity and biosafety requirements must be respected, according to the activity carried out and the area in which it takes place.

Minimum requirements for ASFV laboratories must be fulfilled when analyses using inactivated antigens are conducted, whereas maximum requirements must be fulfilled in any laboratory where the ASFV is amplified by in vitro replication into cell cultures.

The developed and adopted recommendations by the Italian NRL included specific requirements for ASF laboratories and animal facilities. Such requirements included those already listed in the repealed Commission Decision 2003/422/EC [18] which were related to the following sections: "General environment", "Laboratory clothing", "Control of personnel", "Equipment" for diagnostic laboratories, and "General environment", "Laboratory clothing", "Control of personnel", "Equipment" and "Animals" for the experimental animal rooms with the addition of new ones (i.e., "Specific disinfection procedures" and "Handling procedures with respect to infected material" for the diagnostic laboratories, and "Specific disinfection procedures" for the experimental animal rooms). In addition, for each section, additional new and/or updated requirements were described. Therefore, Tables S1 and S2 of Commission Decision 2003/422/EC [18] reporting the requirements for ASF laboratories and animal facility were modified as reported in Tables 1 and 2.

Table 1. Biosafety requirements for Level 3 (BSL-3) laboratories with respect to ASFV. In bold are sections present in the repealed Commission Decision 2003/422/EC.

<table>
<tr><th colspan="4">Principles of ASFV Biological Containment Appropriate for Diagnostic Laboratories</th></tr>
<tr><th>Section</th><th>Containment Measures</th><th>Minimum Requirements</th><th>Additional Requirements</th></tr>
<tr><td rowspan="10">General environment</td><td>Working area</td><td>The working area must be separated from any other activity in the same building. Dedicated rooms limited to defined procedures.</td><td>The working area must be separated from any other activity in the same building. Exclusively ASFV-dedicated rooms limited to defined procedures.</td></tr>
<tr><td>Atmospheric pressure</td><td>Normal atmospheric pressure.</td><td>Assisted negative pressure ventilation is recommended.
A visual indicator that displays real-time pressure differentials should be available inside and outside the containment facility to confirm that personnel can enter safely.
Audible alarms should be considered to notify personnel of air flow disruption.</td></tr>
<tr><td>Air filtering system</td><td colspan="2">Vacuum lines must be protected with at least one HEPA [1] filter.</td></tr>
<tr><td>Working area</td><td colspan="2">The working area must be air-locked to allow for disinfection.</td></tr>
<tr><td>Vector control</td><td colspan="2">Efficient vector control (rodents and insects).</td></tr>
<tr><td>Surfaces</td><td colspan="2">• Surfaces must be water repellent and easy to clean. Carpets and rugs are not permitted. Seams, floors, walls, windows, and ceiling surfaces should be sealed. Spaces around doors and ventilation openings should suitable for sealing to facilitate space decontamination.
• Surfaces must be resistant to acids, alkalis, solvents, and disinfectants.</td></tr>
<tr><td>ASFV repository</td><td colspan="2">• All stocks of the ASF virus must be kept in secure storage, whether frozen or freeze-dried. All individual ampoules must be clearly labelled, and comprehensive records maintained of virus stocks together with dates and results of quality control checks. Records must also be kept of viruses added to stock, with details of the source, and of viruses issued to other laboratories.</td></tr>
<tr><td>Windows inspection and personnel safety</td><td colspan="2">An inspection window or other device which allows for the unobstructed view of its occupants and for the entry of at least two people in the diagnostic laboratory is recommended.</td></tr>
<tr><td>Facilities organization</td><td colspan="2">• The facility must be arranged so that personnel ingress and egress are only possible through a series of rooms consisting of:
- A ventilated vestibule with compressible gaskets on the two doors.
- A "clean" changing room outside the containment area.
- A shower room at the non-containment/containment boundary (filter area).
- A "dirty" changing room within the containment area.
• Room within the laboratory for ASFV stock with controlled access. Storage devices must be kept closed and should be accessible to authorized personnel only.
• Toilets for staff must be available outside the containment barrier in the "clean" changing room.
• An emergency eye washer and hand-washing basins must be available in the laboratory.
• A shower must be available in the access filter area. It must be used in case of emergency.
• A biohazard sign must be present on the door.</td></tr>
</table>

Table 1. *Cont.*

Principles of ASFV Biological Containment Appropriate for Diagnostic Laboratories			
Section	**Containment Measures**	**Minimum Requirements**	**Additional Requirements**
Equipment	Biological safety cabinet	Biological safety cabinet (class II) used for all manipulations of live virus. Cabinet should have double HEPA [1] filtration of exhaust air.	
	Autoclave	An autoclave must be available in the facility, preferably within the laboratory.	An autoclave must be available within the laboratory.
	Other equipment	All equipment needed for laboratory procedures to be available within the dedicated laboratory suite.	
Personnel	Entry	Entry to unit limited to named, trained personnel.	Entry to unit limited to named, trained personnel by clearing house.
	Personal dispositions	• Wash and disinfect hands upon leaving the unit. • Personnel not permitted to visit premises with pigs for a minimum of 48 h [2] after leaving the unit.	
	Personnel training	• Laboratory personnel has specific technical and scientific skills. • Personnel to receive updates and additional training when equipment, procedures, or policies change. • An institutional policy regarding visitor training, occupational health requirements, and safety communication is considered.	
Laboratory clothing		• Dedicated outer clothing to be used only in the ASF virus unit. • Disposable gloves to be used for all manipulations of infected material. • Outer clothing to be sterilized before removal from the unit or washed at a high temperature within the unit.	• Complete change of clothes upon entry. • Laboratory clothing to be used only in the ASF virus unit. • Disposable gloves to be used for all manipulations of infected material. • Clothing to be sterilized before removal from the unit or washed at a high temperature within the unit.
Specific disinfection procedures	Potentially or effective contaminated liquid effluents.	Liquid waste effluents must be treated to inactivate the ASF virus (heat or chemical). A wastewater treatment is required.	
	Potentially or effective contaminated solid waste.	Solid waste must be treated to inactivate the ASF virus (heat/incineration or chemical).	
Handling procedures of infected material	Infectious material.	Infectious material must be handled in safety cabinets.	
	Inactivated antigens	The processing of tissues for direct immunofluorescence test (DIFT), molecular biology (PCR), or serology using inactivated antigens may be conducted at a lower containment level, provided that the minimum requirements reported in the present table are fulfilled, basic hygiene is maintained, and post-operational disinfection with the safe disposal of carcasses, tissues, and sera is conducted.	

[1] High Efficiency Particulate Air (HEPA) filter. [2] The duration of the ban for personnel to visit premises with pigs is based on the indications reported in the FAO document [26], proposed as a valid minimum requirements for biosafety measures in the containment working area as well.

Table 2. Biosafety requirements for Level 3 (ABSL-3) experimental animal rooms with respect to ASFV. In bold are sections present in the repealed Commission Decision 2003/422/EC.

Biosafety Requirements for ASF Experimental Animal Rooms		
Section	**Containment Measures**	**Requirements**
General environment	Working area	The working area must be separated from any other activity in the same building. Exclusively ASFV-dedicated rooms limited to defined procedures.
	Atmospheric pressure	Assisted negative pressure ventilation. A visual indicator that displays real-time pressure differentials should be available inside and outside the containment facility to confirm that personnel can enter safely. Audible alarms should be considered to notify personnel of air flow disruption.
	Air filtering system	Vacuum lines must be protected with at least one HEPA [1] filter.
	Working area	The working area must be air-locked to allow for disinfection at end of experiment.
	Vector control	Efficient vector control (rodents and insects).
	Surfaces	• Surfaces must be water repellent and easy to clean. Carpets and rugs are not permitted. Seams, floors, walls, windows, and ceiling surfaces should be sealed. Spaces around doors and ventilation openings should be suitable for sealing to facilitate space decontamination. Surfaces must be resistant to acids, alkalis, solvents, and disinfectants.
	Windows inspection and personnel safety	An inspection window or other device which allows for the unobstructed view of its occupants and for the entry of at least two people in the experimental animal room is recommended.
	Facilities organization	• The facility must be arranged so that personnel ingress and egress are only possible through a series of rooms consisting of: - A ventilated vestibule with compressible gaskets on the two doors. - A "clean" changing room outside the containment area. - A shower room at the non-containment/containment boundary (filter area). - A "dirty" changing room within the containment area. • Animal facilities must be separated by lockable doors. • Toilets for staff must be available outside the containment barrier in the "clean" changing room. An emergency eye washer and hand-washing basins must be available in the laboratory. • A shower must be available in the access filter area. • A biohazard sign must be present on the door.
Equipment	Approved mean for safe disposal	Pathological incinerators, or other approved means, are recommended for the safe disposal of the large carcasses of infected animals and other solid waste.
	Other equipment	All equipment needed for laboratory procedures to be available within the dedicated laboratory suite.

Table 2. *Cont.*

Biosafety Requirements for ASF Experimental Animal Rooms		
Section	**Containment Measures**	**Requirements**
Personnel	Entry	Entry to unit limited to named personnel by clearing house.
	Personal dispositions	Personnel not permitted to visit premises with pigs for a minimum of 48 h [2] after leaving the unit.
	Personnel training	• Laboratory personnel has specific technical and scientific skills. • Personnel to receive updates and additional training when equipment, procedures, or policies change. • An institutional policy regarding visitor training, occupational health requirements, and safety communication is considered.
Laboratory clothing	Personnel change on entry	Complete change of clothes upon entry. Complete animal facility clothing (including undergarments, pants and shirts or jump suits, and shoes and gloves) must be provided in the "dirty" changing room or, alternatively, in the "clean" changing room and put on by personnel before entering the research areas.
	Personnel change on exit	Complete change of clothes upon exit. When leaving an ABSL-3 animal space that acts as the primary barrier and that contains large volumes of aerosols containing highly infectious agents (an animal room, necropsy room, carcass disposal area, contaminated corridor, etc.), personnel must: • Remove "dirty" lab clothing. • Take a shower. • Wear "clean" clothing immediately after leaving this high-risk animal space and before going to any other part of the ABSL-3 facility. Soiled clothing worn in an ABSL-3 space must be eliminated or autoclaved before being laundered.
Specific disinfection procedures	Potentially or effective contaminated liquid effluents	Liquid waste effluents must be treated to inactivate the ASF virus (heat or chemical). A wastewater treatment is required.
	Potentially or effective contaminated solid waste	Solid waste must be treated to inactivate the ASF virus (heat/incineration or chemical) or stored in containers that must be chemically and externally decontaminated before proper disposal as per Regulation 1069/2009 of the European Parliament and of the Council.
Animals		All animals must be slaughtered before leaving the unit, and post-mortem examinations are to be completed within the biosafe area. Carcasses must be treated to inactivate the ASF virus (heat/incineration) or cut and stored in containers that must be chemically and externally decontaminated before proper disposal as per Regulation 1069/2009 of the European Parliament and of the Council.

[1] High Efficiency Particulate Air (HEPA) filter. [2] The duration of the ban for personnel to visit premises with pigs is based on the indications reported in the FAO document [26], proposed as valid minimum requirements for biosafety measures in the containment working area as well.

3.3. Application of Internal Audit Activity at the NRL for ASF

The internal audit occurred annually to ensure the proper application and the efficacy of the biosafety/biosecurity system and avoid every accidental or intentional release of ASFV infected material outside the BSL-3 containment area. Considering the complexity of the checklist, an internal audit was performed each time with different queries never investigated before. This approach allowed for every aspect of the biosafety/biosecurity system to be checked, gradually. The audit report included findings and observations. The audit findings highlighted significant security holes that required immediate intervention, while observations included suggestions aimed at improving the biosafety/biosecurity system further. As a demonstration, we report some significant audit findings and one observation that arose over time during the internal audit process in Table 3.

Table 3. Internal audit results: gaps and actions whose implementation has been necessary for the biosafety and biosecurity of the containment area at the NRL.

Activity	Requirements	GAP	Action
Findings			
1. Possibility to use ABSL-3 animal facility for trials with biological agents requiring a BSL lower than three when activities with ASFV are suspended.	• Secondary independent controlled access to the ABSL-3 animal facility dedicated exclusively to the authorized personnel working with biological agents different from ASFV. • SOPs to access the ABSL-3 animal facility for personnel working with biological agents different from ASFV.	Lack of SOPs to access the ABSL-3 animal facility for personnel working with biological agents different from ASFV.	Writing ad hoc SOPs
2. Stream of samples between the laboratory and animal facility.	• Physical separation between the animal facilities of the BSL-3 and the laboratory. • Passage of authorized, named personnel only from the laboratory to the animal facilities and vice versa. • SOPs on sample exchange between BSL-3 diagnostic laboratory and ABSL-3 animal facility.	• Lack of lockable doors between animal facilities of the BSL-3 and laboratory. • Lack of SOPs on sample exchange between BSL-3 diagnostic laboratory and ABSL-3 animal facility.	• Implementation of lockable doors between animal facilities of the BSL-3 and laboratory • Writing ad hoc SOPs.
Observations			
Activity	**Requirements**	**GAP**	**Action**
3. Proper storage and preservation of biological material.	ASF virus repository.	Presence of a virus repository managed by complex, cumbersome, and not completely safe software for the loading/unloading of aliquots.	Implementation of the new updated software-based system for the ASFV repository.

4. Discussion

Health biotechnology and bioengineering have recently undergone major advances with respect to both human and animal medicine. Since microorganisms may be harmful or pathogenic to animals and humans, to guarantee public health, every possible scenario of an outbreak occurring as a result of the release of a biohazard into the environment cannot be neglected [27]. This implies putting specific biosafety and biosecurity procedures in place. The ever-growing handling of ASFV in laboratories, necessary for either diagnostic or research in this period, requires particular attention to avoid the release of the virus from the laboratories, with consequent detrimental effects on animal health and the pork industry. WHO, FAO, WOAH, national governments, other international agencies, and biosafety associations have published broad biosafety manuals and guidelines in collaboration with expert groups over time. The specific requirements for ASFV laboratories and animal facilities were reported in the Commission Decision of 26 May 2003 n. 422 [18]. However, recently, the Commission Delegated Regulation (EU) 2020/689 of 17 December 2019 [17] repealed the previous Commission Decision, and specific ASF provisions became suddenly unavailable.

Since standards on biosafety and biosecurity may vary according to the laboratory, institution, or even among countries [28,29], a harmonization of criteria is necessary to avoid discrepancies and provide greater protection of the environment and public health. For this purpose, the biosafety group assisted by the NRL drafted specific recommendations for ASF laboratories and animal facilities that aim to bridge the present regulatory gap. The developed specific requirements were based on provisions reported in the repealed Commission Decision 2003/422/EC [18], in the Directive 2000/54/EC of the European Parliament and of the Council still in force [19], and in the biosafety guidelines of international agencies, as well as on the experience of the Italian NRL in the field.

The developed requirements specific for ASF laboratories and animal facilities exhibit the same scheme adopted by the Commission Decision 2003/422/EC [18] to aid consultation by users. However, "Specific disinfection procedures", "Handling procedures of infected material", and "Animals" (only for the ASF ABSL-3 experimental animal rooms) were introduced as new sections. In addition, sub-sections were used to further aid the reader during the consultation of the requirements and to emphasize newly proposed issues, in particular related to the correct functioning of the infrastructure.

The information was reported succinctly, and the words "must", "should", and "recommended" were utilized in association with mandatory or optional provisions. Standard practices and recommendations necessary to ensure the biosafety of the laboratories, such as bans on eating, drinking, smoking, etc., were not included, unlike other biosafety manuals previously proposed by international agencies. Herein, we report only the structural, equipment, and procedural requirements specific to the handling of ASFV by laboratories and animal facilities. More specifically, instructions on structural characteristics of the working area of both laboratories and animal facilities BSL-3 for ASF were added. In particular, the requirement that the working area be air-locked to allow for disinfection was introduced as mandatory. Indeed, periodic disinfections, useful in the ABSL-3 to decontaminate facility rooms, require a sealed environment. In the "General environment" section, the "Atmospheric pressure" subsection of laboratories BSL-3 for ASF entailing assisted negative pressure ventilation was added under the supplementary requirements as a recommended feature. This characteristic was a mandatory requirement for ASF ABSL-3; in the present version, it was considered an added element of biosafety and, therefore, introduced as a suggestion for BSL-3 laboratories with respect to ASF. In the same subsection, a visual indicator displaying real-time pressure differentials and audible alarms were indicated as optional. These recommendations were also included in relation to the ABSL-3 facilities for ASF. An "ASFV repository" subsection was introduced as a requirement for BSL-3 laboratories for ASF. Indeed, generally, in diagnostic and research laboratories dealing with ASF, virus banking is utilized. This practice requires the implementation of precautions to ensure accurate retrieval, proper storage, and disposal. Therefore, in this subsection, any information regarding the proper storage of the ASFV was outlined for compliance with high biosecurity and biosafety levels. In addition, a specific section dedicated to personnel safety was introduced. It was entitled "Windows inspection and personnel safety" and includes some recommended structural requirements and procedures, such as the inspection of the BSL-3 laboratory and animal facility and the recommendation to require a minimum of two individuals when entering the containment area. This section was introduced in relation to both laboratories and animal facilities BSL-3 dealing with ASF. Moreover, a "Facilities organization" subsection was added to both BSL-3 laboratory and animal facility to provide some standard guidelines useful for the design of laboratory facilities that deal with ASFV. Thus, infrastructure endowment, a proper subdivision of the premises, and service requirements were clearly reported. The section "Equipment", reporting the requirements for BSL-3 laboratories in relation to ASFV, was separated into three subsections. With respect to "Biological safety cabinet" and "Other equipment", the original guidelines (2003/422/EC) [18] were maintained. A section entitled "Autoclave" was added to emphasize the necessity to include such instrumentation in ASFV-dedicated facilities. Similarly, with respect to animal facilities, the "Equipment" section was divided into two subsections, one for the safe disposal and the other for all the necessary equipment. Therefore, "Approved means for safe disposal" was introduced to provide guidelines on the proper and safe disposal of large, infected animal carcasses and other type of waste (solid and liquid). More specifically, an incinerator or other approved piece of equipment was suggested. However, to allow for the proper functioning of the animal facility without an incinerator, strict provisions necessary for the safe disposal of waste and carcasses were introduced.

Regarding the "Personnel" section, guidelines on personnel training were introduced in relation to both laboratories and animal facilities, as specific and regular technical and scientific training activities were considered essential. A further novelty concerned the introduction of the "Specific disinfection procedures". The section was split into two subsections for laboratories and animal facilities: "Potentially or effectively contaminated liquid effluents" and "Potentially or effectively contaminated solid waste", where specific guidelines were included. Both heat and chemical inactivation treatments for ASFV were mentioned, and specific wastewater treatment was indicated as a requirement. Under the requirements for animal facilities, more specific guidelines were reported for the disposal of solid waste. Indeed, to allow for the safe disposal of solid waste, even from the animal facilities that do not have an incinerator, information on proper storage of waste was added. More specifically, the solid and liquid waste can be stored in containers that must be chemically and externally decontaminated before proper disposal, as per Regulation 1069/2009 [30]. Similarly, under the section "Animals" referring to the requirements for animal facilities, information on proper disposal of carcasses resulting from in vivo experimentation was included also for facilities without incinerators. For a well-functioning biosecurity system, the proper implementation of the requirements must be accompanied by an effective risk assessment and an internal audit system. The risk assessment has to be considered as a dynamic system necessary from the beginning of the BSL-3 containment area activity and throughout its lifetime when structural and process changes occur and new risks might, therefore, be identified.

The developed risk assessment, based on the biological risk analysis process flowchart described in the WOAH Terrestrial Animal Health Manual [24], identifies the level of risk associated with each hazard. In order to reduce each risk to acceptable levels, specific operating and management modalities must be adopted, and specific procedures were, therefore, drawn up. It is not a static process, but rather a constantly evolving one based on a few, simple yet fundamental steps: (1) information gathering, (2) risk evaluation, (3) risk control strategy development, (4) risk control measure selection and implementation, (5) risks and risk control measures review [24].

On the other hand, the internal audit system is a tool to ensure that the biological containment measures, internal procedures, and effective risk assessment are properly implemented, ensuring the best functioning of the biosafety/biosecurity system and avoiding any accidental or intentional release of ASFV-infected material outside the BSL-3 containment area. The internal audit represents an essential tool to highlight critical aspects of the BSL-3 containment area of the Italian NRL in relation to ASFV. All critical audit findings emerged during the first period of the BSL-3 containment area activity. They were related to "Infrastructure management", a macro-area checking procedural issues, disposal and release of materials, access to the working area, and management of premises and equipment. The ABSL-3 animal facility constitutes a valuable containment premise for the conduction of in vivo trial activities and, therefore, usable also for in vivo studies relying on other biological agents that require a containment level lower than three. However, to ensure a high biosafety level, strict provisions must be indicated and fulfilled when this situation occurs.

The internal audit performed during the first years of activity of the BSL-3 containment area highlighted the absence of a specific procedure for the regulation of the access to and of the activities occurring within the ABSL-3 animal facilities for personnel dedicated to research activities with biological agents different from ASFV. This quality system gap was bridged by making a change to the pre-existing procedures consisting of the addition of specific provisions for activities with biological agents different from ASFV.

An infrastructural gap detected by the internal audit was related to the stream of samples between the laboratory and the animal facility. The laboratory and the animal facility of the BSL-3 containment area must be clearly separated. Despite the existence of a door with a key, the passage from the laboratory to the animal facility and vice versa by named personnel only could not be ensured. Therefore, the installation of an electronic door lock with a code and numeric keypad to regulate the access between the two premises for the BSL-3 containment area was introduced. In addition, a procedural gap was identified. Indeed, a stream of samples can be generated during the experimental trials (samples coming from the animal facility towards the laboratory for diagnostic activities and viruses coming from the laboratory toward the animal facilities for trials). The internal audit showed that no clear provisions were in place for the passage of samples between the laboratory and the animal facilities. Therefore, specific changes to the procedure that regulates access to and activities in the containment area were requested, and instructions on methodologies for the transfer of biological material from animal facilities to the laboratory and vice versa were introduced.

Moreover, the internal audit activity suggested the development of an ASFV repository. A software-based system to archive ASFV biological material, virus strains, and related information was considered necessary and, therefore, introduced.

5. Conclusions

Developing high-quality public standard procedures represents the right way to ensure a high biosafety and biosecurity system, improving global public health and protecting the agro-livestock sector. Although aspects of biosafety, biosecurity, and responsible conduct are common in the literature, the lack of integration leaves space for improvement [31]. The requirements for ASF laboratories and animal facilities, the risk assessment, and the internal audit developed by Italian NRL has led to the drafting of a main document (a biosafety and biosecurity manual) with several linked SOPs.

The provisions reported in the present work, alongside the risk assessment and internal audit system resulting from the efforts and substantial experience of the Italian NRL in relation to ASF, is an example of updated and more comprehensive specific guidelines. These may ensure high biosafety and biosecurity standards for laboratories working with ASFV, as well as representing a useful tool for those who have to design laboratory facilities for the handling of ASFV.

Thanks to this complex biological containment system, along with a major shift in mindset, resting on transversal capabilities, a great improvement of the biological containment system was obtained in the Italian NRL ASF BSL-3 containment area.

Supplementary Materials: The following supporting information can be downloaded at: https://www.mdpi.com/article/10.3390/ani14030454/s1, Supplementary materials report the original tables present in the repealed Commission Decision 2003/422/EC. Table S1. Principles of biological containment appropriate for diagnostic laboratories; Table S2. Biosafety requirements for experimental animal rooms.

Author Contributions: Conceptualization, S.P. (Silvia Pavone), S.C. and F.F.; methodology, S.P. (Silvia Pavone), C.I., M.G., M.S.B., C.R., S.P. (Stefano Petrini), S.C. and F.F.; writing—original draft preparation, S.P. (Silvia Pavone), S.C. and F.F.; writing—review and editing, S.P. (Silvia Pavone), S.C., M.S.B., F.F., C.I. and M.G.; visualization, F.F., S.P. (Silvia Pavone) and S.C.; supervision, F.F. and S.C.; funding acquisition, F.F., M.S.B. and S.C. All authors have read and agreed to the published version of the manuscript.

Funding: This research was funded by a grant from the Italian Ministry of Health: grant number D63C22000820001 (Peste Suina Africana: ruolo dei fattori ambientali nella trasmissione diretta e indiretta della malattia, PSA-RESIST project, RC0052022).

Institutional Review Board Statement: Not applicable.

Informed Consent Statement: Not applicable.

Data Availability Statement: Data are contained within the article and Supplementary Materials.

Conflicts of Interest: The funders had no role in the design of the study, in the collection, analyses, or interpretation of the data, in the writing of the manuscript, or in the decision to publish the results.

References

1. World Health Organization (WHO). *Manual of Diagnostic Tests and Vaccines for Terrestrial Animals of the World Organisation for Animal Health (OIE)*; World Organisation for Animal Health: Pairs, France, 2018; pp. 48–63.
2. World Health Organization (WHO). Biosafety. Available online: https://www.emro.who.int/health-topics/biosafety/index.html (accessed on 12 November 2023).
3. Environmental Health & Safety (Ed.) *Biosafety and Biosecurity Manual*; Michigan State University: Lansing, MI, USA, 2022.
4. World Organisation for Animal Health. *Biorisk Management Laboratory Biosecurity Guidance*; World Organisation for Animal Health: Pairs, France, 2006.
5. Lathers, C.M. Risk Assessment in Regulatory Policy Making for Human and Veterinary Public Health. *J. Clin. Pharmacol.* **2002**, *42*, 846–866. [CrossRef] [PubMed]
6. Centers for Disease Control and Prevention, N.I. of H. *Biosafety in Microbiological and Biomedical Laboratories (BMBL-6)*, 6th ed.; Centers for Disease Control and Prevention: Atlanta, GA, USA, 2020.
7. Ta, L.; Gosa, L.; Nathanson, D.A. *Biosafety and Biohazards: Understanding Biosafety Levels and Meeting Safety Requirements of a Biobank*; Springer: Berlin/Heidelberg, Germany, 2019; pp. 213–225.
8. Zhiming, Y. Current Status and Future Challenges of High-Level Biosafety Laboratories in China. *J. Biosaf. Biosecurity* **2019**, *1*, 123–127. [CrossRef] [PubMed]
9. Gaudreault, N.N.; Madden, D.W.; Wilson, W.C.; Trujillo, J.D.; Richt, J.A. African Swine Fever Virus: An Emerging DNA Arbovirus. *Front. Vet. Sci.* **2020**, *7*, 215. [CrossRef] [PubMed]
10. Alonso, C.; Borca, M.; Dixon, L.; Revilla, Y.; Rodriguez, F.; Escribano, J.M. ICTV Virus Taxonomy Profile: Asfarviridae. *J. Gen. Virol.* **2018**, *99*, 613–614. [CrossRef] [PubMed]
11. Chenais, E.; Ståhl, K.; Guberti, V.; Depner, K. Identification of Wild Boar–Habitat Epidemiologic Cycle in African Swine Fever Epizootic. *Emerg. Infect. Dis.* **2018**, *24*, 810–812. [CrossRef] [PubMed]
12. Mazur-Panasiuk, N.; Żmudzki, J.; Woźniakowski, G. African Swine Fever Virus—Persistence in Different Environmental Conditions and the Possibility of Its Indirect Transmission. *J. Vet. Res.* **2019**, *63*, 303–310. [CrossRef] [PubMed]
13. Carlson, J.; Fischer, M.; Zani, L.; Eschbaumer, M.; Fuchs, W.; Mettenleiter, T.; Beer, M.; Blome, S. Stability of African Swine Fever Virus in Soil and Options to Mitigate the Potential Transmission Risk. *Pathogens* **2020**, *9*, 977. [CrossRef] [PubMed]
14. Boinas, F.S.; Wilson, A.J.; Hutchings, G.H.; Martins, C.; Dixon, L.J. The Persistence of African Swine Fever Virus in Field-Infected Ornithodoros Erraticus during the ASF Endemic Period in Portugal. *PLoS ONE* **2011**, *6*, e20383. [CrossRef] [PubMed]
15. Brown, V.R.; Bevins, S.N. A Review of African Swine Fever and the Potential for Introduction into the United States and the Possibility of Subsequent Establishment in Feral Swine and Native Ticks. *Front. Vet. Sci.* **2018**, *5*, 11. [CrossRef] [PubMed]
16. Li, Z.; Chen, W.; Qiu, Z.; Li, Y.; Fan, J.; Wu, K.; Li, X.; Zhao, M.; Ding, H.; Fan, S.; et al. African Swine Fever Virus: A Review. *Life* **2022**, *12*, 1255. [CrossRef] [PubMed]
17. European Commission. *Commission Delegated Regulation (EU) 2020/689 of 17 December 2019 Supplementing Regulation (EU) 2016/429 of the European Parliament and of the Council*; European Commission: Brussels, Belgium, 2020.
18. European Commission. *2003/422/EC: Commission Decision of 26 May 2003 Approving an African Swine Fever Diagnostic Manual*; European Commission: Brussels, Belgium, 2003.
19. European Commission. *Directive 2000/54/EC of the European Parliament and of the Council of 18 September 2000 on the Protection of Workers from Risks Related to Exposure to Biological Agents at Work*; European Commission: Brussels, Belgium, 2000.
20. Pavone, S.; Iscaro, C.; Dettori, A.; Feliziani, F. African Swine Fever: The State of the Art in Italy. *Animals* **2023**, *13*, 2998. [CrossRef] [PubMed]
21. Ståhl, K.; Boklund, A.; Podgórski, T.; Vergne, T.; Abrahantes, J.C.; Papanikolaou, A.; Zancanaro, G.; Mur, L. Epidemiological Analysis of African Swine Fever in the European Union during 2022. *EFSA J.* **2023**, *21*, e08016. [CrossRef] [PubMed]
22. Sánchez-Vizcaíno, J.M.; Mur, L.; Martínez-López, B. African Swine Fever: An Epidemiological Update. *Transbound. Emerg. Dis.* **2012**, *59*, 27–35. [CrossRef] [PubMed]
23. Muñoz-Pérez, C.; Jurado, C.; Sánchez-Vizcaíno, J.M. African Swine Fever Vaccine: Turning a Dream into Reality. *Transbound. Emerg. Dis.* **2021**, *68*, 2657–2668. [CrossRef] [PubMed]
24. World Organisation for Animal Health Chapter 1.1.4 Biosafety and Biosecurity: Standard for Managing Biological Risk in the Veterinary Laboratory and Animal Facilities. In *Manual of Diagnostic Tests and Vaccines for Terrestrial Animals*; World Organisation for Animal Health: Pairs, France, 2018.
25. World Health Organization. *Laboratory Biosafety Manual*, 4th ed.; World Health Organization: Geneva, Switzerland, 2020; ISBN 9789240011311.
26. Guberti, V.; Khomenko, S.; Masiulis, M.; Kerba, S. *African Swine Fever in Wild Boar*; FAO: World Organisation for Animal Health (WOAH) (founded as OIE); European Commission: Brussels, Belgium, 2022; ISBN 978-92-5-136565-6.
27. Nordmann, B.D. Issues in Biosecurity and Biosafety. *Int. J. Antimicrob. Agents* **2010**, *36*, S66–S69. [CrossRef] [PubMed]

28. Le Duc, J.W.; Anderson, K.; Bloom, M.E.; Estep, J.E.; Feldmann, H.; Geisbert, J.B.; Geisbert, T.W.; Hensley, L.; Holbrook, M.; Jahrling, P.B.; et al. Framework for Leadership and Training of Biosafety Level 4 Laboratory Workers. *Emerg. Infect. Dis.* **2008**, *14*, 1685–1688. [CrossRef] [PubMed]
29. Lipsitch, M.; Bloom, B.R. Rethinking Biosafety in Research on Potential Pandemic Pathogens. *mBio* **2012**, *3*, 1110–1128. [CrossRef] [PubMed]
30. European Commission. *Regulation (EC) 1069/2009 of the European Parliament and of the Council of 21 October 2009 Laying down Health Rules as Regards Animal by-Products and Derived Products Not Intended for Human Consumption and Repealing Regulation (EC) No 1774/2002*; European Commission: Brussels, Belgium, 2009.
31. Perkins, D.; Danskin, K.; Rowe, A.E.; Livinski, A.A. The Culture of Biosafety, Biosecurity, and Responsible Conduct in the Life Sciences: A Comprehensive Literature Review. *Appl. Biosaf.* **2019**, *24*, 34–45. [CrossRef] [PubMed]

Disclaimer/Publisher's Note: The statements, opinions and data contained in all publications are solely those of the individual author(s) and contributor(s) and not of MDPI and/or the editor(s). MDPI and/or the editor(s) disclaim responsibility for any injury to people or property resulting from any ideas, methods, instructions or products referred to in the content.

animals

MDPI

Article

Evaluation of Extended Storage of Swine Complete Feed for Inactivation of Viral Contamination and Effect on Nutritional, Microbiological, and Toxicological Profiles

Jordan T. Gebhardt [1,*], Scott A. Dee [2], Erin Little [2], Brittney N. Scales [3] and Doug R. Kern [3]

1 Department of Diagnostic Medicine/Pathobiology, College of Veterinary Medicine, Kansas State University, Manhattan, KS 66506, USA
2 Pipestone Applied Research, Pipestone Veterinary Services, Pipestone, MN 56164, USA; scottdee7255@gmail.com (S.A.D.); erin.little@pipestone.com (E.L.)
3 eGenesis, Cambridge, MA 02140, USA; brittney.scales@egenesisbio.com (B.N.S.); doug@kernmail.org (D.R.K.)
* Correspondence: jgebhardt@vet.k-state.edu

Simple Summary: The extended storage of feed ingredients has been suggested as a method to mitigate the risk of viral transmission through contaminated ingredients. To validate the approach of extended storage of complete swine feed for inactivation of swine viruses, an experiment was conducted to evaluate the ability of extended storage to inactivate the porcine reproductive and respiratory syndrome virus (PRRSV), porcine epidemic diarrhea virus (PEDV), and Senecavirus A (SVA) when stored for 58 d at 23.9 °C using a self-contained intermodal shipping container. Minimal detrimental impacts were observed when storing feed for 58 d on feed quality measurements including mycotoxin detection, identification and quantification of bacteria, and stability of vitamins and phytase. No infectivity was observed for PRRSV, PEDV, or SVA following 58 d storage at 23.9 °C with the exception of a documented biocontainment breach for PRRSV. The results indicate that the extended storage of complete swine feed can be a method to reduce risks associated with pathogen transmission through feed while having minimal effects on measures of nutritional quality.

Abstract: The extended storage of feed ingredients has been suggested as a method to mitigate the risk of pathogen transmission through contaminated ingredients. To validate the approach of extended storage of complete swine feed for the inactivation of swine viruses, an experiment was conducted wherein swine feed was inoculated with 10 mL of 1×10^5 $TCID_{50}$/mL of porcine reproductive and respiratory syndrome virus (PRRSV), porcine epidemic diarrhea virus (PEDV), and Senecavirus A (SVA) and stored for 58 d at 23.9 °C. Measures of feed quality were also evaluated at the initiation and conclusion of the storage period including screening for mycotoxins, characterization of select microbiological measures, and stability of phytase and dietary vitamins. Storing feed for 58 d under either ambient or anaerobic and temperature-controlled storage conditions did not result in substantial concerns related to microbiological profiles. Upon exposure to the feed following 58 d of storage in a swine bioassay, previously confirmed naïve pigs showed no signs of PEDV or SVA replication as detected by the PCR screening of oral fluids and serum antibody screening. Infection with SVA was documented in the positive control room through diagnostic testing through the State of Minnesota. For PRRSV, the positive control room demonstrated infection. For rooms consuming inoculated feed stored for 58 d, there was no evidence of PRRSV infection with the exception of unintentional aerosol transmission via a documented biocontainment breach. In summary, storing feed for 58 d at anaerobic and temperature-controlled environmental conditions of 23.9 °C validates that the extended storage of complete swine feed can be a method to reduce risks associated with pathogen transmission through feed while having minimal effects on measures of nutritional quality.

Keywords: feed; PEDV; PRRSV; SVA; quality; swine; virus

1. Introduction

Contamination of feed ingredients or complete feed has been shown to be a potential route of pathogen transmission [1,2]. To control this potential route of disease transmission, efforts can focus on the prevention of the contamination step from occurring or implementing strategies to reduce the survival of the bacteria or virus in the material, and thus reduce the likelihood of causing clinical disease when presented to a naïve animal. Prevention strategies focus on biosecurity practices such as ingredient sourcing, controlling points of entry within the feed manufacturing system such as the ingredient-receiving pit, and implementing measures to reduce the likelihood of personnel or other fomites unintentionally introducing the virus into an ingredient or complete feed.

In the scope of feed biosecurity, after prevention practices have been implemented, a common follow-up step is the incorporation of active mitigation techniques to reduce pathogen survival and/or ability to cause infection when exposed to susceptible animals. Multiple practices have been shown to reduce the survival of pathogens including bacteria and viruses in ingredients and/or complete feed including thermal processing [3], chemical feed additives [4–10], and irradiation [8]. While these processes can all reduce viral survival, their use in different practical settings is largely dependent on a number of factors including cost, availability, level of risk aversion desired by the customer, and impact on nutritional components such as vitamins and enzymes. Historically, irradiation has been shown to be a highly effective method to reduce the contamination of feed ingredients or complete feed [8]. Given the high level of efficacy for feed sanitation, irradiation has been commonly used in biomedical research settings to ensure that the feed provided to animals is free of pathogenic agents. However, potential negative consequences of irradiation include significant financial cost, degradation in nutrients including vitamins, and limited throughput based on current infrastructure. Furthermore, research has explored the extended storage of inoculated feed ingredients to reduce viral detection and infectivity. These studies have demonstrated the viral inactivation characteristics of multiple viruses in a variety of feed ingredients and complete feed under different conditions [11,12].

Research has extended beyond the laboratory and has used real-world conditions to evaluate the stability of SVA in soybean meal using 1 metric ton volumes using a temperature of 23.9 °C and a storage of 30 d [13]. However, since feed has the potential for contamination during the milling and delivery phases of the supply chain, additional validation is needed for the extended storage of complete swine feed to inactivate viruses on a practical scale. Thus, exploring potential strategies to achieve high levels of efficacy using alternative methods to irradiation would be useful for a number of industries, including biomedical research using pigs as a research model, given their higher feed intake relative to smaller research animals such as rodents. Previous research has established that a 30 d holding period was effective to inactivate SVA when stored at 23.9 °C [13], and within this experiment it was decided to evaluate viral inactivation and feed quality with a storage period approximately twice as long as the previously established holding guidelines. Our hypothesis is that the extended storage of complete swine feed will result in viral inactivation and will maintain satisfactory levels of nutrient stability. Thus, the objective of this experiment was to validate a 58 d anaerobic and temperature-controlled extended storage protocol to determine the effect on the nutritional quality of complete feed and for inactivation of three swine viruses (PEDV, PRRSV, SVA) when inoculated into complete swine feed.

2. Materials and Methods

2.1. General

To store swine feed during the extended storage period, custom center-flow solid polyethylene containers (dimensions: 1.4 × 1.0 × 0.8 m; RPP Containers, Cincinnati, OH, USA) were used, each having a capacity of approximately 450 kg per container. A climate-controlled intermodal shipping container was used to create the specified environmental storage conditions (A&M Cold Storage and Trailer Leasing, Suwanee, GA, USA; 12.2 m long × 2.3 m wide × 2.6 m tall). The intermodal shipping container was fitted with a self-contained heating and refrigeration unit that operated off of a 480 volt, 3-phase power supply. The dimensions of the feed storage containers were designed so that the intermodal shipping container could store 28 feed totes (14 per layer; two layers high).

This study used a commercial formulation of swine finishing diet (Table 1). The dietary formulation was a standard corn–soybean meal-dried distillers grains with solubles-based diet. A subset of the feed contained a commercially available blend of formic and propionic acid (0.3% inclusion; pHorce, Anpario, Nottinghamshire, UK). The dietary acidifier was included in virus-inoculated feed totes and feed totes used to assess microbiological, toxicological, and nutritional profiles. The remainder of the feed which was used to feed pigs within the bioassay following consumption of the virus-inoculated feed did not contain feed additives and corn was added at the expense of the dietary acidifier. Diets were manufactured at a commercial feed mill (Chandler Feeds, Woodstock, MN, USA) and transported to the biosafety level 2 (BSL-2) Pipestone Research facility. Upon arrival at the facility, a small amount of feed was transferred into the bottom of each container; then, a 2.26 kg block of dry ice was added to the storage container, and the container was then filled to approximately 450 kg of feed per container. A total of 29 feed storage containers were used, each containing approximately 450 kg of swine feed (Figure 1). Methods used in the experiment are shown in Figure 2.

Table 1. Diet composition (as-fed basis) [1].

Item:	Swine Finishing Diet
Ingredient, %	
Corn	84.81
Soybean meal, 46.5% CP	7.35
Corn DDGS	5.00
Calcium carbonate	1.00
Salt	0.50
L-Lys HCl	0.37
Monocalcium phosphate, 21% P	0.25
Threonine [1]	0.13
Vitamin premix [2]	0.13
Trace mineral premix [3]	0.13
L-Trp	0.02
Phytase [4]	0.02
DL-Met	0.01
Dietary acidifier [5]	0.30
Total	100
Calculated analysis, %	
CP	12.00
Ether extract	3.42
Ca	0.48
Available P	0.24

[1] Thr-PRO (CJ, Downers Grove, IL, USA). [2] Each kilogram contains 1,653,468 IU vitamin A, 661,387 IU vitamin D3, 17,637 IU vitamin E, 1,323 mg menadione, 3,307 mg riboflavin, 11,023 mg pantothenic acid, 19,842 mg niacin, 13.2 mg vitamin B12. [3] Each kilogram contains 22 g Mn, 73 g Fe, 73 g Zn, 11 g Cu, 198 mg I, and 198 mg Se. [4] HiPhos 2700, DSM Nutritional Products, Parsippany, NJ. [5] Commercial blend of formic and propionic acid (pHorce, Anpario, Nottinghamshire, United Kingdom). Feed additive was included in virus-inoculated feed totes and feed totes used to assess microbiological, toxicological, and nutritional profiles. Remainder of feed did not contain feed additives and corn was added at the expense of the dietary acidifier.

2.2. Inoculum Preparation and Tote Inoculation

Following transfer of feed into feed storage containers, 5 containers were inoculated with viral inoculum prior to being sealed. Viral inoculum consisted of PRRSV-144 L1C variant, PEDV, and SVA and was confirmed to be viable through the propagation and titration procedures at South Dakota State University. Selection of these viruses was based on the recent emergence of the PRRSV-144 L1C variant, history of PEDV in feed biosecurity research, and stability of SVA in feed ingredients and complete feed. Inoculation procedures were identical to those previously described [13], with the addition of PEDV in the current experiment. Briefly, each of the 3 viruses was prepared into a 10 mL ice cube containing a mixture of each virus with a total dose of 1×10^5 $TCID_{50}$ per virus. Each virus was diluted in 30 mL of minimum essential medium (MEM, Sigma-Aldrich, St. Louis, MO, USA). Ice cubes were prepared by freezing 10 mL aliquots of the mixture in 50 mL conical centrifuge tubes (Corning Inc., Corning, NY, USA) at −80 °C.

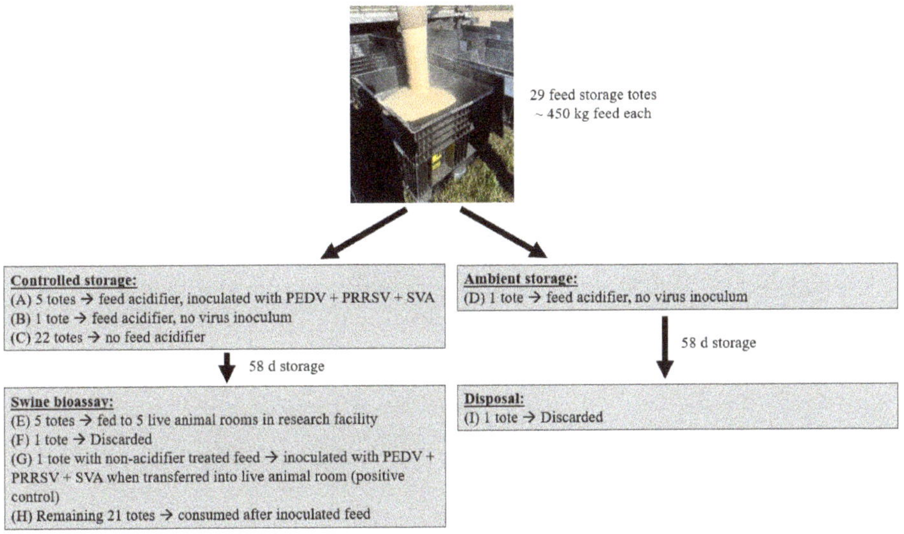

Figure 1. Experimental design.

To inoculate totes, a solid metal rod (1.2 m length × 4.45 cm diameter) was inserted into the lumen of a PVC pipe (1.2 m length × 5.08 cm diameter). The combined rod and pipe "inoculation instrument" was manually forced into the middle of each tote. Once the instrument was in the proper place, the internal metal rod was removed, providing a clear path for the ice cube to travel into the interior of the tote, unimpeded by the presence of meal. Following the removal of the rod, the 10 mL ice cube was dropped into the PVC tube, the PVC tube was removed, resulting in the cube being buried by the meal and creation of the "hot spot" centralized in the tote.

2.3. Storage Procedure

The containers had solid plastic lids attached and were stored outside overnight to allow excess CO_2 to sublimate for human health considerations when transferring into a confined airspace of the intermodal shipping container. Following overnight storage in open air, feed storage containers were transferred to their appropriate storage location. A total of 28 totes were transferred into the intermodal shipping container to fill to full capacity. Five totes contained feed treated with the commercial feed acidifier and treated with the viral inoculum, one tote was transferred and contained feed treated with the commercial feed acidifier and no inoculation with virus to serve as a source of feed to be used to assess stability of nutrients including vitamins and enzyme stability. This tote was used so no known viral-inoculated feed would be sent to laboratories for nutritional analysis. Additionally, 22 totes were transferred into the intermodal shipping container which contained feed not treated with the commercial feed acidifier and this feed was used to maintain pigs during the bioassay following consumption of the viral-inoculated feed. Following placement of the feed in the intermodal shipping container, approximately 45 kg of dry ice was placed between the feed storage containers, the door was sealed, and a vent located at the top of the intermodal shipping container was kept open for 12 additional hours to allow any pressure to escape as the dry ice continued to undergo sublimation and the denser CO_2 displaced air out of the discharge vent. After this period of venting excess pressure, the container was fully sealed, and the integrated refrigeration unit was set to a target temperature of 23.9 °C based on previous research [13], which would be automatically maintained over the storage period.

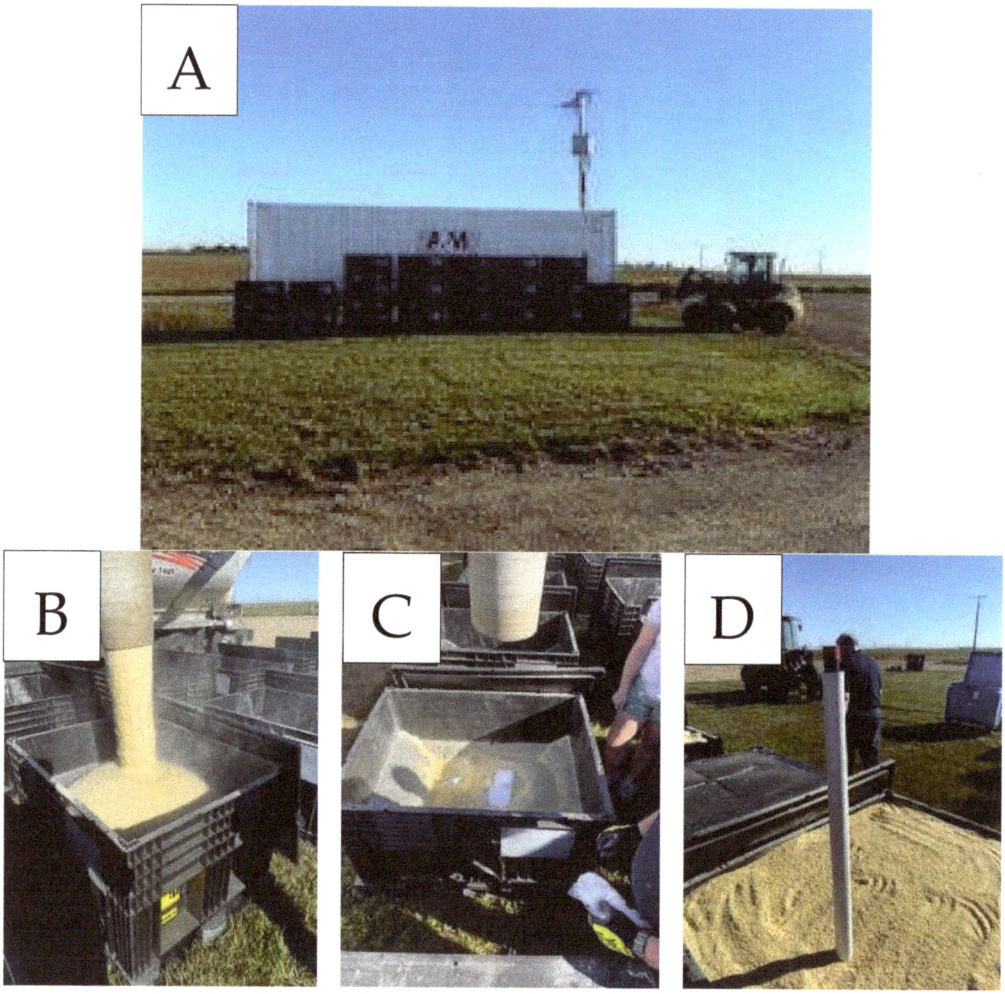

Figure 2. Equipment and processes. (**A**) Intermodal shipping container along with feed storage containers; (**B**) filling of feed storage containers; (**C**) adding dry ice to feed storage container after bottom surface has been covered with base layer of feed; and (**D**) viral inoculation technique once feed storage totes have been filled.

In addition to the feed storage totes under anaerobic and temperature-controlled storage conditions in the intermodal shipping container, one tote was placed in an uninsulated machinery storage shed to characterize the nutritional stability in uncontrolled environmental storage conditions. This study was conducted in southwest Minnesota from approximately September through October 2022, which resulted in significant fluctuation in ambient conditions.

2.4. Environmental Monitoring during Storage

During the process of filling feed storage containers, temperature and relative humidity (% RH) data loggers (RC-51H waterproof USB temperature data logger, Elitech, San Jose, CA, USA) were placed to characterize storage conditions with data recorded every 10 min continuously throughout the storage period. A total of 11 data loggers were used, with six being placed within feed storage totes stored in controlled conditions and one within the feed storage tote stored in ambient conditions to characterize the conditions within the feed. In addition to monitoring conditions within feed, three data loggers were placed in the intermodal shipping container approximately 11.5 cm from the floor and distributed evenly within the intermodal shipping container to characterize air temperature and humidity over the storage period. And a final data logger was placed in the machinery storage shed to characterize the conditions for the tote stored in ambient environmental conditions.

2.5. Feed Sampling and Quality Measurements

Feed samples were collected on d 0 from the feed storage tote which was stored at ambient conditions and from the feed storage tote to be stored in controlled environmental conditions. Feed samples were collected from totes containing feed treated with the commercial blend of formic and propionic acids. Feed was sampled using a feed probe to collect 10 samples per feed storage container and were mixed to create a composite sample for each container. This sample process was repeated following the 58 d storage period. Feed samples from d 0 and d 58 were analyzed at the North Dakota State University Veterinary Diagnostic Laboratory using liquid chromatography–tandem mass spectrometry (LC/MS/MS). Feed samples were also analyzed to characterize microbiological profiles by assessing for concentration of yeast and mold (FDA BAM Chapter 18), aerobic plate count (AOAC 966.23), *Salmonella* (AOAC-RI 121501), *Escherichia coli* O157:H7 (AOAC-RI 031002), *Staphylococcus aureus* (FDA BAM Chapter 12), and *Clostridium perfringens* (ISO 7937). Feed samples collected on d 0 as well as feed samples stored for 58 d at ambient, controlled, and refrigerated (4 °C) conditions were analyzed at the DSM Nutritional Products Technical Marketing Analytical Services Laboratory (Belvidere, NJ, USA) for vitamins A, B5, D3, E, and phytase activity.

2.6. Virus Infectivity Assessment

Following the 58 d storage period, assessment of viral infectivity was performed using a BSL-2 live animal research facility. This facility had six rooms, each with its own feed bin, allowing for an individual tote to be allocated to a specific bin and be fed to a specific group of pigs. These rooms were organized into separate airspaces using filtration of incoming and outgoing air, with separate shower in/out, designated clean and dirty areas with a clear line of separation, along with the use of separate clothing, footwear, disposable gloves, and equipment per room to prevent room-to-room cross-contamination.

At the conclusion of the 58 d storage period, five sets of totes were transferred into five separate feed bins. A sixth room was used as a positive control where a feed tote was transferred into the feed bin with feed that was identical in formulation with the exception of not containing the commercial feed additive blend containing formic and propionic acid. The same viral inoculum was then added directly to the feed auger at the bottom of the bin and transferred into the live animal room using the automated feeding system.

Each room was filled with approximately 30 pigs originating from a PRRSV, SVA, and PEDV-negative farm. Following consumption of the test feed, the pigs were fed diets that did not contain the formic/propionic acid commercial feed additive for the remainder of the study. During the 28-d bioassay, all six rooms of pigs were assessed for evidence of SVA and PRRSV 144 L1C infection through the collection of oral fluids (one rope per pen, six ropes per room) at d 0 (arrival), 14, and 30 of each replicate. In addition, daily observations were conducted, looking for clinical signs of SVA infection (vesicles and lameness) and signs of PRRSV 144 L1C variant infection (pyrexia, hyperemia, dyspnea, and weight loss). Selected cases of mortality were necropsied, lymphoid tissues collected, and tested by PCR for the presence of SVA and PRRSV RNA. Serum samples were analyzed for the presence of PEDV (fluorescent focus neutralization assay, PEDV USA/Colorado/2013 used as indicator virus), PRRSV (X3 Elisa), and SVA (IFA) antibodies. Oral fluid samples were analyzed using RT-PCR for detection of viral RNA. All diagnostic testing was performed at the Iowa State University Veterinary Diagnostic Laboratory.

2.7. Statistical Analysis

Data were reported using appropriate statistical tests when possible and using descriptive statistics when sufficient variability was not present to satisfy the assumptions of ANOVA. Visualization of the temperature and humidity data over the storage period was performed using the ggplot2 package in R using the ggplot function (version 4.2.1; 23 June 2022). For vitamin stability analysis, data were analyzed as a one-way ANOVA using the lmer function from the lme4 package in R with sample storage condition as a fixed effect. An individual feed sample was considered the experimental unit. Degrees of freedom were approximated using the Kenward–Roger approach and pairwise comparisons were performed using a Tukey multiple comparison adjustment. Results are considered significant at $p \leq 0.05$ and marginally significant from $p > 0.05$ to $p \leq 0.10$.

3. Results

3.1. Storage Conditions

The environmental conditions for the ambient and controlled storage conditions are shown in Figure 3. Under controlled storage, the thermostat was set to 23.9 °C and the temperature remained relatively consistent over the duration of storage. There were differences observed based on the location of the data logger within the intermodal shipping container based on restricted airflow as the shipping container was completely full of feed storage totes, which likely limited air circulation. Relative humidity also differed based on the data logger location within the intermodal shipping container and there was a general reduction in relative humidity over time. Under ambient storage conditions, there was a high degree of variation in temperature and humidity within day as well as over the entire duration of the extended storage. Drastic differences in temperature and relative humidity were observed, as would be expected with no environmental control, and the temperature decreased over time, as would be expected based on geographic region and time of the year progressing through the fall months.

Temperature and relative humidity were also measured within the feed storage totes. Under controlled storage, there was variability in temperature and humidity between the feed storage totes evaluated, and this is likely due to the tote position within the intermodal shipping container. However, the feed remained at a more consistent temperature and relative humidity over time in the controlled storage conditions compared to the ambient storage conditions, as would be expected based on the environmental data. Over the course of the experiment, the temperature of the feed stored under ambient conditions decreased reflecting the environment, whereas the feed in the controlled storage remained consistent in terms of temperature and humidity.

3.2. Mycotoxin Analysis

To validate the safety of the feed following extended storage, a series of assays were conducted to look at mycotoxin concentrations as well as to evaluate the stability of key nutritional components including vitamins and the phytase enzyme. Samples collected at the beginning of the experiment and following 58 d of storage in either ambient or anaerobic and temperature-controlled conditions did not contain any detectable mycotoxins (Table 2). On d 0, there were detectable levels of yeast in one of two feed samples (120 cfu/g), mold in both feed samples (2700–2800 cfu/g), as well as aerobic plate counts ranging from 2000 to 5300 cfu/g. These levels are within the range of expected parameters for commercial swine feed based on the author's experience with this diagnostic testing. Following the 58 d storage at ambient conditions, there was a slight increase in the detection of yeast (490 cfu/g) relative to d 0 samples, similar aerobic plate count (2200 cfu/g), and reduced mold concentration (320 cfu/g). Following 58 d of controlled storage, there was no detectable yeast (<10 cfu/g), and the concentration of mold (90 cfu/g) and aerobic plate count (260 cfu/g) were both lower than samples collected on d 0. Thus, it did not appear that storage of complete feed for 58 d under either ambient or controlled anaerobic and temperature-controlled storage conditions resulted in substantial concerns related to microbiological profiles.

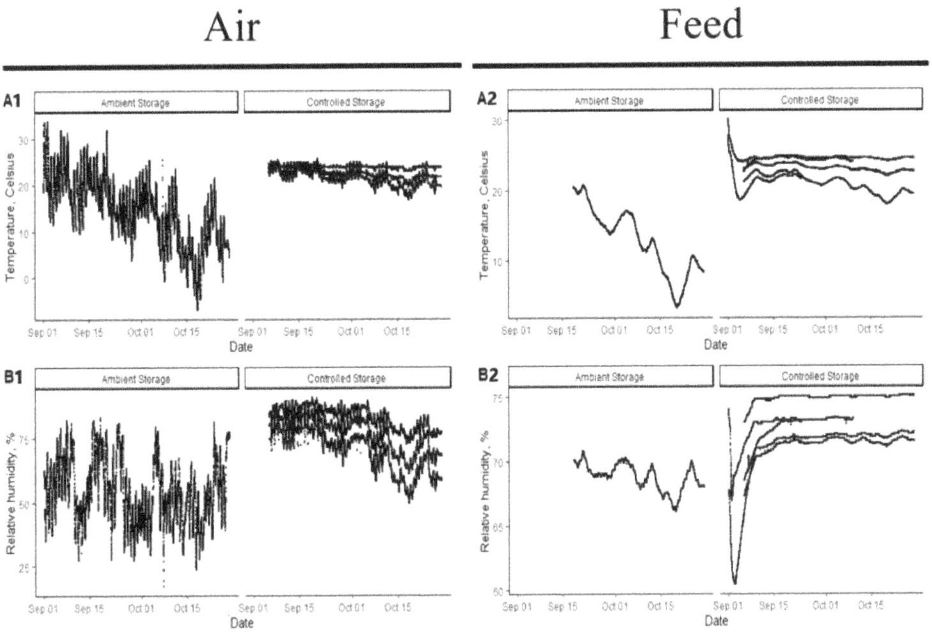

Figure 3. Temperature (**A1**) and relative humidity (**B1**) of air in ambient or controlled storage conditions. Temperature (**A2**) and relative humidity (**B2**) of feed stored in ambient or controlled storage conditions.

3.3. Nutritional Analysis

For analysis of the concentration of the phytase enzyme, the analyzed content was 175% of the calculated claim of concentration on d 0 (Table 3). Following 58 d storage either at refrigerated temperatures, anaerobic and temperature-controlled storage in the intermodal shipping container, or under ambient conditions, there was a reduction ($p < 0.05$) in phytase activity compared to d 0. However, there were no differences between the three storage conditions ($p > 0.05$) for phytase activity when stored for 58 d. For vitamin A, the analyzed concentration on d 58 for feed samples stored in controlled or ambient conditions had greater ($p < 0.05$) vitamin A concentration compared to the d 0 baseline, and feed stored for 58 d under ambient conditions had greater ($p < 0.05$) vitamin A concentration compared to feed stored at 4 °C for 58 d.

There is substantial variation associated with vitamin A analysis. This variation can be caused by variation due to sample collection, as well as laboratory analytical variation. Following 58 d storage at either controlled or ambient conditions, the concentration of vitamin A was still near the calculated claim for vitamin A concentration, indicating there was not a significant reduction in the concentration over time. For vitamin B5, there was a reduction in concentration following 58 d of storage at 4 °C, ambient, and controlled conditions compared to the d 0 baseline ($p < 0.05$), but concentrations were still near formulated claim levels following this period of storage. There was no evidence of changes

($p > 0.05$) in vitamin D3 or vitamin E concentration following 58 d of storage at any of the conditions evaluated.

Table 2. Effect of complete feed storage on detection of mycotoxins and microbial profile.

Item		D 0 [1]		d 58	
	Limit of Detection [2]	Sample 1	Sample 2	Controlled Storage	Ambient Storage
Mycotoxin analysis					
Aflatoxin B1	20	<20	<20	<20	<20
Aflatoxin B2	20	<20	<20	<20	<20
Aflatoxin G1	20	<20	<20	<20	<20
Aflatoxin G2	20	<200	<200	<200	<200
Fumonisin B1	200	<200	<200	<200	<200
Fumonisin B2	200	<200	<200	<200	<200
Fumonisin B3	200	<200	<200	<200	<200
HT-2	200	<20	<20	<20	<20
T-2	20	<200	<200	<200	<200
DAS	200	<20	<20	<20	<20
Ochratoxin A	20	<20	<20	<20	<20
Sterigmatocystin	20	<100	<100	<100	<100
Zearalenone	100	<200	<200	<200	<200
DON	200	<200	<200	<200	<200
15-ADON	200	<200	<200	<200	<200
3-ADON	200	<20	<20	<20	<20
Microbiology analysis					
Yeast, cfu/g	10 cfu/g	120	<10	<10	490
Mold, cfu/g	10 cfu/g	2800	2700	90	320
Salmonella, cfu/25 g	ND per 25 g	ND	ND	ND	ND
E. coli 0157:H7, cfu/25 g	ND per 25 g	ND	ND	ND	ND
Staphylococcus aureus, cfu/g	10 cfu/g	<10	<10	<10	<10
Aerobic plate count, cfu/g	10 cfu/g	5300	2000	260	2200
Clostridium perfringens, cfu/g	10 cfu/g	<10	<10	<10	<10

[1] Two samples analyzed on d 0 representing two composite samples of the manufactured feed. Samples for d 58 controlled storage and d 58 ambient storage are the result of 1 sample collected from each storage condition. [2] For mycotoxin testing, limit of detection if practical quantification limit (PQL, μg/kg) as reported by North Dakota State University Veterinary Diagnostic Laboratory (Fargo, ND, USA). For microbiology testing, limit of detection as reported by Eurofins Microbiology Laboratories (Garden Grove, CA, USA).

Table 3. Effect of complete feed storage on quantification of dietary vitamins and phytase [1].

Item	d 0	d 58		
	Baseline	4 °C	Controlled Storage	Ambient Storage
n=	6	6	3	3
Phytase				
Claim, FYT/kg	500	500	500	500
Analyzed, FYT/kg	877 ± 113.5^{a}	578 ± 54.7^{b}	584 ± 85.2^{b}	497 ± 43.8^{b}
Percent of claim	175 ± 22.7^{a}	116 ± 10.9^{b}	117 ± 17.0^{b}	99 ± 8.8^{b}
Vitamin A				
Claim, IU/kg	4135	4135	4135	4135
Analyzed, IU/kg	2425 ± 634.0^{c}	2702 ± 788.6^{bc}	4174 ± 984.5^{ab}	4571 ± 1015.4^{a}
Percent of claim	59 ± 15.3^{c}	65 ± 19.1^{bc}	101 ± 23.8^{ab}	111 ± 24.6^{a}

Table 3. *Cont.*

Item	d 0	d 58		
	Baseline	4 °C	Controlled Storage	Ambient Storage
Vitamin B5				
Claim, mg/kg	28	28	28	28
Analyzed, mg/kg	91.5 ± 10.7 [a]	24.8 ± 12.6 [b]	27.3 ± 4.2 [b]	25.0 ± 2.6 [b]
Percent of claim	327 ± 38.2 [a]	89 ± 44.9 [b]	98 ± 14.9 [b]	89 ± 9.4 [b]
Vitamin D3				
Claim, IU/kg	1654	1654	1654	1654
Analyzed, IU/kg	1459 ± 411.2	1539 ± 222.5	1587 ± 119.2	1659 ± 180.4
Percent of claim	88 ± 24.9	93 ± 13.5	96 ± 7.2	100 ± 10.9
Vitamin E				
Claim, IU/kg	44	44	44	44
Analyzed, IU/kg	48.8 ± 5.7	50.5 ± 2.1	45.0 ± 0.0	43.7 ± 2.5
Percent of claim	111 ± 13.0	115 ± 4.7	102 ± 0.0	99 ± 5.7

[1] Samples analyzed by DSM Nutritional Products Technical Marketing Analytical Services Laboratory (Belvidere, NJ, USA). Differences between sample storage conditions for analyzed phytase or vitamin concentration and percentage of recovery relative to calculated claim was evaluated using a linear model. If results of the one-way ANOVA were $p < 0.05$, pairwise comparisons between storage conditions were performed and means lacking common superscript within analysis row differ ($p < 0.05$). Mean ± standard deviation.

3.4. Viral Stability during Storage

Following extended storage, inoculated feed samples were fed to pigs to determine whether infectious PRRSV, PEDV, or SVA was present. On d 0, there was no evidence of serum antibodies for any of the three viruses (Table 4). On d 27 of the bioassay, antibodies were detected for PRRSV in the positive control room (virus inoculated into feed immediately prior to feeding) as well as the adjacent room. There were no serum antibodies detected on d 27 for PEDV or SVA. Over the course of the experiment, there was no PEDV or SVA RNA detected in oral fluids on any sampling day (0, 6, 14, 20, and 27 dpi). For PRRSV, RNA was detected in three of six rooms (rooms 2, 3, and 5) on 0 dpi, indicating the presence of fragments of viral RNA within the feed and thus oral fluids of pigs sampled. However, in rooms 2 and 3, no subsequent oral fluid samples and the sample collected from room 5 on 6 dpi did not contain detectable PRRSV RNA, indicating no evidence of infection and confirming that the virus detected on 0 dpi was not infectious. For the positive control room, RNA was detected in oral fluids beginning on 6 dpi and all subsequent samples collected at 14, 20, and 27 dpi contained PRRSV RNA, indicating that the infectious virus was present in the inoculated feed sample. On 14 dpi, PRRSV RNA was detected in room 5, which was adjacent to the positive control room. This detection was attributed to a biosecurity breach within the facility and unintentional airflow was discovered between those two rooms. Thus, based on the lack of PRRSV RNA detection in room 5 on 6 dpi and subsequent detection beginning on 14 dpi, it appears that the infectious virus had spread from the positive control room into room 5 via this air leak.

Table 4. Effect of complete feed storage on serum antibody testing [1].

Item	0 dpi			27 dpl		
	PEDV	PRRSV	SVA	PEDV	PRRSV	SVA
Test room 1	0/10	0/10	0/10	0/10	0/10	0/10
Test room 2	0/9	0/9	0/9	0/9	0/9	0/9
Test room 3	0/10	0/10	0/10	0/10	0/10	0/10
Test room 4	0/9	0/9	0/9	0/9	0/9	0/9
Test room 5	0/11	0/11	0/11	0/11	10/11	0/11
Positive control	0/10	0/10	0/10	0/10	10/10	0/10

[1] Samples analyzed by Iowa State Veterinary Diagnostic Laboratory (Ames, IA, USA). PEDV = porcine epidemic diarrhea virus, PRRSV = porcine reproductive and respiratory syndrome virus, SVA = senecavirus A, dpi = day post-inoculation.

No SVA RNA or antibodies to SVA were detected in any of the experimental rooms including the positive control room. However, clinical signs were present in the positive control room and through diagnostic testing through the State of Minnesota confirmed the presence of lesions consistent with SVA and detection of viral RNA in clinical samples from the positive control room, indicating that the infectious virus was present in the inoculated positive control feed.

4. Discussion

Contamination of feed ingredients before milling or complete feed during or post milling and delivery has been shown to be a potential route of pathogen transmission [1,2]. To control this potential route of disease transmission, efforts can focus on the prevention of the contamination step from occurring or implementing strategies to reduce the survival of the bacteria or virus in the material, and thus reduce the likelihood of causing clinical disease when presented to a naïve animal. Prevention of pathogen introduction and strategies that can be implemented have been previously described [14]. Strategies to reduce the survival of pathogens in feed ingredients or complete feed include thermal processing through strategies such as pelleting of complete diets [3], chemical feed additives [4–10], irradiation [8], and extended storage [13]. While these processes all can reduce viral survival, their use in different practical settings is largely dependent on several factors including cost, availability, and the level of risk aversion desired by the customer.

Historically, the irradiation of feed ingredients and complete diets has been shown to reduce the concentration of bacteria within feed and improves growth performance [15,16]. Additionally, irradiation has been shown to inactivate swine viruses including PEDV [8,17]. Given the high degree of efficacy for feed sanitation, irradiation has been commonly used in biomedical research settings to ensure that feed provided to animals is free of pathogenic agents. However, potential negative consequences of irradiation include significant financial cost, degradation of nutrients including vitamins [18,19], and limited throughout based on current infrastructure.

Furthermore, research has explored the extended storage of inoculated feed ingredients to reduce viral detection and infectivity. These studies have characterized the viral inactivation characteristics of multiple viruses in a variety of feed ingredients and complete feed under different conditions [11,12]. Research has extended beyond the laboratory and has used real-world conditions to evaluate the stability of SVA in soybean meal using 1 metric ton volumes [13]. However, additional validation is needed for extended storage of complete swine feed to inactivate viruses on a practical scale. In the current experiment, our goal was to build upon the previous literature and validate the extended storage of swine complete feed for the inactivation of swine viruses for potential use in settings such as biomedical research to provide alternative efficacious techniques to maintain feed safety using techniques other than irradiation.

As expected, storing complete swine feed under anaerobic and temperature-controlled environmental conditions reduced the variation in temperature and humidity compared to storage in ambient environmental conditions. It is hypothesized that maintaining a constant temperature and humidity would preserve feed quality compared to fluctuations in temperature and humidity, although little evidence is available currently. In the current experiment, there was no evidence of mycotoxin production over the 58 d storage period in either ambient or controlled environmental conditions. Furthermore, there was limited evidence of an increase in detrimental microbiological measures over the 58 d storage in either ambient or controlled storage conditions. The yeast concentration increased over the 58 d storage period for feed stored in ambient environmental conditions, but the measured concentration of 490 cfu/g is still within common values observed in commercial swine feed, and yeast is also commonly added as a feed additive in commercial swine diets [20]. The observed levels would not be expected to cause any detrimental health outcomes.

Phytase is a common enzyme used in swine production to increase the utilization of plant-origin phosphorous in the form of phytase-bound P. Phytase stability is a critical measurement and its recovery decreases after extended storage [21] and following thermal processes such as pelleting [22,23]. It is common for the measured concentration of phytase to be well-above formulated levels as a margin of safety given these decay properties. In the current study, the analyzed concentration of phytase on d 0 was approximately 175% of the formulated value. There was a decay in the phytase activity following the 58 d storage period in all storage conditions, but the concentration of phytase was still approximately 100% of formulated claim following this period of extended storage. Thus, the results observed in the current study are consistent with the previous literature and illustrate that the phytase enzyme does lose activity following extended periods of storage which must be considered when implementing extended storage practices.

There is significant variation when measuring dietary vitamin concentration in finished feed attributed to both sampling and analytical variation. In the current experiment, there was evidence of decay over time for vitamin B5, but no evidence of reduced vitamin activity over time for vitamin A, D3, or E under any storage conditions.

Previous research evaluated the stability of vitamins in vitamin premixes and observed that vitamin A is one of the most unstable vitamins in premixes, while vitamins D3, B5, and E were rather stable [24]. Similar results were observed by other researchers who observed that vitamin B5 was rather stable when stored in vitamin and vitamin-trace mineral premixes [25]. In the current experiment, vitamin B5 demonstrated a greater loss over the 58 d storage period compared to the other vitamins evaluated. Previous research also evaluated the stability of vitamin A in vitamin premixes and observed a high degree of stability over time up to 1 year of storage [26], which is similar to the results of the observed study albeit it a much shorter storage duration was investigated, and storage form was different (vitamin premix vs. complete swine feed). Data also exist indicating that vitamin form can have a significant impact on stability during storage. Particularly for vitamin A, there are multiple forms available including a cross-linked beadlet form and non-cross-linked forms. The cross-linked form has greater stability [27], and this was the form that was used in the current experiment. Although there are data evaluating vitamin stability over extended storage periods, much of the research has focused on the practical storage of vitamin and vitamin-trace mineral premixes and not the extended storage of complete feed.

From the perspective of measurements of feed quality, we did not observe significant detrimental effects of storing complete feed for 58 d at controlled or ambient conditions. It is important to consider, however, that the current experiment was aimed to validate practical conditions which included the use of a commercial feed additive containing a blend of formic and propionic acids as well as the creation of a semi-anaerobic environment using dry ice (solid CO_2) when filling feed storage totes. Additional research is needed to fully understand the impact of feed storage on nutrient stability and microbiological profiles.

Following this period of extended anaerobic and temperature-controlled storage, an important goal was to validate inactivation of PEDV, PRRSV, and SVA. For PRRSV, evidence of infection was observed in the positive control room as expected, validating that including viral inoculum into feed immediately prior to feeding caused infection. There was a detection of viral RNA in oral fluids on 0 dpi, which indicates the presence of viral RNA in the feed but did not lead to infection, as documented by the lack of antibody production over time in those rooms as well as the lack of further detection of PRRSV RNA in oral fluids at later sampling points. There was detection of PRRSV infection in a room which was fed feed inoculated with all three viruses and stored for 58 d at 23.9 °C. Following closer investigation, a biosecurity gap was identified as a faulty seal between the positive control room and that test room which allowed unintentional airflow between the rooms. Due to the lack of infection documented on 6 dpi as measured using oral fluid samples, it appears that this biosecurity breach was the cause for the infection in that test room. This represents a limitation of this study. However, the four other test rooms (rooms 1–4) did not display any evidence of infection, giving a reasonable amount of confidence that the identified biosecurity breach was the cause of infection in test room 5 and not the extended storage of the feed. Nonetheless, this presents a limitation to the study and viral survival over the 58 d holding period cannot be completely ruled out.

A limitation of the current experiment was that there was no detection of viral RNA for PEDV in any of the test rooms along with no detectable levels of antibodies throughout the experiment. The viruses used in the current experiment were similar to those used in previous research [10]. In the current experiment, a total inoculum dosage of 1×10^5 $TCID_{50}$ per virus was inoculated into each feed tote and directly into the feed bin for the positive control room (approximately 0.1 $TCID_{50}$ per gram of feed). In the previous work by Dee et al. [10], a total of 1×10^7 $TCID_{50}$ of PEDV, PRRSV, and SVA was inoculated into 1300 kg of complete feed (7.69 $TCID^{50}$ per gram of feed for each virus). This difference in initial viral concentration may partially explain the lack of infectivity with PEDV in the current experiment, as the work by Schumacher et al. [28] which demonstrated that a concentration of 5.6×10^1 $TCID_{50}$ was required to cause infection under the conditions used. For SVA, however, clinical signs were present in the positive control room and through diagnostic testing through the State of Minnesota confirmed the presence of lesions consistent with SVA and the detection of viral RNA in clinical samples from the positive control room, indicating that the infectious virus was present in the inoculated positive control feed. For SVA, Buckley et al. [29] reported that approximately 6.3×10^2 $TCID_{50}$ was required to cause infection when the dose was provided via liquid inoculum, and Caserta et al. [30] reported that a dose of 1×10^5 $TCID_{50}$, 1×10^6 $TCID_{50}$, and 1×10^7 $TCID_{50}$ all caused infection through inoculated feed. The dose used in the current experiment was 1×10^5 $TCID_{50}$, which was sufficient to cause clinical infection, but did not stimulate the production of antibodies or result in shedding via oral fluids. While the current study did not successfully result in infection with PEDV in the positive control room, the results confirm that pigs were infected with PRRSV and SVA when viral inoculum was provided through feed.

The study herein was designed to validate the principle of extended storage for the inactivation of swine viruses, which has been repeated with multiple individual ingredients but has not been fully characterized for complete feed post-feed manufacture. A limitation of the current study is a limited sample size to fully characterize the effects of extended feed storage on nutritional, microbiological, and toxicological outcomes. Based on facility limitations and storage logistics, we were only able to include a limited amount of replication for evaluation of infectivity characteristics and microbiological/toxicological profiles. Follow-up research should be conducted to expand the base of knowledge of the effect of this practice on these key measurements of feed safety and quality. Based on the current data, there does not appear to be any significant negative consequences of 58 d storage at controlled anaerobic environmental conditions.

This represents a significant advancement in the body of knowledge, as nearly all previous research with viral contaminants in swine feed have focused on ingredient storage. This experiment is the first to describe how real-world scale storage of complete swine feed can inactivate viral pathogens. While extended storage of complete feed is not practical in all situations, there is opportunity for its implementation in situations wherein a high degree of risk aversion is needed, such as biomedical research and/or high genetic value animals. Further research should be conducted to explore the stability of other pathogens which may be present in feed such as bacterial contamination or additional viral pathogens beyond those evaluated herein. Our study evaluated extended storage in anaerobic conditions established through the sublimation of dry ice and subsequent displacement of air from the storage containers and intermodal shipping containers. Additional research should be conducted to characterize the efficacy of this procedure compared to extended storage in non-anaerobic conditions as well as whether other compounds such as nitrogen could be used to create these anaerobic conditions. Furthermore, while we have established that a 58 d storage period is effective under the conditions evaluated, further research should be conducted to establish the optimum storage time based on pathogen inactivation and nutrient stability and feed quality characteristics. Nonetheless, the results provide the first evidence characterizing the potential applicability of this technique for the mitigation of viral contamination in swine complete feed.

5. Conclusions

In summary, storing feed for 58 d in controlled anaerobic environmental conditions of 23.9 °C did not result in microbiological or toxicological concerns and only resulted in minimal detrimental effects on nutrient stability. No evidence of infection with PEDV, PRRSV, or SVA was detected in the swine bioassay rooms fed inoculated feed following this 58 d storage period at 23.9 °C with the exception of the aerosol transmission of PRRSV via a documented biocontainment breach. Furthermore, if using extended storage for complete feed, it is recommended for this to be performed at controlled temperatures to maximize the consistency in viral inactivation. Ambient storage may be effective under certain conditions, but inactivation would be expected to be greatly reduced if the feed is stored at temperatures near or below freezing. Additional research should focus on evaluating the time necessary to store complete swine feed for inactivation of swine viruses with consideration given to shorter storage durations. Together, these results validate that the extended storage of complete swine feed can be a method to reduce risk associated with pathogen transmission through feed while having minimal effects on measures of nutritional quality. While this practice may not be applicable in commercial swine production, it may be considered as a means of improving the safety of swine feed for applications such as biomedical research and/or feeding of high genetic value animals wherein a high degree of risk aversion is required to maintain animal health.

Author Contributions: Conceptualization, D.R.K., B.N.S. and J.T.G.; methodology, S.A.D., E.L., J.T.G. and D.R.K.; validation, E.L.; formal analysis, J.T.G.; investigation, E.L., S.A.D. and D.R.K.; writing—original draft preparation, J.T.G.; writing—review and editing, S.A.D., E.L., B.N.S. and D.R.K.; project administration, S.A.D. and D.R.K.; funding acquisition, D.R.K. and B.N.S. All authors have read and agreed to the published version of the manuscript.

Funding: Project funding provided by eGenesis (Cambridge, MA, USA).

Institutional Review Board Statement: All animals used in this experiment were managed in accordance with the institutional animal care and use guidelines observed by the Pipestone Applied Research Institutional Animal Care and Use Committee, #2022-2.

Informed Consent Statement: Not applicable.

Data Availability Statement: The data presented in this study are available on request from the corresponding author. The data are not publicly available due to [All data generated or analyzed during this study are included in this published article].

Acknowledgments: Contribution no. 23-293-J from the Kansas Agric. Exp. Stn., Manhattan KS, 66506-0210 USA. Appreciation is expressed to the Pipestone Applied Research team for project coordination and to Jon Bergstrom and DSM Nutritional Products for technical support and for analysis of phytase and vitamins.

Conflicts of Interest: Brittney Scales and Doug Kern are present and past employees of eGenesis who provided project funding for this research. The authors declare no further conflicts of interest.

References

1. Dee, S.; Clement, T.; Schelkopf, A.; Nerem, J.; Knudsen, D.; Christopher-Hennings, J.; Nelson, E. An evaluation of contaminated complete feed as a vehicle for porcine epidemic diarrhea virus infection of naive pigs following consumption via natural feeding behavior: Proof of concept. *BMC Vet. Res.* **2014**, *10*, 176. [CrossRef]
2. Harrison, O.L.; Gebhardt, J.T.; Paulk, C.B.; Plattner, B.L.; Woodworth, J.C.; Rensing, S.; Jones, C.K.; Trinetta, V. Inoculation of weaned pigs by feed, water, and airborne transmission of *Salmonella enterica* Serotype 4,[5],12:i:-. *J. Food Prot.* **2022**, *85*, 693–700. [CrossRef]
3. Cochrane, R.A.; Schumacher, L.L.; Dritz, S.S.; Woodworth, J.C.; Huss, A.R.; Stark, C.R.; DeRouchey, J.M.; Tokach, M.D.; Goodband, R.D.; Bia, J.; et al. Effect of pelleting on survival of porcine epidemic diarrhea virus-contaminated feed. *J. Anim. Sci.* **2017**, *95*, 1170–1178. [CrossRef] [PubMed]

4. Dee, S.; Neill, C.; Clement, T.; Singrey, A.; Christopher-Hennings, J.; Nelson, E. An evaluation of porcine epidemic diarrhea virus survival in individual feed ingredients in the presence or absence of a liquid antimicrobial. *Porc. Health Manag.* **2015**, *1*, 9. [CrossRef]
5. Cochrane, R.A.; Woodworth, J.C.; Dritz, S.S.; Huss, A.R.; Stark, C.R.; Hesse, R.A.; Tokach, M.D.; Bai, J.F.; Jones, C.K. Evaluating chemical mitigation of porcine epidemic diarrhea virus in swine feed and ingredients. *J. Anim. Sci.* **2015**, *93* (Suppl. 2), 90. [CrossRef]
6. Dee, S.; Neill, C.; Singrey, A.; Clement, T.; Cochrane, R.; Jones, C.; Patterson, G.; Spronk, G.; Christopher-Hennings, J.; Nelson, E. Modeling the transboundary risk of feed ingredients contaminated with porcine epidemic diarrhea virus. *BMC Vet. Res.* **2016**, *12*, 51. [CrossRef] [PubMed]
7. Cochrane, R.A.; Saensukjaroenphon, M.; Dritz, S.S.; Woodworth, J.C.; Huss, A.R.; Stark, C.R.; DeRouchey, J.M.; Tokach, M.D.; Goodband, R.D.; Bai, J.F.; et al. Evaluating the inclusion level of medium chain fatty acids to reduce the risk of PEDV in feed and spray-dried animal plasma. *J. Anim. Sci.* **2016**, *94* (Suppl. 2), 50. [CrossRef]
8. Trudeau, M.P.; Verma, H.; Sampedro, F.; Urriola, P.E.; Shurson, G.C.; McKelvey, J.; Pillai, S.D.; Goyal, S.M. Comparison of thermal and non-thermal processing of swine feed and the use of selected feed additives on inactivation of porcine epidemic diarrhea virus (PEDV). *PLoS ONE* **2016**, *11*, e0158128. [CrossRef] [PubMed]
9. Gebhardt, J.T.; Woodworth, J.C.; Jones, C.K.; Tokach, M.D.; Gauger, P.C.; Main, R.G.; Zhang, J.; Chen, Q.; DeRouchey, J.M.; Goodband, R.D.; et al. Determining the impact of commercial feed additives as potential porcine epidemic diarrhea virus mitigation strategies as determined by polymerase chain reaction analysis and bioassay. *Transl. Anim. Sci.* **2019**, *3*, 93–102. [CrossRef]
10. Dee, S.A.; Niederwerder, M.C.; Edler, R.; Hanson, D.; Singrey, A.; Cochrane, R.; Spronk, G.; Nelson, E. An evaluation of additives for mitigating the risk of virus-contaminated feed using an ice-block challenge model. *Transbound. Emerg. Dis.* **2021**, *68*, 833–845. [CrossRef]
11. Stoian, A.M.M.; Zimmerman, J.; Ji, J.; Hefley, T.J.; Dee, S.; Diel, D.G.; Rowland, R.R.R.; Niederwerder, M.C. Half-life of African swine fever virus in shipped feed. *Emerg. Infect. Dis.* **2019**, *25*, 2261–2263. [CrossRef]
12. Niederwerder, M.C.; Khanal, P.; Foland, T.; Constance, L.A.; Stoian, A.M.M.; Deavours, A.; Haase, K.; Cino-Ozuna, A.G. Stability of African swine fever virus in feed during environmental storage. *Transbound. Emerg. Dis.* **2022**, *69*, 3216–3224. [CrossRef]
13. Dee, N.; Havas, K.; Shah, A.; Singrey, A.; Spronk, G.; Niederwerder, M.; Nelson, E.; Dee, S. Evaluating the effect of temperature on viral survival in plant-based feed during storage. *Transbound. Emerg. Dis.* **2022**, *69*, e2105–e2110. [CrossRef]
14. Cochrane, R.A.; Dritz, S.S.; Woodworth, J.C.; Stark, C.R.; Huss, A.R.; Cano, J.P.; Thompson, R.W.; Fahrenholz, A.C.; Jones, C.K. Feed mill biosecurity plans: A systematic approach to prevent biological pathogens in swine feed. *J. Swine Health Prod.* **2016**, *24*, 154–164.
15. DeRouchey, J.M.; Tokach, M.D.; Nelssen, J.L.; Goodband, R.D.; Dritz, S.S.; Woodworth, J.C.; James, B.W.; Real, D.E. Effect of irradiation of individual feed ingredients and the complete diet on nursery pig performance. *J. Anim. Sci.* **2003**, *81*, 1799–1805. [CrossRef] [PubMed]
16. Groesbeck, C.N.; DeRouchey, J.M.; Tokach, M.D.; Goodband, R.D.; Dritz, S.S.; Nelssen, J.L. Effects of irradiation of feed ingredients added to meal or pelleted diets on growth performance of weanling pigs. *J. Anim. Sci.* **2009**, *87*, 3997–4002. [CrossRef] [PubMed]
17. Liu, Y.; Shao, Y.; Wang, L.; Lu, W.; Li, S.; Xu, D.; Fu, Y.V. Inactivation of porcine epidemic diarrhea virus with electron beam irradiation under cold chain conditions. *Environ. Technol. Innov.* **2022**, *27*, 102715. [CrossRef] [PubMed]
18. Eggum, B.O. Effect of radiation treatment on protein quality and vitamin content of animal feeds. In *Decontamination of Animal Feeds by Irradiation*; International Atomic Energy Agency: Vienna, Austria, 1979.
19. Kilcast, D. Effect of irradiation on vitamins. *Food Chem.* **1994**, *49*, 157–164. [CrossRef]
20. Namted, S.; Poungpong, K.; Loongyai, W.; Rakangthong, C.; Bunchasak, C. A review: Using yeast extract as feed additive in pig diets. *Adv. Anim. Vet. Sci.* **2022**, *10*, 2384–2395. [CrossRef]
21. De Jong, J.A.; DeRouchey, J.M.; Tokach, M.D.; Dritz, S.S.; Goodband, R.D.; Woodworth, J.C.; Jones, C.K.; Stark, C.R. Stability of commercial phytase sources under different environmental conditions. *J. Anim. Sci.* **2016**, *94*, 4259–4266. [CrossRef]
22. De Jong, J.A.; Woodworth, J.C.; DeRouchey, J.M.; Goodband, R.D.; Tokach, M.D.; Dritz, S.S.; Stark, C.R.; Jones, C.K. Stability of four commercial phytase products under increasing thermal conditioning temperatures. *Transl. Anim. Sci.* **2017**, *1*, 255–260. [CrossRef]
23. Truelock, C.; Yoder, A.; Evans, C.; Stark, C.; Paulk, C. The effects of pelleting process parameters and phytase source on the in-feed stability of phytase. *Anim. Feed Sci. Technol.* **2022**, *294*, 115407. [CrossRef]
24. Shurson, G.C.; Salzer, T.M.; Koehler, D.D.; Whitney, M.H. Effect of metal specific amino acid complexes and inorganic trace minerals on vitamin stability in premixes. *Anim. Feed Sci. Technol.* **2011**, *163*, 200–206. [CrossRef]
25. Yang, P.; Wang, H.K.; Zhu, M.; Li, L.X.; Ma, Y.X. Degradation kinetics of vitamins in premixes for pig: Effects of choline, high concentrations of copper and zinc, and storage time. *Anim. Biosci.* **2021**, *34*, 701–713. [CrossRef] [PubMed]
26. Yang, P.; Wang, H.; Li, L.; Zhang, N.; Ma, Y. The stability of vitamin A from different sources in vitamin premixes and vitamin-trace mineral premixes. *Appl. Sci.* **2021**, *11*, 3657. [CrossRef]
27. Coelho, M. Vitamin Stability in premixes and feeds. A practical approach in ruminant diets. In Proceedings of the 13th Annual Florida Ruminant Nutrition Symposium, Gainesville, FL, USA, 10–11 January 2002; pp. 127–145.

28. Schumacher, L.L.; Woodworth, J.C.; Jones, C.K.; Chen, Q.; Zhang, J.; Gauger, P.C.; Stark, C.R.; Main, R.G.; Hesse, R.A.; Tokach, M.D.; et al. Evaluation of the minimum infectious dose of porcine epidemic diarrhea virus in virus-inoculated feed. *Am. J. Vet. Res.* **2016**, *77*, 1108–1113. [CrossRef] [PubMed]
29. Buckley, A.; Lager, K. Infectious dose of Senecavirus A in market weight and neonatal pigs. *PLoS ONE* **2022**, *17*, e0267145. [CrossRef] [PubMed]
30. Caserta, L.C.; Noll, J.C.G.; Singrey, A.; Niederwerder, M.C.; Dee, S.; Nelson, E.A.; Diel, D.G. Stability of Senecavirus A in animal feed ingredients and infection following consumption of contaminated feed. *Transbound. Emerg. Dis.* **2022**, *69*, 88–96. [CrossRef] [PubMed]

Disclaimer/Publisher's Note: The statements, opinions and data contained in all publications are solely those of the individual author(s) and contributor(s) and not of MDPI and/or the editor(s). MDPI and/or the editor(s) disclaim responsibility for any injury to people or property resulting from any ideas, methods, instructions or products referred to in the content.

animals

MDPI

Article

Evaluation of Truck Cab Decontamination Procedures following Inoculation with Porcine Epidemic Diarrhea Virus and Porcine Reproductive and Respiratory Syndrome Virus

Grace E. Houston [1], Cassandra K. Jones [2], Caitlin Evans [3], Haley K. Otott [3], Charles R. Stark [3], Jianfa Bai [1], Elizabeth G. Poulsen Porter [1], Marcelo N. de Almeida [4], Jianqiang Zhang [4], Phillip C. Gauger [4], Allison K. Blomme [3], Jason C. Woodworth [2], Chad B. Paulk [3] and Jordan T. Gebhardt [1,*]

1 Department of Diagnostic Medicine/Pathobiology, College of Veterinary Medicine, Kansas State University, Manhattan, KS 66506-0201, USA

2 Department of Animal Sciences and Industry, College of Agriculture, Kansas State University, Manhattan, KS 66506-0201, USA; jonesc@ksu.edu (C.K.J.); jwoodworth@ksu.edu (J.C.W.)

3 Department of Grain Science and Industry, College of Agriculture, Kansas State University, Manhattan, KS 66506-0201, USA; ceevans@ksu.edu (C.E.); haley27@ksu.edu (H.K.O.); ablomme@ksu.edu (A.K.B.); cpaulk@ksu.edu (C.B.P.)

4 Department of Veterinary Diagnostic and Production Animal Medicine, College of Veterinary Medicine, Iowa State University, Ames, IA 50011-1134, USA; malmeida@iastate.edu (M.N.d.A.); jqzhang@iastate.edu (J.Z.); pcgauger@iastate.edu (P.C.G.)

* Correspondence: jgebhardt@vet.k-state.edu

Simple Summary: It is widely understood that transportation vehicles can serve as potential vectors of pathogen transmission within swine production systems for pathogens of concern such as porcine epidemic diarrhea virus (PEDV) and porcine reproductive and respiratory syndrome virus (PRRSV). The objective of this experiment was to evaluate several decontamination methods for the mitigation of PEDV and PRRSV within truck cabs. A total of three full-sized truck cabs were modified for use in a BSL-2 research facility with multiple surface types including fabric, rubber, and plastic. Surfaces were inoculated with either PEDV alone, PRRSV alone, PEDV + an organic matter mixture of feces and dirt, or PRRSV + organic mixture. Practical decontamination methods were then applied using a pump sprayer or Hurricane fogger system, or a commercially available gaseous chlorine dioxide system. Several differences were observed within the different combinations of disinfectants and surfaces, indicating that under the conditions of this study, the ability of different disinfectants to reduce the detection of PEDV and PRRSV genetic materials differed depending on the surface being evaluated. In general, most disinfectant applications were only able to reduce the quantity of detectable virus but not completely eliminate it from the surface.

Abstract: This experiment aimed to evaluate commercially available disinfectants and their application methods against porcine epidemic diarrhea virus (PEDV) and porcine reproductive and respiratory syndrome virus (PRRSV) on truck cab surfaces. Plastic, fabric, and rubber surfaces inoculated with PEDV or PRRSV were placed in a full-scale truck cab and then treated with one of eight randomly assigned disinfectant treatments. After application, surfaces were environmentally sampled with cotton gauze and tested for PEDV and PRRSV using qPCR duplex analysis. There was a disinfectant × surface interaction ($p < 0.0001$), indicating a detectable amount of PEDV or PRRSV RNA was impacted by disinfectant treatment and surface material. For rubber surfaces, 10% bleach application had lower detectable amounts of RNA compared to all other treatments ($p < 0.05$) except Intervention via misting fumigation, which was intermediate. In both fabric and plastic surfaces, there was no evidence ($p > 0.05$) of a difference in detectable RNA between disinfectant treatments. For disinfectant treatments, fabric surfaces with no chemical treatment had less detectable viral RNA compared to the corresponding plastic and rubber ($p < 0.05$). Intervention applied via pump sprayer to fabric surfaces had less detectable viral RNA than plastic ($p < 0.05$). Furthermore, 10% bleach applied via pump sprayer to fabric and rubber surfaces had less detectable viral RNA than plastic ($p < 0.05$). Also, a 10 h downtime, with no chemical application or gaseous fumigation for 10 h, applied to fabric surfaces had less detectable viral RNA than other surfaces ($p < 0.05$). Sixteen treatments were evaluated via swine bioassay, but all samples failed to produce infectivity. In summary, commercially available disinfectants successfully reduced detectable viral RNA on surfaces but did not eliminate viral genetic material, highlighting the importance of bioexclusion of pathogens of interest.

Keywords: disinfectants; porcine epidemic diarrhea virus; porcine reproductive and respiratory syndrome virus; truck cabs; swine

1. Introduction

Biosecurity and maintaining healthy populations of animals are critical for modern swine production. Key viral pathogens affecting commercial swine production include porcine reproductive and respiratory syndrome virus (PRRSV) and porcine epidemic diarrhea virus (PEDV). While PRRSV has been present in North America for decades, PEDV was first detected in North America in 2013. Specific estimates of economic impact are challenging to summarize, and the estimated economic impact of PRRSV in North America was USD 664 million as per a 2013 publication by Holtkamp et al. [1]. Additional research has been conducted to estimate the economic impact at the farm level in the United States [2], China [3], and Germany [4], and additional information is being generated regarding the impact of PRRSV on wean-finish mortality in commercial settings [5]. While PRRSV is one of the costliest diseases affecting swine production in North America, the rapid distribution of PEDV in the United States beginning in 2013 resulted in substantial industry losses [6]. Following the initial introduction into the United States in 2013, the incidence of PEDV on sow farms has been lower compared to the initial epidemic [7]; however, it remains a challenging virus to control especially in growing and finishing pig populations due to gaps in biosecurity practices.

Both viruses are enveloped, positive-stranded RNA viruses [8,9]. The transmission of PRRSV occurs through exposure via the respiratory and/or oral routes as well as through vertical transmission in gestating females [10]. The transmission of PEDV occurs through fecal–oral transmission [9]. The control of these pathogens involves attention to several different potential vectors including animal movement, supply entry into farms, personnel movement, and other fomites such as transportation vehicles. Research suggests that transportation vehicles can play a role in the transmission of bacterial [11] and viral pathogens [12–15]. Cleaning and disinfecting metal surfaces has been shown to reduce detectable PRRSV and PEDV [16–21]. These methods are successful at reducing detectable viral RNA, and techniques including power washing, disinfecting, and heating semi-truck trailers responsible for live animal transportation are commonplace in the current United States swine industry as biosecurity measures. However, when considering the role of feed delivery in pathogen introduction, the likelihood of detecting PEDV or porcine deltacoronavirus RNA was highest within the truck cab of the feed delivery truck as opposed to the trailer [22].

To date, truck cabs have been an under-evaluated potential source of pathogen transmission, and very little is known regarding the best practices for their decontamination. A challenge associated with the cab of feed delivery trucks is that multiple surface types are present within a single space. When considering the efficacy of disinfectants and surface types, a previous study has found that surface type can influence the detection of viral RNA after disinfectant treatment [23]. However, most disinfection protocols are developed by extrapolations from laboratory settings or are otherwise lacking for applying laboratory bench data to real-world settings. Therefore, the objective of this study was to evaluate commercially available disinfectants and application methods on different surfaces present within truck cabs for efficacy in reducing detectable PEDV and PRRSV RNA and subsequent infectivity of surface environmental samples.

2. Materials and Methods

The surface inoculation and sample collection for this study were conducted at the Cargill Feed Science Research Center (FSRC) at the O.H. Kruse Feed Mill of Kansas State University (KSU) in Manhattan, KS, USA with approval from the KSU Institutional Biosafety Committee (Project Approval #1511). The inoculation of surfaces was performed within a biosafety cabinet (BSC) housed within the BSL-2 space of the FSRC. This study was set up in an 8 × 3 × 2 factorial with eight disinfectant methods, three different surfaces, and two viruses, with each combination of factors repeated three times.

2.1. Preparation of Inoculum

Viral inoculum was propagated and titrated at the South Dakota State University Animal Disease Research & Diagnostic Laboratory using common virological techniques. Briefly, stock PEDV was cultured using Vero-76 cells, and for PRRSV, MARC-145 cells were used. Following propagation and virus titration, viral inoculum was prepared by placing 25 mL of PEDV (USA/CO/2013 isolate with a titer of 1.33×10^6 $TCID_{50}$/mL; GenBank accession number: KF272920) and 25 mL of PRRSV (1-7-4 isolate with a titer of 1.33×10^6 $TCID_{50}$/mL) into separate containers. In each container, 225 mL of phosphate-buffered solution (PBS) was added to achieve a final concentration of 10^5 $TCID_{50}$/mL. Each virus was divided into five containers containing approximately 45 mL volume, sealed and stored at −80 °C for two weeks. When virus was needed for the study, the containers of each virus were defrosted in a lukewarm water bath (approximately 36 °C to 40 °C) and used immediately. Inoculum samples were taken during the study to evaluate Ct values over time. When evaluating the Ct values of pure virus stock, the PRRSV stock was always one to two Ct values higher (Ct values ranging from 23.3 to 24.0) than the virus stock of PEDV (Ct values ranging from 22.1 to 22.3).

2.2. Preparation of Surfaces and Disinfectant

Plastic (0.32 × 10.16 × 20.32 cm white high-density polyethylene panel; Menards, Eau Claire, WI, USA), rubber (Heavy-Duty 45.72 × 45.72 × 0.64 cm Rubber Gym Tiles; BCG, Boston, MA, USA), and fabric surfaces (upholstery fabric, no water resistance treatment; Joann's Fabrics, Hudson, OH, USA) were cut into 10 cm × 10 cm squares for the creation of surface coupons. There were a total of 48 coupons for each surface type (48 plastic, 48 rubber, 48 fabric). Velcro strips were applied to the back of the surface coupons prior to inoculation, and surface coupons were placed into a transportation container (Promoze Food Storage Containers, Seattle, WA, USA), and they remained in these storage containers until ready for placement into the truck cabs. Three truck cabs of similar make and model were sourced from a local salvage yard (Beloit, KS, USA). The truck cabs were removed from their frame at the salvage yard, transported to the FSRC for the study, and placed on wheels for easy mobility. Truck cabs were visually and physically inspected for any holes or inadequate seals and sealed as necessary using either silicone sealant (DAP; Menards, Eau Claire, WI, USA) or super glue (Gorilla Glue; Gorilla Glue Co, Sharonville, OH, USA). At the start of the study, truck cabs were wheeled into the BSL-2 space, their wheels were locked, and they remained in the same location for the duration of the study. Each truck had eight disinfectant treatments applied in random order. Disinfectant treatments included the following:

1. No disinfectant.
2. 1:256 dilution of Synergize (Neogen Corp, Lexington, KY, USA) applied through misting fumigation (Hurricane Ultra II Portable Electric Fogger, Curtis Dyna-Fog Ltd., Westfield, IN, USA).
3. 1:256 dilution of Synergize applied through pump sprayer (Chapin Sure Spray 1 Gallon Tank Sprayer, Menards, Eau Claire, WI, USA).
4. 1:64 dilution of Intervention (Virox, Oakville, ON, USA) applied through misting fumigation.
5. 1:64 dilution of Intervention applied through pump sprayer.

6. 10% bleach (7.55% sodium hypochlorite germicidal bleach; Clorox, Oakland, CA, USA) solution applied through pump sprayer.
7. No chemical treatment, 10 h holding time.
8. Gaseous fumigation over 10 h with chlorine dioxide (ProKure G; ProKure Solutions, Phoenix, AZ, USA).

The designated concentration of wet disinfectants was prepared daily. Once prepared, solutions were poured into their respective application method tool: pump sprayer or misting fumigation. The water-based chlorine dioxide pouch was utilized following manufacturer-labeled instructions for the gaseous fumigation application.

2.3. *Surface Inoculation and Disinfectant Application*

Surfaces were inoculated with either 1 mL of PRRSV (titer of 10^5 $TCID_{50}$/mL) or PEDV (titer of 10^5 $TCID_{50}$/mL). Surfaces were allowed to dry for 1 h prior to placement within an individual truck cab. When surfaces were ready for placement, surface coupons were placed within the truck cab with nitrile gloves changed in-between new surface coupons. Coupon placement was predetermined prior to the start of the study to ensure a consistent placement of surface coupons with each treatment and corresponded to the location within truck cabs from which the surfaces would naturally be located. Plastic coupon surfaces were placed on the dashboard, rubber surface coupons were placed on the floorboard, and fabric surface coupons were placed on the driver's seat as shown in Figure 1. Surfaces shown include both surfaces with viral inoculum applied directly, or surfaces containing the presence of organic matter and viral inoculum. The results herein represent the results for clean surfaces only.

After surfaces were placed in the truck cab, a randomly assigned disinfectant treatment application was conducted. Independent disinfectant treatments were applied in each truck cab within each run, although each disinfectant treatment was performed in each truck cab for a total of 3 replicates for each disinfectant treatment. For the pump sprayer application, the applicator was positioned outside the truck cab on the driver's side, and the liquid was applied in a snake-like application going from the front to the back of the cab, resulting in 50–90 g of disinfectant solution being used per treatment application. For the misting fogger application, the head of the fogger was angled and secured at 90° (parallel with the ground), placed in the passenger side seat, and aimed for the driver side of the truck cab. Once set in location, the flow rate was set to 2, turned on, the passenger door was closed, and the fogger was allowed to run for 5 min. The amount of disinfection used for this application ranged from 220 to 340 g. Once the application of the pump sprayer and the misting fogger was completed, wet application methods were allowed to dry for 15 min prior to environmental sample collection.

For the gaseous fumigation treatment, the plastic container was placed in the passenger's seat, doors were closed, and fumigation was allowed to occur for 10 h after which the driver and passenger doors were opened, and the truck cab was allowed to dissipate for 1 h before environmental samples were collected. For the no chemical treatment and the 10 h downtime treatment, surface coupons were inoculated with virus, placed in truck cabs as previously mentioned, and they remained in the truck cab for approximately 15 min or 10 h, respectively, before sampling. Each combination of virus, surface, and disinfectant treatment was replicated three times.

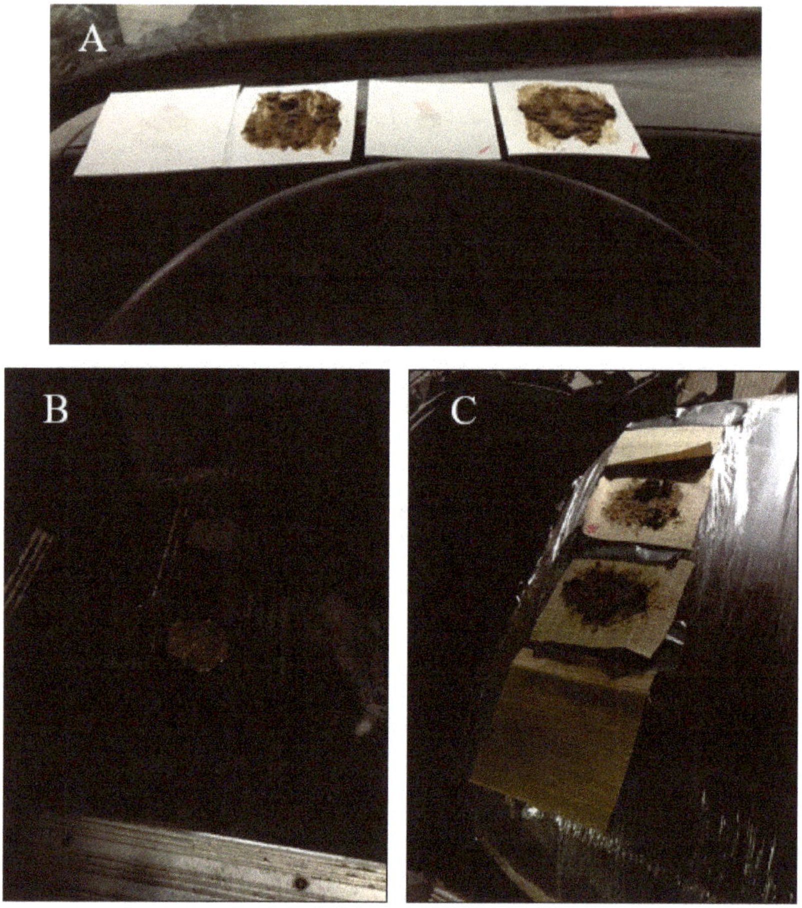

Figure 1. Surface inoculation for plastic (**A**), rubber (**B**), and fabric (**C**) coupons within full-sized truck cabs. Surfaces inoculated with organic material shown here are not reported in the current manuscript.

2.4. Environment Sampling

Environmental samples were collected with cotton gauze as previously described [24]. Briefly, a 10 cm × 10 cm cotton gauze, pre-moistened with 5 mL PBS, was utilized to swab the surface coupon after treatment application. Environmental samples were transferred to the BSC, 20 mL of PBS was added to them, and they were inverted for 5–10 s and allowed to sit at room temperature for 1 h. The samples were then vortexed for 15 s, and the supernatant was transferred into 1.75 mL cryovials and 15 mL conical tubes. Samples were stored at −80 °C until PCR analysis. Once all samples were collected, surface coupons were discarded, and cabs were cleaned, sprayed with 1:256 glutaraldehyde, and allowed to dry for 20 min to prevent virus accumulation within the truck cabs.

2.5. Reverse Transcription Real-Time PCR Analysis

Reverse transcription real-time PCR (qRT-PCR) was conducted at the Molecular Research and Development Laboratory within the Kansas State Veterinary Diagnostic Laboratory. Fifty µL of supernatant from each sample was loaded into a deep-well plate and extracted using a Kingfisher Flex magnetic particle processor (Fisher Scientific, Pittsburg, PA, USA) and the MagMAX-96 Viral RNA Isolation kit (Life Technologies, Grand Island, NY, USA) according to manufacturer's instructions with one modification, reducing the final elution volume to 60 µL. One negative extraction control consisting of all reagents except the sample (replaced with PBS buffer) was included in each extraction. Positive controls of each stock virus were also included with each extraction. Extracted RNA was frozen at −80 °C until assayed by qRT-PCR using a multiplex qRT-PCR assay developed by the KSU Molecular Research and Development Laboratory to simultaneously detect and quantify PRRSV and PEDV RNA. Analyzed values represent cycle threshold (Ct) at which virus was detected. If a sample had no detectable PRRSV or PEDV RNA, a sample was assigned a value of 45 as a total of 45 cycles were run for each sample. For result interpretation, greater Ct values indicate less detectable viral RNA.

2.6. Bioassay Analysis

A total of 16 treatments were selected for pig bioassay to assess the infectivity of virus present in them (Table 1). The bioassay procedure was conducted 13 months after surface inoculation and environmental sample collection. The project was approved through Iowa State University's institutional animal care and use committee (Project #IACUC-22-082) and institutional biosafety committee (Project #22-045). A total of 48 crossbred, 10-day-old pigs of mixed sex were sourced from a single commercial, a crossbred farrow-to-wean herd with no prior exposure to PEDV and PRRSV. The Iowa State University Veterinary Diagnostic Laboratory confirmed all pigs to be negative for PRRSV using blood samples and PEDV, porcine deltacoronavirus, and transmissible gastroenteritis virus using fecal swabs via the respective real-time RT-PCR. Blood serum testing further confirmed that pigs were negative for PRRSV antibody using nucleocapsid protein-based ELISA and negative for PEDV antibody using an indirect fluorescent antibody assay. Pigs were allowed 3 days of acclimation prior to beginning of the bioassay. Three pigs were housed in a room, and each pig was challenged via an oral gavage of PEDV inoculum, an intramuscular (IM) injection, and an intranasal application of PRRSV. The oral gavage method was modeled similar to the described in the previous research [25], utilizing a 10 French feed tube and 60 mL syringe (10 mL/pig). For the PRRSV inoculum, 3 mL was administered to the muscle of the cervical region. For PRRSV intranasal inoculation, the tip of a 3 mL Luer slip syringe was inserted inside one of the pig's nostrils, and 1 mL of inoculum was allowed to drip directly into the pig's nostril. Rectal swabs were collected on days −3, 0, 4, and 7 post inoculation (dpi) from all pigs and tested for PEDV RNA using qRT-PCR. Serum samples were collected on −3, 0, 4, and 7 dpi from all pigs and tested for PRRSV RNA using qRT-PCR. Following humane euthanasia at 7 dpi, the small intestine, the cecum, an aliquot of cecal contents as described by Schumacher et al. [26], and lung samples were collected at necropsy. A negative bioassay was concluded if all rectal swabs, serum samples, lung tissue, and cecum contents had non-detectable levels of PEDV or PRRSV. If any samples had detectable RNA, the result would be considered a positive bioassay.

Table 1. Summary of the 16 treatments chosen for live pig bioassay [1].

Surface Type	Disinfectant Treatment
Rubber (n = 8)	No disinfectant
	No chemical, 10 h downtime
	Gaseous fumigation, 10 h downtime
	Misting fumigation, Intervention
	Misting fumigation, Synergize
	Pump sprayer, Intervention
	Pump sprayer, Synergize
	Pump sprayer, 10% bleach
Plastic (n = 4)	No disinfectant
	No chemical, 10 h downtime
	Gaseous fumigation, 10 h downtime
	Pump sprayer, 10% bleach

Table 1. *Cont.*

Surface Type	Disinfectant Treatment
Fabric (n = 4)	No disinfectant No chemical, 10 h downtime Gaseous fumigation, 10 h downtime Pump sprayer, 10% bleach

[1] Supernatant from environmental samples taken after disinfectant application to a truck cab were used as the inoculum for the live pig bioassay. A total of 48 crossbred, 10-day-old pigs of mixed sex were administered 10 mL oral gavage of porcine epidemic diarrhea virus (PEDV) and 3 mL intramuscular (IM) injection, and 1 mL/nostril intranasal application of porcine reproductive and respiratory syndrome virus (PRRSV) after a three-day acclimation period.

2.7. Statistical Analysis

Data were analyzed in a split-plot design with truck cab as the experimental unit for disinfectant treatment and surface coupons as the experimental unit for surface type (fabric, plastic, or rubber) and virus (PEDV or PRRSV). There were three replications per treatment. The Ct value of each sample was analyzed with ANOVA and *F*-test through the aov function in R programming language (R Foundation for Statistical Computing, Vienna, Austria). Fixed effects considered the disinfectant treatment, surface treatment, and virus type, while the random effect was truck cab defining it as the experimental unit for disinfectant treatment to account for the split-plot design. Results of Ct data are reported as least squares means ± standard error of the mean. All statistical models were evaluated using the visual assessment of studentized residuals, and assumptions appeared to be reasonably met. A Tukey multiple comparison adjustment was utilized to control Type I error rate. Results were considered significant at $p \leq 0.05$ and marginally significant between $p > 0.05$ and $p \leq 0.10$.

3. Results and Discussion

Both PEDV and PRRSV are considered endemic, or present at normal levels, within the swine-dense regions of the United States. Control of these and other pathogens are critical to maintain animal health and welfare. Given the significant economic impact of PRRSV and PEDV, continued efforts must be placed on the control of these pathogens. Both viruses share a similar structure being enveloped, positive-sense, single-stranded RNA viruses [9,10]. Importantly, because of their viral envelope, disinfection and inactivation is readily possible if correct procedures are implemented.

The transmission of PRRSV occurs through exposure via the respiratory and/or oral routes as well as through vertical transmission in gestating females [10], while the transmission of PEDV occurs through fecal–oral transmission [9]. Thus, for an effective control program, biosecurity efforts must be implemented to reduce the risk of these transmission events from occurring by focusing on controlling vectors including animal movement, supply entry into farms, personnel movement, and other fomites such as transportation vehicles.

The control of these viral pathogens requires a complex strategy addressing all possible routes of entry including animal movement, supply entry into farms, personnel movement, and other fomites such as transportation vehicles. Significant research efforts have focused on diagnostic testing to properly determine the health status of animals prior to movement, the decontamination of incoming supplies [27], the filtration of incoming air [28], and multiple investigations to address feed biosecurity and reduce the likelihood of pathogen introduction via incoming feed as reviewed by Stewart et al. [29].

It has been documented that transportation vehicles can contribute to the spread of both pathogens in addition to the direct contact routes of transmission [12,13,30,31]. Given this potential risk of viral transmission, there has been research documenting the benefits of power washing, disinfecting, and then heating livestock trailers to reduce the amount of potential infective material to limit the spread of disease to naïve pigs due to both PEDV and PRRSV [16–19,32]. However, most of this research has focused on how to reduce transmission in transportation with live animals and not on transportation vehicles that can frequently visit production sites without live animals, including feed delivery trucks. Furthermore, most research has focused on the inactivation of viruses on surfaces in an experimental setting, but little research has focused on the practical application of decontamination treatments in real-world settings using complex surfaces such as those found in the cab of semi-trucks that contain multiple surfaces as well as many surfaces with different orientations, making the practical application of disinfectants challenging.

There is evidence to suggest that when sampling for viruses or bacteria within a feed mill, a majority of the samples containing detectable pathogens were from the cab of the delivery truck, most likely due to the driver getting in and out of the cab for deliveries at production sites and potentially "tracking" virus back into the cab [11,14,15,22]. There are ways to mitigate this risk, including the use of shoe covers for each delivery or the application of disinfectant to the truck cab since the application of disinfectant has been successful for PEDV and PRRSV reduction in terms of livestock trailers; however, challenges exist when trying to transfer these procedures to truck cabs. The problems to consider when applying disinfectant to the truck cab are how to apply the disinfectant, the type of disinfectant (in terms of wet, dry, or fumigation), and the implications of disinfectant on the surfaces since there are various surface types present throughout the truck cab, unlike livestock trailers that are primarily made of metal. There is minimal research looking at different application methods and how these applications impact the amount of detectable virus within truck cabs on different surface types. Therefore, this study aimed to quantify the impact of commercially available disinfectants applied through different application methods on detectable viral RNA on fabric, plastic, and rubber surfaces inoculated with either PEDV or PRRSV and if detectable viral RNA could cause infection in susceptible pigs via swine bioassay.

There was no evidence of a disinfectant × surface × virus interaction ($p = 0.959$; Table 2), surface × virus interaction ($p = 0.926$), or disinfectant × virus interaction ($p = 0.508$).

Table 2. Effect of surface type, disinfectant, and virus on the detection of viral RNA during truck cab decontamination [1].

Item	Surface Type					
	Fabric		Plastic		Rubber	
	PEDV	PRRSV	PEDV	PRRSV	PEDV	PRRSV
Proportion PCR positive						
No disinfectant [2]	3/3	3/3	3/3	3/3	3/3	3/3
Misting fumigation [3]						
Intervention [4]	3/3	3/3	3/3	3/3	2/3	2/3
Synergize [5]	2/3	3/3	3/3	3/3	3/3	3/3
Pump sprayer [6]						
Intervention	3/3	3/3	3/3	3/3	3/3	3/3
Synergize	2/3	3/3	2/3	3/3	3/3	3/3
10% Bleach [7]	2/3	0/3	3/3	3/3	1/3	0/3
10 h downtime [8]						
No chemical	3/3	2/3	3/3	3/3	3/3	3/3
Gaseous treatment [9]	3/3	1/3	3/3	3/3	3/3	3/3
Cycle threshold [10]						
No disinfectant	34.6	37.2	26.7	30.6	26.7	31.4
Misting fumigation						
Intervention	33.4	38.2	28.1	31.6	34.2	36.7
Synergize	36.2	36.4	29.7	34.2	30.3	33.3

Table 2. *Cont.*

Item	Surface Type					
	Fabric		Plastic		Rubber	
	PEDV	PRRSV	PEDV	PRRSV	PEDV	PRRSV
Pump sprayer						
Intervention	34.8	37.5	28.3	31.3	28.8	32.0
Synergize	37.3	38.7	33.0	32.6	30.6	33.5
10% Bleach	40.7	45.0	26.7	31.2	41.2	45.0
10 h downtime						
No chemical	36.4	40.3	27.8	29.8	29.7	30.2
Gaseous treatment	36.8	44.4	28.3	31.9	28.6	33.2

[1] Surfaces were inoculated with 1 mL of porcine epidemic diarrhea virus (PEDV) or porcine reproductive and respiratory syndrome virus (PRRSV), randomly placed within the truck cab, and subjected to a randomly assigned disinfectant treatment. Samples with no detectable RNA were assigned a value of 45. [2] Surfaces were inoculated with pure virus and allowed to sit within the truck cab for 15 min. These surfaces were not treated with a disinfectant application. [3] Truck cabs had a hurricane fumigation system placed in the passenger seat and directed toward the driver's side. The hurricane fumigation system was filled with respective disinfectant and was allowed to run 5 min for each treatment. [4] 1:64 dilution of Intervention (Virox, Oakville, ON, USA). [5] 1:256 dilution of Synergize (Neogen Corp, Lexington, KY, USA). [6] Truck cabs had disinfectant applied with a conventional pump sprayer with the designated disinfectant. [7] Household bleach (10% dilution; The Clorox Company, Oakland, CA; 7.55% sodium hypochlorite). [8] Surfaces were inoculated with pure virus and allowed to sit within the truck cab for 10 h. These surfaces were not treated with a disinfectant application. [9] Truck cabs had gaseous chlorine dioxide (ProKure G; ProKure Solutions, Phoenix, AZ, USA) placed on their passenger side seats, and the truck cabs were allowed to fumigate for 10 h. [10] Disinfectant × surface × virus interaction, $p = 0.959$; SEM = 1.84.

There was a significant disinfectant × surface interaction ($p < 0.0001$; Table 3), indicating that the quantity of PEDV or PRRSV RNA detected on surfaces differed based on disinfectant treatment.

Table 3. Effect of surface type and disinfectant on the detection of viral RNA during truck cab decontamination [1].

Item	Proportion PCR Positive			Ct Value [2]		
	Fabric	Plastic	Rubber	Fabric	Plastic	Rubber
No disinfectant [3]	6/6	6/6	6/6	35.9 ^{c,d,e,f,g,h}	28.6 ^{a,b}	29.0 ^{a,b}
Misting fumigation [4]						
Intervention [5]	6/6	6/6	5/6	35.8 ^{a,b,c,d,e,f,g,h}	29.8 ^{a,b,c,d}	35.4 ^{a,b,c,d,e,f,g,h}
Synergize [6]	5/6	6/6	6/6	36.6 ^{a,b,c,d,e,f,g,h}	31.9 ^{a,b,c,d,e,f}	31.8 ^{a,b,c,d,e,f}
Pump sprayer [7]						
Intervention	6/6	6/6	6/6	36.1 ^{b,d,e,f,g,h}	29.8 ^{a,c}	30.4 ^{a,b,c,d,e,f}
Synergize	5/6	5/6	6/6	38.0 ^{e,f,g,h}	32.8 ^{a,b,c,d,e,f,g}	32.0 ^{a,b,c,d,e,f}
10% Bleach [8]	2/6	6/6	1/6	42.9 ^{h}	29.0 ^{a,b,c,d}	43.1 ^{h}
10 h downtime [9]						
No chemical	5/6	6/6	6/6	38.4 ^{f,g,h}	28.8 ^{a,b,c,d}	30.0 ^{a,b,c,d}
Gaseous treatment [10]	4/6	6/6	6/6	40.6 ^{g,h}	30.1 ^{a,b,c,d,e}	30.9 ^{a,b,c,d,e,f}

[1] Surfaces were inoculated with 1 mL of porcine epidemic diarrhea virus (PEDV) or porcine reproductive and respiratory syndrome virus (PRRSV), randomly placed within the truck cab, and subjected to a randomly assigned disinfectant treatment. Samples with no detectable RNA were assigned a value of 45. [2] Disinfectant × surface, $p < 0.0001$; SEM = 1.45. [3] Surfaces were inoculated with pure virus and allowed to sit within the truck cab for 15 min. These surfaces were not treated with a disinfectant application. [4] Truck cabs had a hurricane fumigation system placed in the passenger seat and directed toward the driver's side. The hurricane fumigation system was filled with respective disinfectant and was allowed to run 5 min for each treatment. [5] 1:64 dilution of Intervention (Virox, Oakville, ON, USA). [6] 1:256 dilution of Synergize (Neogen Corp, Lexington, KY, USA). [7] Truck cabs had disinfectant applied with a conventional pump sprayer with the designated disinfectant. [8] Household bleach (10% dilution; The Clorox Company, Oakland, CA, USA; 7.55% sodium hypochlorite). [9] Surfaces were inoculated with pure virus and allowed to sit within the truck cab for 10 h. These surfaces were not treated with a disinfectant application. [10] Truck cabs had gaseous chlorine dioxide (ProKure G; ProKure Solutions, Phoenix, AZ, USA) placed on their passenger side seat, and the truck cabs were allowed to fumigate for 10 h. $^{a–h}$ Means lacking common superscripts differ, $p < 0.05$.

For rubber surfaces, environmental samples from the 10% bleach application had the least ($p < 0.05$) amount of PEDV or PRRSV RNA detected compared to the other disinfectant methods with Intervention via misting fumigation application as intermediate. For plastic and fabric surfaces, there was no evidence of a statistical difference between decontamination methods ($p > 0.05$). Furthermore, for the no disinfectant treatment, environmental swabs from fabric surfaces detected less PEDV or PRRSV RNA than that from plastic or rubber surfaces ($p < 0.05$). For the pump sprayer with Intervention disinfectant treatment, environmental swabs from fabric surfaces detected less ($p < 0.05$) PEDV or PRRSV RNA compared to that from rubber surfaces with plastic surfaces as intermediate. For the pump sprayer with 10% bleach disinfectant treatment, environmental swabs from fabric and rubber surfaces detected less PEDV or PRRSV RNA than that from plastic surfaces ($p < 0.05$). For the no chemical 10 h downtime disinfectant treatment, environmental swabs from fabric surfaces detected less PEDV or PRRSV RNA when compared to that from plastic or rubber surfaces ($p < 0.05$). Lastly, for the gaseous treatment 10 h downtime disinfectant application, environmental swabs from fabric surfaces detected less PEDV or PRRSV RNA when compared to that from plastic or rubber surfaces ($p < 0.05$). These results are similar to Muckey et al. [23], where there was also a treatment × surface interaction, indicating that when implementing disinfectant applications, the surface should be taken into consideration since surface material has the potential to influence the amount of detectable viral RNA. Given this commonality between research studies, there is some suggestion that surface type can influence the amount of detectable viral RNA recovered within sampling techniques.

The main effect of disinfection treatment was statistically significant ($p = 0.016$; Table 4), with environmental swabs collected after 10% bleach treatment, detecting less PEDV or PRRSV RNA ($p < 0.05$) when compared to environmental swabs from no disinfectant, Intervention via pump sprayers, and no chemical application 10 h downtime with all other disinfectant treatment applications being intermediate.

Disinfectants have been shown to reduce the amount of detectable PRRSV or PEDV RNA [16–18,33]. However, research suggests that the application of sodium hypochlorite, or bleach, results in the greatest reduction of PEDV, PRRSV, and other enteric viruses or bacteria, sometimes producing non-detectable test results for pathogens compared to other disinfectants or soaps used based on the surface material, which is consistent with the findings from this study [16–18,33–35].

The main effect of the surface was statistically significant ($p < 0.0001$), with environmental swabs from plastic surfaces detecting greater amounts of PEDV or PRRSV RNA when compared to those from rubber, and the samples from rubber detecting greater amounts of PEDV or PRRSV RNA when compared to those from fabric ($p < 0.05$). Rabuza et al. [36] reported that when sampling with cotton gauze, a methodology similar to that used in this experiment, the cotton gauze method extracted lower quantities of *Klebsiella pneumoniae* and *Staphylococcus aureus* from inoculated fabric surfaces when compared to other methods like contact plating, destructive elution methodology, or nondestructive elution methodology. For contact plating, the fabric sample had RODAC agar poured over it, so it could stick to the medium, while destructive or nondestructive elution methodologies are wash-off methods where microorganisms are eluted from the fabrics by shaking the fabrics for a specific time in an elution medium. Then, if the sample was assigned the nondestructive elution methodology, the sample was subjected to forced desorption by pressing the microorganism through the fabric without destroying the fabric. While contact plating, destructive elution methodology, or nondestructive elution methodology are employed to quantify bacteria on fabric surfaces, all methodologies require a fabric sample. These methodologies have their strengths, but there is minimal real-world application since pieces of a seat cover cannot be cut off from a seat within a truck cab. The advantage of the sampling methodology used in the current experiment is the ease of application in a practical field setting. Fabric surfaces have the potential to absorb and hold wet disinfectants for longer periods when compared to other non-absorptive surfaces like plastic or rubber, indicating that inherent properties of the surface could lend to longer disinfectant contact times. When consulting hospital procedures for disinfectant application, spraying disinfectants on various surfaces is a common solution for disinfection. However, excessive wetting of fabric surfaces is highly discouraged due to patient discomfort, and disinfectants will continuously leak out of fabric surfaces for extended periods [37]. It is uncertain which factor, either sampling method or the inherent properties of fabric, may have potentially impacted the Ct results from this study. These findings highlight the need for more research to be conducted on disinfectant application on fabric surfaces and applicable sampling methodology. The main effect of virus ($p < 0.0001$) was significant, with environmental swabs detecting greater amounts of PEDV RNA compared to the detection of PRRSV RNA ($p < 0.05$).

Table 4. Main effects of disinfectant, surface type, and virus on the detection of viral RNA during truck cab decontamination [1].

Item	Proportion PCR Positive	Ct Value	SEM	p=
Disinfectant			1.11	0.016
No disinfectant [2]	18/18	31.2 [a]		
Hurricane fumigation [3]				
Intervention [4]	16/18	33.7 [a,b]		
Synergize [5]	17/18	33.4 [a,b]		
Pump sprayer [6]				
Intervention	18/18	32.1 [a]		
Synergize	16/18	34.3 [a,b]		
10% Bleach [7]	9/18	38.3 [b]		
10 h Downtime				
No chemical [8]	17/18	32.4 [a]		
Gaseous treatment [9]	16/18	33.9 [a,b]		
Surface Type			0.53	<0.0001
Fabric	39/48	38.0 [c]		
Plastic	47/48	30.1 [a]		
Rubber	41/48	32.8 [b]		
Virus			0.48	<0.0001
PEDV	62/72	32.0 [a]		
PRRSV	62/72	35.3 [b]		

[1] Surfaces were inoculated with 1 mL of porcine epidemic diarrhea virus (PEDV) or porcine reproductive and respiratory syndrome virus (PRRSV), randomly placed within the truck cab, and subjected to a randomly assigned disinfectant treatment. Samples with no detectable RNA were assigned a value of 45. [2] Surfaces were inoculated with pure virus and allowed to sit within the truck cab for 15 min. These surfaces were not treated with a disinfectant application. [3] Truck cabs had a hurricane fumigation system placed in the passenger seat and directed toward the driver's side. The hurricane fumigation system was filled with respective disinfectant and was allowed to run 5 min for each treatment. [4] 1:64 dilution of Intervention (Virox, Oakville, ON, USA). [5] 1:256 dilution of Synergize (Neogen Corp, Lexington, KY, USA). [6] Truck cabs had disinfectant applied with a conventional pump sprayer with the designated disinfectant. [7] Household bleach (10% dilution; The Clorox Company, Oakland, CA, USA; 7.55% sodium hypochlorite). [8] Surfaces were inoculated with pure virus and allowed to sit within the truck cab for 10 h. These surfaces were not treated with a disinfectant application. [9] Truck cabs had gaseous chlorine dioxide (ProKure G; ProKure Solutions, Phoenix, AZ, USA) placed on their passenger side seat, and the truck cabs were allowed to fumigate for 10 h. [a,b,c] Means within main effect lacking common superscripts differ, $p < 0.05$.

Supernatant from environmental samples of surfaces inoculated with PEDV and PRRSV after disinfectant treatment application were utilized in bioassay and failed to produce infectivity. To our knowledge, this is the first published study utilizing inoculum from environmental samples in which the positive control treatment groups (with Ct values ranging 26.4–37.2) did not cause clinical infection in pigs via bioassay. A potential explanation for the failure to produce infectivity is the time from sample collection and processing to the time of bioassay, potentially impacting infectivity factors for PEDV and PRRSV. For example, the length of time from sample collection to bioassay for this study was 13 months, while for the dust sample study, the length of time from sample collection to bioassay was 11 months [38]. More research is needed to understand the impacts of long-term storage on virus infectivity.

This study aimed to test disinfectant applications that could be implemented between feed deliveries or while feed trucks are loaded at the feed mill. Additionally, the data presented here provide value to a broader segment of the industry in comparing disinfectants through different applications that could be quickly implemented through a variety of truck cabs like those used in live-animal delivery, veterinary services, and other trucks associated with the swine and animal industries.

4. Conclusions

In conclusion, the disinfection of truck cabs has its challenges given the variety of surfaces present. The disinfectant treatments tested in this trial could only reduce the detectable amount of viral RNA and did not completely eliminate the viral genetic material across surface types. Based on this study and methodology utilized for bioassay, this amount of detectable viral RNA was not sufficient to cause infectivity in a swine bioassay. Additional research is warranted to refine methods of detecting infectious viruses across varying surfaces.

Author Contributions: Conceptualization, G.E.H., J.T.G., C.K.J., J.C.W. and C.B.P.; methodology, J.T.G., C.K.J., J.C.W., C.B.P., M.N.d.A., J.Z. and P.C.G.; validation, G.E.H., J.T.G., M.N.d.A., J.Z. and P.C.G.; formal analysis, G.E.H. and J.T.G.; investigation, G.E.H., C.E., H.K.O., C.R.S., J.B., E.G.P.P., M.N.d.A., J.Z., P.C.G., A.K.B., C.B.P. and J.T.G.; resources, C.E., H.K.O., C.R.S. and C.B.P.; data curation, G.E.H. and J.T.G.; writing—original draft preparation, G.E.H.; writing—review and editing, C.K.J., C.E., H.K.O., C.R.S., J.B., E.G.P.P., M.N.d.A., J.Z., P.C.G., A.K.B., J.C.W., C.B.P. and J.T.G.; supervision, C.K.J. and J.T.G.; project administration, J.T.G., M.N.d.A. and P.C.G.; funding acquisition, J.T.G., C.K.J., J.C.W. and C.B.P. All authors have read and agreed to the published version of the manuscript.

Funding: Funding was, wholly or in part, provided by the National Pork Board through grant nos. 20-084 and 22-060.

Institutional Review Board Statement: Surface inoculation and sampling procedures were reviewed and approved by the Kansas State University Institutional Biosafety Committee. Animal use procedures were reviewed and approved by the Iowa State University Institutional Animal Care and Use Committee (IACUC number 22-082, April 2022).

Informed Consent Statement: Not applicable.

Data Availability Statement: All data generated or analyzed during this study are included in this published article.

Acknowledgments: Contribution no. 24-099-J from the Kansas Agric. Exp. Stn., Manhattan KS, 66506-0210 USA. Special thanks to Roman Pogranichniy and Joe Anderson for technical support.

Conflicts of Interest: The authors declare no conflict of interest.

References

1. Holtkamp, D.J.; Kliebenstein, J.B.; Neumann, E.J.; Zimmerman, J.J.; Rotto, H.F.; Yoder, T.K.; Wang, C.; Yeske, P.E.; Mowrer, C.L.; Haley, C.A. Assessment of the economic impact of porcine reproductive and respiratory syndrome virus on United States pork producers. *J. Swine Health Prod.* **2013**, *21*, 72–84.
2. Valdes-Donoso, P.; Alvarez, J.; Jarvis, L.S.; Morrison, R.B.; Perez, A.M. Production losses from an endemic animal disease: Porcine reproductive and respiratory syndrome (PRRS) in selected midwest US sow farms. *Front. Vet. Sci.* **2018**, *5*, 102. [CrossRef] [PubMed]
3. Zhang, Z.; Li, Z.; Li, H.; Yang, S.; Ren, F.; Bian, T.; Sun, L.; Zhou, B.; Zhou, L.; Qu, X. The economic impact of porcine reproductive and respiratory syndrome outbreak in four Chinese farms: Based on cost and revenue analysis. *Front. Vet. Sci.* **2022**, *9*, 1024720. [CrossRef]
4. Renken, C.; Nathues, C.; Swam, H.; Fiebig, K.; Weiss, C.; Eddicks, M.; Ritzmann, M.; Natheus, H. Application of an economic calculator to determine the cost of porcine reproductive and respiratory syndrome at farm-level in 21 pig herds in Germany. *Porcine Health Manag.* **2021**, *7*, 3. [CrossRef] [PubMed]
5. Magalhaes, E.S.; Zimmerman, J.J.; Thomas, P.; Moura, C.A.A.; Trevisan, G.; Holtkamp, D.J.; Wang, C.; Rademacher, C.; Silva, G.S.; Linhares, D.C.L. Whole-herd risk factors associated with wean-to-finish mortality under the conditions of a midwestern USA swine production system. *Prev. Vet. Med.* **2022**, *198*, 105545. [CrossRef] [PubMed]

6. Niederwerder, M.C.; Hesse, R.A. Swine enteric coronavirus disease: A review of 4 years with porcine epidemic diarrhoea virus and porcine deltacoronavirus in the United States and Canada. *Transbound. Emerg. Dis.* **2018**, *65*, 660–675. [CrossRef] [PubMed]
7. Morrison Swine Health Monitoring Project. (6 October 2023). PED Cumulative Incidence (Last Updated October 2023). Morrison Swine Health Monitoring Project. Available online: https://mshmp.umn.edu/reports#Charts (accessed on 7 January 2024).
8. Lunney, J.K.; Fang, Y.; Ladinig, A.; Chen, N.; Li, Y.; Rowland, B.; Renukaradhya, G.J. Porcine reproductive and respiratory syndrome virus (PRRSV): Pathogenesis and interaction with the immune system. *Annu. Rev. Anim. Biosci.* **2015**, *4*, 129–154. [CrossRef] [PubMed]
9. Jung, K.; Saif, L.J.; Wang, Q. Porcine epidemic diarrhea virus (PEDV): An update on etiology, transmission, pathogenesis, and prevention and control. *Virus Res.* **2020**, *286*, 198045. [CrossRef]
10. Pileri, E.; Mateu, E. Review on the transmission porcine reproductive and respiratory syndrome virus between pigs and farms and impact on vaccination. *Vet. Res.* **2016**, *47*, 108. [CrossRef]
11. Fedorka-Cray, P.; Hogg, A.; Gray, J.T.; Lorenzen, K.; Velasquez, J.; Von Behren, P. Feed and feed trucks as sources of Salmonella contamination in swine. *J. Swine Health Prod.* **1997**, *5*, 189–193.
12. Dee, S.; Deen, J.; Otake, S.; Pijoan, C. An experimental model to evaluate the role of transport vehicles as a source of transmission of porcine reproductive and respiratory syndrome virus to susceptible pigs. *Can. J. Vet. Res.* **2004**, *68*, 208–214. [PubMed]
13. Lowe, J.; Gauger, P.; Harmon, K.; Zhang, J.; Connor, J.; Yeske, P.; Loula, T.; Levis, I.; Dufresne, L.; Main, R. Role of transportation in spread of porcine epidemic diarrhea virus infection, United States. *Emerg. Infect. Dis.* **2014**, *20*, 872–874. [CrossRef] [PubMed]
14. Elijah, C.G.; Harrison, O.L.; Blomme, A.K.; Woodworth, J.C.; Jones, C.K.; Paulk, C.B.; Gebhardt, J.T. Understanding the role of feed manufacturing and delivery within a series of porcine deltacoronavirus investigations. *J. Swine Health Prod.* **2022**, *30*, 17–23. [CrossRef]
15. Gebhardt, J.T.; Dritz, S.S.; Elijah, C.G.; Jones, C.K.; Paulk, C.B.; Woodworth, J.C. Sampling and detection of African swine fever virus within a feed manufacturing and swine production system. *Transbound. Emerg. Dis.* **2022**, *6*, 103–114. [CrossRef]
16. Dee, S.; Deen, J.; Burns, D.; Douthit, G.; Pijoan, C. An assessment of sanitation protocols for commercial transport vehicles contaminated with porcine reproductive and respiratory syndrome virus. *Can. J. Vet. Res.* **2004**, *68*, 208–214.
17. Dee, S.; Deen, J.; Pijoan, C. Evaluation of 4 intervention strategies to prevent the mechanical transmission of porcine reproductive and respiratory syndrome virus. *Can. J. Vet. Res.* **2004**, *68*, 19–26.
18. Dee, S.; Deen, J.; Burns, D.; Douthit, G.; Pijoan, C. An evaluation of disinfectants for the sanitation of porcine reproductive and respiratory syndrome virus-contaminated transport vehicles at cold temperatures. *Can. J. Vet. Res.* **2005**, *69*, 64–70.
19. Baker, K.L.; Thomas, P.R.; Karriker, L.A.; Ramirez, A.; Zhang, J.; Wang, C.; Holtkamp, D.J. Evaluation of an accelerated hydrogen peroxide disinfectant to inactivate porcine epidemic diarrhea virus in swine feces on aluminum surfaces under freezing conditions. *BMC Vet. Res.* **2017**, *13*, 372. [CrossRef]
20. Holtkamp, D.J.; Myers, J.; Thomas, P.R.; Karriker, L.A.; Ramirez, A.; Zhang, J.; Wang, C. Efficacy of accelerated hydrogen peroxide disinfectant to inactivate porcine epidemic diarrhea virus in swine feces on metal surfaces. *Can. J. Vet. Res.* **2017**, *81*, 100–107.
21. Baker, K.L.; Mowrer, C.L.; Zhang, J.; Chen, Q.; Ramirez, A.; Wang, C.; Karriker, L.A.; Holtkamp, D.J. Evaluation of a peroxygen-based disinfectant for inactivation of porcine epidemic diarrhea virus at low temperatures on metal surfaces. *Vet. Microbiol.* **2018**, *214*, 99–107. [CrossRef]
22. Greiner, L.L. Evaluation of the likelihood of detection of porcine epidemic diarrhea virus or porcine deltacoronavirus ribonucleic acid in areas within feed mills. *J. Swine Health Prod.* **2016**, *24*, 198–204.
23. Muckey, M.B.; Jones, C.K.; Woodworth, J.C.; Paulk, C.B.; Dritz, S.S.; Gebhardt, J.T. Using environmental sampling to evaluate the effectiveness of decontamination methods to reduce detection of porcine epidemic diarrhea virus RNA on feed manufacturing surfaces. *Transl. Anim. Sci.* **2021**, *5*, txab121. [CrossRef]
24. Elijah, C.G.; Trujillo, J.D.; Jones, C.K.; Gaudreault, N.N.; Stark, C.R.; Cool, K.R.; Paulk, C.B.; Kwon, T.; Woodworth, J.C.; Morozov, I.; et al. Evaluating the distribution of African swine fever virus within a feed mill environment following manufacture of inoculated feed. *PLoS ONE* **2021**, *16*, e0256138. [CrossRef]
25. Thomas, J.T.; Chen, Q.; Gauger, P.C.; Giménez-Lirola, L.G.; Sinha, A.; Harmon, K.M.; Madson, D.M.; Burrough, E.R.; Magstadt, D.R.; Salzbrenner, H.M.; et al. Effect of porcine epidemic diarrhea virus infectious doses on infection outcomes in naïve conventional neonatal and weaned pigs. *PLoS ONE* **2015**, *10*, e0139266. [CrossRef]
26. Schumacher, L.L.; A Cochrane, R.; Huss, A.R.; Gebhardt, J.T.; Woodworth, J.C.; Stark, C.R.; Jones, C.K.; Bai, J.; Main, R.G.; Chen, Q.; et al. Feed batch sequencing to decrease the risk of porcine epidemic diarrhea virus (PEDV) cross-contamination during feed manufacturing. *J. Anim. Sci.* **2018**, *96*, 4562–4570. [CrossRef]
27. Ruston, C.; Holtkamp, D.; Li, P.; Koziel, J.A.; Stephan, A.; Loesekann, T.; Torremorell, M.; Murray, D.; Wedel, K.; Johnson, C.; et al. Effectively using ultraviolet-C light for supply decontamination on swine farms. *J. Swine Health Prod.* **2022**, *30*, 101–106. [CrossRef]
28. Dee, S.; Batista, L.; Deen, J.; Pijoan, C. Evaluation of an air-filtration system for preventing aerosol transmission of porcine reproductive and respiratory syndrome virus. *Can J. Vet. Res.* **2005**, *69*, 293–298. [PubMed]
29. Stewart, S.C.; Dritz, S.S.; Woodworth, J.C.; Paulk, C.; Jones, C.K. A review of strategies to impact swine feed biosecurity. *Anim. Health Res. Rev.* **2020**, *21*, 61–68. [CrossRef]
30. Dee, S.; Deen, J.; Rossow, K.; Wiese, C.; Otake, S.; Joo, H.S.; Pijoan, C. Mechanical transmission of porcine reproductive and respiratory syndrome virus throughout a coordinated sequence of events during cold weather. *Can. J. Vet. Res.* **2002**, *66*, 232–239.

31. Dee, S.; Deen, J.; Rossow, K.; Wiese, C.; Eliason, R.; Otake, S.; Joo, H.S.; Pijoan, C. Mechanical transmission of porcine reproductive and respiratory syndrome virus throughout a coordinated sequence of events during warm weather. *Can. J. Vet. Res.* **2003**, *67*, 12–19.
32. Neat, R.A.; Zhang, J.; Hoang, H.; McKeen, L.; Mowrer, C.L.; Holtkamp, D.J. Disinfection and conditions associated with thermo-assisted drying and decontamination inconsistently produce negative PRRSV rRT-PCR results on metal surfaces. *Vet. Microbiol.* **2021**, *262*, 109240. [CrossRef] [PubMed]
33. Bowman, A.S.; Nolting, J.M.; Nelson, S.W.; Bliss, N.; Stull, J.W.; Wang, Q.; Premanandan, C. Effects of disinfection on the molecular detection of porcine epidemic diarrhea virus. *J. Vet. Mic.* **2015**, *179*, 213–218. [CrossRef] [PubMed]
34. Gerba, C.P.; Kennedy, D. Enteric Virus Survival during household laundering and impact of disinfection with Ssodium hypochlorite. *Appl. Environ. Microbiol.* **2007**, *73*, 4425–4428. [CrossRef] [PubMed]
35. Honisch, M.; Stamminger, R.; Bockmühl, D. Impact of wash cycle time, temperature, and detergent formulation on the hygiene effectiveness of domestic laundering. *J. Appl. Microbiol.* **2014**, *117*, 1787–1797. [CrossRef]
36. Rabuza, U.; Sostar-Turk, S.; Fijan, S. Efficiency of four sampling methods used to detect two common nosocomial pathogens on textile. *Text. Res. J.* **2012**, *82*, 2099–2105. [CrossRef]
37. Creamer, E.; Humphreys, H. The contribution of beds to healthcare-associated infection: The importance of adequate decontamination. *J. Hosp. Infect.* **2008**, *69*, 8–23. [CrossRef]
38. Schumacher, L.L.; Huss, A.R.; Cochrane, R.A.; Stark, C.R.; Woodworth, J.C.; Bai, J.; Poulsen, E.G.; Chen, Q.; Main, R.G.; Zhang, J.; et al. Characterizing the rapid spread of porcine epidemic diarrhea virus (PEDV) through an animal food manufacturing facility. *PLoS ONE* **2017**, *12*, e0187309. [CrossRef]

Disclaimer/Publisher's Note: The statements, opinions and data contained in all publications are solely those of the individual author(s) and contributor(s) and not of MDPI and/or the editor(s). MDPI and/or the editor(s) disclaim responsibility for any injury to people or property resulting from any ideas, methods, instructions or products referred to in the content.

animals MDPI

Article

Situation Analysis and Recommendations for the Biosecurity Status of Dairy Farms in Punjab, India: A Cross-Sectional Survey

Pankaj Dhaka [1,2,*], Ilias Chantziaras [1], Deepthi Vijay [2,3], Manmeet Singh [2], Jasbir Singh Bedi [2], Nele Caekebeke [1] and Jeroen Dewulf [1]

[1] Veterinary Epidemiology Unit, Faculty of Veterinary Medicine, Ghent University, Salisburylaan 133, 9820 Merelbeke, Belgium; ilias.chantziaras@ugent.be (I.C.); nele.caekebeke@biocheckgent.com (N.C.); jeroen.dewulf@ugent.be (J.D.)

[2] Centre for One Health, College of Veterinary Science, Guru Angad Dev Veterinary and Animal Sciences University, Ludhiana 141004, India; deepthivijay@kvasu.ac.in (D.V.); meet.mehn@gmail.com (M.S.); bedijasbir78@gmail.com (J.S.B.)

[3] Department of Veterinary Public Health, College of Veterinary and Animal Sciences, Kerala Veterinary and Animal Sciences University, Mannuthy, Thrissur 680651, India

* Correspondence: pankaj.dhaka2@gmail.com; Tel.: +91-7983-816-228

Simple Summary: Farm biosecurity is crucial for preventing infections and maintaining animal health. The present study assessed biosecurity practices on dairy farms in Punjab, India, using a standardized scoring system. The study found that the mean external and internal biosecurity scores for the selected dairy farms were 45.4% and 43.7%, respectively. Among the different aspects of biosecurity, farms performed best in vermin control and adult cattle management, while they scored lowest in purchase and reproduction, and health management. Interestingly, larger farms tended to have better biosecurity. This study highlights the need for improvement in biosecurity measures on dairy farms, particularly in the areas of purchasing animals and health management. This research underscores the importance of addressing these issues to enhance animal health and reduce disease risk in the region. Further investigations are recommended to better understand the factors influencing biosecurity practices on these farms.

Abstract: Farm biosecurity is an important herd management strategy to assure infection prevention and animal health. The present study aimed to evaluate the implementation of biosecurity measures on dairy farms in Punjab, India, using the standardized Biocheck.UGent™ scoring system. Convenience sampling was used to select 94 dairy farms, comprising both cattle and buffaloes, with a mean herd size of 74.8 animals. The study found that the mean external and internal biosecurity scores for the selected dairy farms were 45.4% and 43.7%, respectively. Among the subcategories, the highest external biosecurity score was observed for 'vermin control and other animals' (63%), and the highest internal biosecurity score was observed for 'adult cattle management' (76.6%). Whereas the lowest score for external biosecurity was observed for 'purchase and reproduction' (30.6%), and the lowest score for internal biosecurity was observed for 'health management' (33.6%). The overall mean biosecurity score of the present study was 44.8%, which was lower than the overall mean global score of 52%. The correlation analysis indicated a positive correlation between herd size and overall biosecurity scores, indicating that larger farms had, on average, higher biosecurity scores. Based on these results, it can be concluded that the implementation of biosecurity measures on dairy farms in Punjab needs improvement, especially in the subcategories of 'purchase and reproduction' and 'health management'. Further research to identify the factors influencing the implementation of biosecurity measures on dairy farms in the region is recommended.

Keywords: biosecurity; buffaloes; cattle; dairy; herd management; India

1. Introduction

Dairy farming is an important animal husbandry sector that contributes significantly to sustainable food production, livelihood, and economies worldwide [1]. Like other livestock sectors, dairy farms face significant risks from the introduction and spillover of infectious diseases, animal welfare concerns, and environmental contamination. These issues pose a threat to food security and safety, as well as cause socioeconomic losses for dairy farmers [2]. A number of infectious agents that affect dairy animals can also be transmitted to humans through direct contact or by consuming contaminated animal food products. This can result in the transmission of various zoonotic diseases [3]. Effective implementation of biosecurity measures is one of the important and widely used strategies to prevent and control the spread of infectious diseases on dairy farms [4].

Farm biosecurity is considered a critical component of One Health, and it involves implementing segregation, hygiene, or management procedures (excluding medically effective feed additives and preventive/curative treatment of animals) to reduce the likelihood of introducing, establishing, surviving, or spreading potential pathogenic agents to, within, or from a farm operation or geographical area [5]. Effective implementation of both external and internal biosecurity measures in dairy farms can significantly reduce the risk of disease transmission from infected animals, contaminated equipment, and people such as farm workers, visitors, and animal health professionals. This, in turn, can improve the health, productivity, reduction in antimicrobial usage, and welfare of the animals [6,7]. Studies have shown that, despite the significant role of biosecurity measures in preventing the spread of infectious diseases and the associated benefits, many dairy farmers lack awareness and knowledge with regard to the benefits and application of biosecurity measures [8].

Dairying is a crucial component of rural livelihood in India, with the Indian dairy sector being arguably the world's largest agri-food sector, providing employment to over 70 million households [9,10]. In India, as per the 20th livestock census, the cattle population is recorded at 193.46 million, and the buffalo population is 109.85 million. These animals, respectively, contribute to 51% and 45% of the total milk production for the year 2021–22 [11]. In the state of Punjab, around 4 million buffaloes and 2.5 million cattle equally share in contributing 50% of the milk production. The region primarily features mixed-type dairy farms (having both buffalo and cattle), with a significant presence in small and medium-sized operations located in rural and peri-urban areas [12]. However, more than half of India's dairy sector remains informal, highlighting the need for strengthening herd health management and techno-managerial approaches to move the sector toward sustainability [13].

India has been identified as a hotspot for emerging infectious diseases [14], and past epidemiological studies have reported the prevalence of various zoonoses associated with farm animals. Some of the important zoonoses include brucellosis [15], anthrax [16], toxoplasmosis [17], Q fever [18], and Crimean–Congo hemorrhagic fever [19]. The concerns have been raised by previous studies conducted in Punjab regarding the potential transmission of zoonotic diseases due to several risk factors and the limited awareness among livestock farmers in the region. Researchers observed that livestock farmers possess a low to medium level of knowledge about prevalent zoonotic diseases in the region [20]. The low level of education and being a cattle farmer were negatively associated with the farmer's knowledge of zoonotic diseases [21]. Furthermore, it was noted that farmers' adoption of biosecurity measures and the utilization of personal protective equipment by veterinary personnel were significantly lacking in the region [22]. This considerably increases the risk of zoonotic disease transmission and underscores the need for comprehensive assessments and awareness initiatives focused on farm biosecurity within the region.

Many times, the animal health sector in India mainly focuses on therapeutic measures and reactive responses toward infectious disease outbreaks [23]. The recent outbreak of lumpy skin disease (LSD) in 15 Indian states in 2022 resulted in significant socioeconomic losses due to bovine morbidity, mortality, and productivity losses [24]. Such outbreaks underscore the need to promote biosecurity management in the dairy sector to control and prevent the spread of emerging animal health threats. Therefore, it is essential to assess and quantify farm biosecurity practices to understand existing management practices and identify areas for improvement in the herd. In this context, the present study aims to assess the level of biosecurity measures in dairy farms in the Punjab state of India and formulate recommendations for adopting biosecurity measures in the study region.

2. Materials and Methods

2.1. Sampling Strategy

A cross-sectional survey was conducted using convenience sampling to recruit volunteer dairy farmers from Punjab. The outreach was facilitated through local contacts and an open call inviting participation in the study. Farm visits were scheduled between July 2021 and November 2022 based on the availability of farmers. A standardized procedure was followed during the visits to ensure uniformity in data collection, and participating farmers completed a farm biosecurity questionnaire. The surveys were carried out on-site, with interviewers simultaneously observing the biosecurity measures in place during the farm visit. The interviewer also observed the ongoing biosecurity measures during the herd visit. Verbal consent was obtained from the farm owners to use the farm data anonymously for academic and research purposes.

2.2. Ethical Statement

The research was conducted in accordance with the Declaration of Helsinki and national standards. All the required ethical considerations were taken into account. The nature of the study was completely voluntary, and informed consent was obtained from study participants. The details of the participants were anonymous, and data confidentiality was properly maintained.

2.3. Data Collection and Biosecurity Quantification

The farm visits were conducted to assess the implementation of biosecurity measures using the Biocheck.UGent™ (Ghent, Belgium) questionnaire. This is a risk-based assessment tool that allows for the quantification and comparison of the biosecurity status of herds within and between countries [25]. The dairy cattle survey includes a maximum of 124 questions, and the survey results in a score from 0 to 100 for both external and internal biosecurity subcategories. In the evaluation process of the survey, responses that described the ideal biosecurity scenario consistently received full points for the question, whereas less favorable answers were assigned partial or no points. In most instances, only two scoring options were applicable, indicating the presence or absence of a specific measure. However, in a few cases, an intermediate score was used. Notably, for certain questions that originally offered three or more response options (e.g., always, sometimes, and never), a simplified scoring approach was adopted, with the intermediate answer (e.g., sometimes) being treated on par with the lowest-scoring response [26]. The final score for internal and external biosecurity is the weighted average of the subcategory scores. A score of 0 represents an 'absolute lack of any biosecurity measures', while a score of 100 indicates 'full application of all assessed biosecurity measures' (Source weblink: https://biocheckgent.com/en/questionnaires/dairy-cattle, accessed on 14 March 2021). Damiaans et al. (2020) provide a more detailed description of the Biocheck.UGent™ system for the dairy cattle [26].

The dairy cattle section of Biocheck.UGent™ consists of five subcategories for external biosecurity ((a) purchase and reproduction, (b) transport and carcass removal, (c) feed and water, (d) visitors and farm workers, and (e) vermin control and other animals) and six subcategories for internal biosecurity ((a) health management, (b) calving management, (c) calf management, (d) dairy management, (e) adult cattle management, and (f) working organization and equipment).

To ensure that the survey questions were easily understood by the dairy farmers, they were translated into the local language (Punjabi or Hindi) as per the requirement. The data collected through the survey were then transferred to the online system of Biocheck.UGent™ (www.BiocheckGent.com) for calculation of the biosecurity scores and further analysis.

2.4. Statistical Analysis

The descriptive statistical analysis for the distribution of responses was carried out using Microsoft® Excel 2019 (Microsoft Corp., Santa Rosa, CA, USA). The mean, median, standard deviation, and minimum and maximum scores were obtained for each subcategory. The normal distribution of the data was assessed by using a Kolmogorov–Smirnov test. The possible associations between herd size and overall biosecurity scores and external and internal biosecurity scores were evaluated by means of correlation analysis. The correlation coefficient was categorized into weak (0–0.25), fair (0.25–0.5), good (0.5–0.75), and excellent (0.75–1), based on the R-value as described by [27]. The statistical significance level was set to $p < 0.05$. Further, to compare biosecurity scores among distinct farm types, namely buffalo farms, cattle farms, and mixed cattle and buffalo farms, Tukey's HSD (honestly significant difference) method was employed for pairwise comparisons. The F statistic initially determined overall mean disparities, and the subsequent application of Tukey's HSD test identified statistically significant differences between various pairs of means, if present.

3. Results

3.1. Herd Characteristics

A total of 94 commercial dairy farms, comprising both cattle and buffaloes, were enrolled in the study. Out of the 94 farms, 44 farms exclusively had cattle, 39 farms had a mix of buffaloes and cattle, and 11 farms solely had buffaloes. The mean herd size of the participating farms was 74.8 bovines (SD = 65.5; range = 12–343). Figure 1 shows the distribution of the selected dairy farms according to their herd size. The median herd size was 52.5 animals, and 75% of the farms (below the upper quartile) had less than 100 animals.

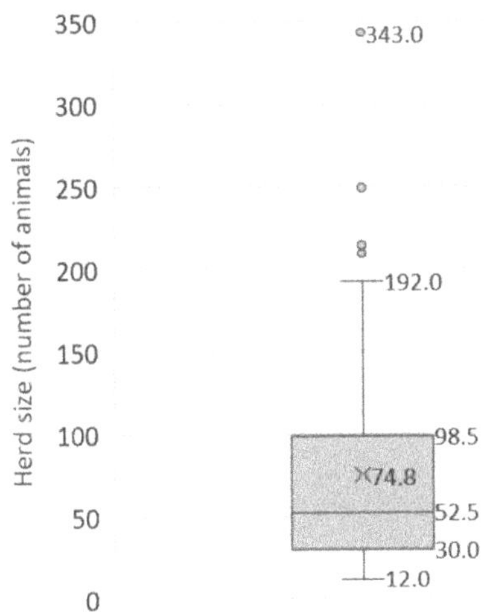

Figure 1. Distribution of the studied farms by number of animals.

3.2. Biosecurity Scoring of Dairy Farms and Comparison with Global Mean Scores

The mean external biosecurity score for the selected dairy farms was observed to be 45.4% (SD = 18), with a median of 42% and a range of 17–84%. Among the different subcategories, the highest mean score was observed for 'vermin control and other animals' (63%), followed by 'feed and water' (61%), 'transport and carcass removal' (52.9%), and 'visitors and farm workers' (47.6%), and the lowest score was observed for 'purchase and reproduction' (30.6%). Comparing the mean scores with the global average (as provided by Biocheck.UGent™), the subcategories 'transport and carcass removal', 'feed and water', and 'vermin control and other animals' had higher mean scores in the present study farms. However, the overall global external biosecurity mean score was higher than the mean score of the present study (67 versus 45.4) (Table 1; Figure 2).

Table 1. Biosecurity scores of the study farms and comparison with the global mean scores.

Biosecurity Parameter	Mean (%)	SD (%)	Median (%)	Min (%)	Max (%)	Global Mean (%) (n = 1848) *
Total external biosecurity score	**45.4**	**18**	**42**	**17**	**84**	**67**
(a) Purchase and reproduction	30.6	17.1	25	4	66	78
(b) Transport and carcass removal	52.9	28.4	52	0	100	47
(c) Feed and water	61	22.4	64	15	100	59
(d) Visitors and farmworkers	47.6	21.1	49	9	92	70
(e) Vermin control and other animals	63	25.2	65.5	0	96	62
Total internal biosecurity score	**43.7**	**22.4**	**33.5**	**14**	**85**	**37**
(a) Health management	33.6	32	18.5	0	93	31
(b) Calving management	42.4	22.1	27	11	91	31
(c) Calf management	41.6	20.2	34.5	10	82	43
(d) Dairy management	55.1	18.6	53	12	85	47
(e) Adult cattle management	76.6	17.2	75	28	100	40
(f) Working organization and equipment	41.9	31.1	17	3	100	38
Overall biosecurity score	**44.8**	**19.6**	**38**	**18**	**83**	**52**

* as of 21 February 2023.

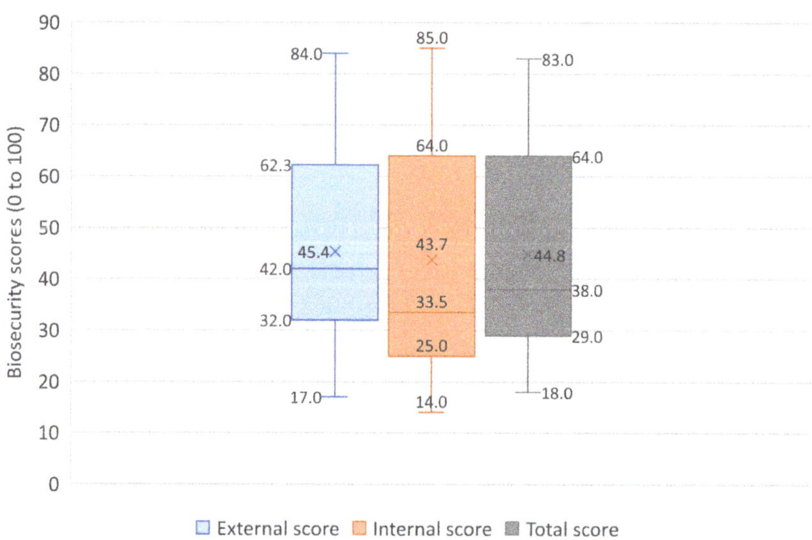

Figure 2. Comparison of external, internal, and total biosecurity scores among the study farms.

The mean internal biosecurity was 43.7% (SD = 22.4), with a median of 33.5% and a range of 14–85%. Among the subcategories, the highest score was observed for 'adult cattle management' (76.6%), followed by 'dairy management' (55.1%), 'calving management' (42.4%), 'working organization and equipment' (41.9%), 'calf management' (41.6%), and the lowest score was observed for 'health management' (33.6%). Comparing the mean scores with the global average, except for 'calf management', the mean scores of all other subcategories were higher in the present study. Additionally, the overall internal biosecurity mean score was higher than the global mean score (43.7 versus 37) (Table 1, Figure 2). However, the overall mean biosecurity score of the present study was 44.8%, which was lower than the overall mean global score of 52%.

3.3. External Biosecurity

The results with regard to purchase and reproduction suggest a lack of focus on animal health assessment during the procurement of new animals. Although farmers usually acquire animals from familiar sources, diagnostic testing for endemic diseases is not commonly conducted. Only a small proportion (2.1%) of farmers reported testing the milk of newly introduced animals, primarily for suspected mastitis cases. Also, only 24.5% (n = 23) of the farmers adhered to the quarantine protocol for newly introduced animals, and there is no uniformity in terms of the duration of quarantine (the average duration observed was 9.7 days). The quarantine protocols are also not being adequately followed during the reintroduction of animals into herds. A total of 27.6% (n = 26) of farmers indicated that their animals leave their farms for animal shows and markets, but over half (53.8%, n = 14) of these farmers neglect to keep their animals in quarantine upon returning to the herd. Many farms (61.7%, n = 58) use artificial insemination as a reproductive tool, whereas 36 farms (38.3%) were still found to rely on a bull service for breeding, but only 22.2% (n = 8) of these farms conduct regular screening of the bulls, mainly for endemic diseases such as brucellosis.

Among the surveyed farms, 31.9% (n = 30) reported performing transport baths, mainly involving the washing of tires for vehicles entering their premises. Only 19.2% (n = 18) of the farms indicated that the transport vehicles are consistently emptied upon arrival. On the other hand, a significant majority of farmers (80.8%, n = 76) acknowledged that animal transportation to their farms often occurs in vehicles not specific to their own farm and that other farm animals are frequently transported together. Moreover, only 16% (n = 15) of the farms indicated they regularly cleaned and disinfected the transport vehicle prior to entry onto the farm.

More than half (55.3%, n = 52) of the farms lacked separate carcass storage space with a hard surface floor. A substantial proportion (30.9%, n = 29) of farmers frequently spread manure from other farms on their farmlands and pastures. While about 60% (n = 56) of farmers protected their feed facilities from pets, over half (56.4%, n = 53) admitted that pet dogs and/or cats could access their stables.

A total of 30.8% (n = 29) of farms required visitors to notify them before visiting. Regular compliance with protocols for farm-specific clothing and boots for workers and visitors was observed in 26.6% of the farms. Furthermore, over half of the farms (53.2%, n = 50) did not provide separate areas for changing clothes, washing hands, and boots.

Only 22.3% (n = 21) of farmers focused on insect control programs (mainly flies and mosquitoes), and only one farm was implementing bird control programs with nets. A total of 24.5% (n = 23) of farms indicated that their animal has access to outside natural water bodies, whereas 30.8% (n = 29) expressed the possibility of contact of their farm animals with animals from other farms.

3.4. Internal Biosecurity

With regard to animal health management, 31.9% (n = 30) of farms routinely isolated sick animals from healthy groups, with the majority (63.8% out of 31.9%) of these farms separating animals only when clinical signs were evident. Individual maternity pens or C-section boxes were provided in 10.6% (n = 10) of farms, while many farms (51.1%, n = 48) used shared maternity pens. Most farms (87.2%, n = 82) implemented proper cleaning and disinfection procedures for used obstetric materials before and after each calving or abortion; however, the remaining farms followed a cleaning protocol and only applied disinfection when handling cases of abortions. The practice of testing animals after abortion was observed to be low, with only 3.2% (n = 3) of farmers conducting the test routinely. However, 93.6% of farmers acknowledged the importance of testing if an animal has repeated abortions or if their veterinarian recommends it. Most of the farms (97.9%, n = 92) disposed of fetal membranes and tissues away from the farm premises (on the manure pile/slurry pit) after a calving or abortion.

A total of 62.8% (n = 59) of farms were found to follow the work line of taking care of sick animals after attending to the healthy animals, whereas 30.8% (n = 29) of the farms were not following any specific order in this context. Only 35.1% (n = 33) of farms maintained animal health records, and 41.5% (n = 39) had written protocols for vaccination, disease treatment, and hygiene procedures. A total of 33% (n = 31) of farmers performed bacteriological testing of farm water. However, they preferred to conduct such testing only when the farm faced adverse health and production incidences, such as calf mortalities, diarrhea, and abortion outbreaks.

Except for nine farms, the others were following the administration of colostrum during the first six hours of birth. However, the assessment of colostrum quality is not commonly practiced. A total of 35.1% (n = 33) of the farms reported housing the calves in individual calf boxes/hutches or separate areas. The majority of the farms (77.7%, n = 73) regularly cleaned the feeding buckets after each feeding session. The concept of frozen colostrum or an artificial reserve of colostrum was not applied by any farm in the study.

The manual milking method was the primary approach utilized for milking, accounting for 85.1% (n = 80) of the farms employing this technique. Among these farms, more than half (51.1%; n = 48) adopted a wet cleaning technique for the teats, followed by drying them with separate towels. A total of 61.7% (n = 58) of farms were examining the milk during fore-stripping. Except for seven farms, the others were keeping the animals upright for 30 min to one hour after milking. Furthermore, the majority of farms (83%; n = 78) followed the practice of milking cows with mastitis and/or a high somatic cell count (SCC) last in the milking order. Half of the farms (50%; n = 47) were clipping udders twice or more per year, whereas 30% (n = 28) were clipping at least once a year. Only 39.4% (n = 37) of the farms were carrying out regular (i.e., minimum once per year) bacterial examination of the udders of all animals. The average frequency for cleaning and disinfection of an adult animal stable was listed to be 3.8 and 3.4 times per year, respectively.

A total of 45.7% (n = 43) of the farms were grouping the animals as per 'age category' in the stable. Only 35.1% (n = 33) of the farms were regularly following hoof disinfection footbath for adult animals. The farm work line in the order of handling youngest to oldest animals was followed by 55.3% (n = 52) of the farms. Most of the farms (96.8%; n = 91) were using their farm-specific equipment and not sharing it with other farms.

3.5. Correlation Analysis and Farm Type Comparisons

Figure 3a depicts the scatter plot between herd size and overall farm biosecurity, for which Spearman's correlation coefficient (r_s) was found to be 0.75 ($p < 0.01$), indicating an 'excellent' positive correlation. Figure 3b depicts the scatter plot between internal and external biosecurity scores, for which r_s was observed as 0.84 ($p < 0.01$), indicating an 'excellent' positive correlation.

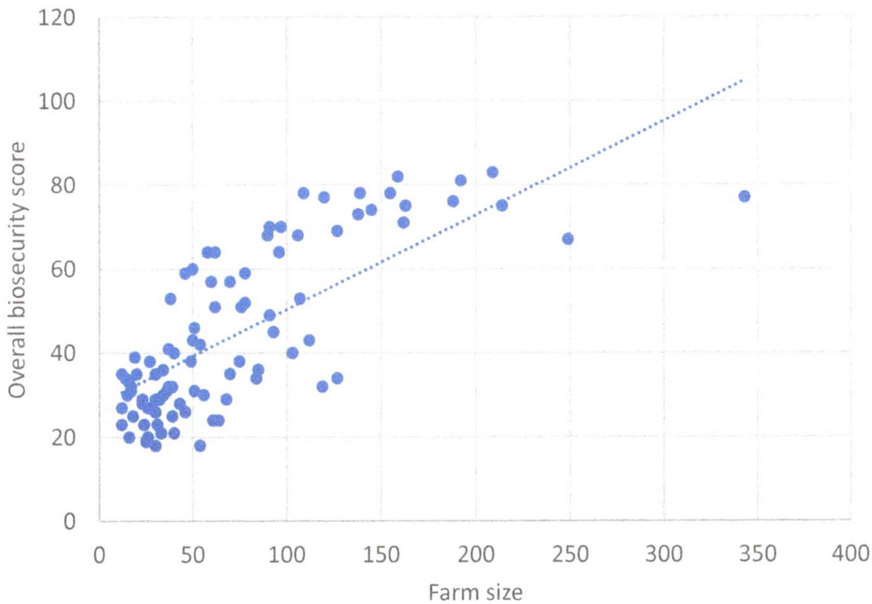

(**a**)

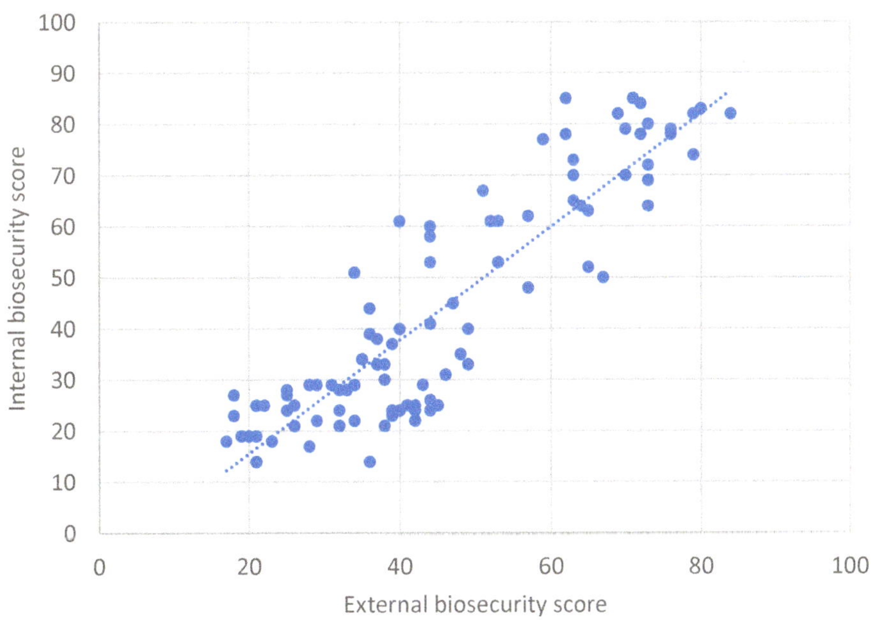

(**b**)

Figure 3. (**a**) Scatter plot between farm size and overall biosecurity scores. (**b**) Scatter plot between external and internal biosecurity scores.

The analysis of biosecurity scores among farm types resulted in an F-ratio of 1.90, with a corresponding p-value of 0.15, indicating no statistical significance at $p < 0.05$. Similarly, pairwise comparisons of biosecurity scores among different farm types also revealed non-significant differences: buffalo farms versus cattle farms (p-value: 0.09), buffalo farms versus mixed farms (p-value: 0.14), and cattle farms versus mixed farms (p-value: 0.97).

4. Discussion

Farm biosecurity plays a crucial role in ensuring sustainable solutions to animal health, production, and welfare challenges [6,7]. However, limited data on dairy farm biosecurity parameters in India exist, and this is the first study in which the Biocheck.UGent™ scoring system is used to quantify these parameters. The Biocheck.UGent™ scoring system is a well-validated system that provides a comprehensive scoring of various farm biosecurity parameters, allowing for comparative analysis both within and between countries [28]. The Biocheck.UGent™ holds great potential for analysis of Indian farm biosecurity scoring and facilitating comparisons with other regions within the country and across the globe. Many of its parameters are well suited for Indian farm biosecurity conditions, enabling a comprehensive evaluation of biosecurity practices. The system can provide valuable insights into the strengths and weaknesses of biosecurity measures implemented in Indian farms, allowing for targeted improvements.

Punjab is the foremost commercial dairy farming region in India. Due to rapid urbanization and rising consumer income levels, there has been a growing demand for high-quality milk and milk products in the country. As a result, there is a noticeable transition occurring from traditional small-scale production to large-scale commercial production to meet these increasing demands [9]. Previous research has indicated that medium and large farmers are more aware and interested in adopting biosecurity interventions [29,30]. This explains the voluntary participation of the majority of the medium and large-scale dairy farmers in the study region.

The study observed that the overall external biosecurity mean score (45.4%) of the farms was lower than the global average (67%). This difference is mainly attributed to the low scores for the subcategories of 'purchase and reproduction' (30.6%) and 'visitors and farmworkers' (47.6%). The reason behind this could be attributed to the prevalent open herd system in most farms, where animals are frequently acquired from known sources without requiring any clinical proof of their health status. As a result, there are limited practices for testing newly purchased animals unless they show clinical signs. In line with this, a study conducted by Denis-Robichaud et al. (2019) revealed a lack of widespread adoption of biosecurity measures among Canadian dairy farms, with 48% of open herds lacking a strategy for introducing new and returning animals [31].

Further, the present study revealed that awareness about quarantine protocols and their implementation was low in the region. Recent studies have emphasized the importance of advocating the implementation of quarantine protocols as part of dairy farm biosecurity measures in order to prevent the introduction and spread of infectious diseases [32–34]. There has been low compliance with testing the maternal immunity of purchased calves, as it is commonly believed that these calves already possess sufficient maternal antibodies against endemic infections. Furthermore, the limited diagnostic facilities for endemic diseases were linked to lower compliance with animal and bull testing. A study by Damiaans et al. (2020) among dairy herds in Belgium found a higher score under the 'purchase and reproduction' category (39%) due to farms not purchasing cattle from external sources [26]. Farms that did carry out external purchases generally took precautions to limit possible disease transmission risks by following testing and quarantine procedures. However, in the present study, there was less emphasis on maintaining closed herds as animal purchase and sale among farmers was common. The low compliance for the use of laboratory diagnosis could be attributed to several factors, such as the unavailability of diagnostic facilities in the region, inadequate emphasis on adopting preventive procedures by veterinarians and/or farmers, high costs associated with diagnostic testing, and insufficient awareness of the benefits of early diagnosis of infectious diseases. Recent studies in the region have underscored the limited availability and adoption of diagnostic tools and preventive measures by animal health service providers. This, in turn, has contributed to the imprudent use of antimicrobial agents in the study region [35,36].

The low score for 'visitors and farmworkers' in terms of compliance with farm biosecurity measures can be attributed to several factors. Firstly, it is important to note that there is a prevailing socio-cultural norm where formal permission to visit farms is not considered important, so there is a general hesitancy toward seeking permission before visiting farms. Secondly, many farm workers are also engaged in work outside of the farm, which increases the risk of introducing diseases to the farm. Furthermore, there is a notable absence of adherence to farm-specific clothing and boots among farm workers, visitors, and animal health professionals. This can be attributed to a lack of awareness among farmers regarding the potential transmission of pathogens from contaminated clothing and boots worn by visitors. A previous study in the Punjab region found low compliance with personal biosecurity practices among the participant farmers, including limited use of personal protective equipment (PPE). Notably, handling the calving and dystocia cases often occurred without sleeves or gloves, and just 4% of surveyed farmers offered PPE to visitors. This highlights a concerning pattern of insufficient personal biosecurity measures within the agricultural community [22]. Studies have observed similar socio-cultural effects on biosecurity compliance in livestock systems across different regions of the world, where cultural beliefs hinder the effective implementation of biosecurity measures [37,38].

The mean internal biosecurity score (43.7%) was found to be higher than the global mean (37%), with scores above 50% observed for 'adult cattle management' and 'dairy management'. This can be attributed to the large experience of the participating farmers in managing dairy farms and their emphasis on the fulfillment of biosecurity parameters associated with their farm boundaries (i.e., internal biosecurity), with less concern for external sources of disease. Moreover, farmers prioritize the health of their milking animals as they hold a high economic value, and the cost of treating diseases such as mastitis and abortions is substantial, in addition to the potential loss of production associated with these diseases. Therefore, they invest their best possible efforts in animal care and hygienic milk production. In a study by Denis-Robichaud et al. (2019) on Canadian dairy farmers' perceptions and adoption of biosecurity practices, limited implementation was observed in both intra- and inter-herd measures aimed at minimizing infection spread. For example, 27% of respondents did not house sick or lame animals in their calving pens. Only 29% consistently ensured cow cleanliness before calving, and 27% regularly sanitized calving pens after each calving event. Additionally, fewer than 15% implemented measures to restrict or manage farm visitors [31].

The overall mean biosecurity score of the present study (44.8%) was lower than the overall mean global score (52%). This can be discussed in light of the fact that the majority of the farms that carried out the survey using Biocheck.UGent™ were from European regions where compliance with farm biosecurity is strongly advocated [26,39].

The positive correlation observed between herd size and overall farm biosecurity (rs: 0.75), as well as between internal and external biosecurity scores (rs: 0.84), reveals valuable insights. This suggests that, in our study context, as the herd size increases, the overall farm biosecurity tends to improve significantly. Past studies observed that herd size could be positively correlated with the economic condition and educational standard of farmers, which might lead to large farms adopting biosecurity measures more efficiently due to greater access to resources and awareness [29,40]. Similarly, the positive correlation observed between internal and external biosecurity scores highlights a strong association between the two aspects of biosecurity. This implies that farms excelling in one dimension of biosecurity are likely to perform well in the other as well, indicating a comprehensive approach to biosecurity implementation [41].

The analysis revealed that farm types (buffalo farms, cattle farms, and mixed cattle and buffalo farms) do not exhibit any significant difference in biosecurity practices in the study region. This uniformity may be attributed to the prevalence of mixed farming practices in the study region, where dairy cattle and buffaloes share similar management and biosecurity requirements. Additionally, regional factors and environmental conditions may contribute to the observed consistency in biosecurity scores across different farm types. Further research can explore the effectiveness of specific biosecurity measures within mixed farming regions to gain deeper insights into these practices.

Although our study did not extensively assess the role of socio-cultural factors and field veterinarians in implementing farm biosecurity, previous research has highlighted their importance. For instance, a study on bovine tuberculosis in New Zealand demonstrated how farmers' livestock purchasing behavior is influenced by the culture of their farm environment. While voluntary disease control schemes have been proposed to drive behavioral change, their success depends on understanding how different triggers work in different situations [42]. Additionally, many veterinarians in certain regions may be more focused on therapeutic approaches rather than preventive approaches. As noted by Preite et al. (2023), there is a need for training and improving biosecurity communication among farm veterinarians, as well as implementing formal regional risk mitigation programs through public–private collaboration to enhance biosecurity measures' acceptance and maintenance on farms [43].

Study recommendations: To optimize biosecurity and mitigate disease risks on dairy farms, it is crucial for animal health professionals to educate farmers about the significance of implementing these biosecurity measures. By emphasizing the long-term benefits, such as improved farm productivity and enhanced animal health, professionals can encourage farmers to prioritize biosecurity practices. Policymakers and institutions should promote awareness and compliance through education and incentives while developing and enforcing regulations. The socio-cultural values of the local community should be considered when designing policies and interventions. The present study reflects the scope of improvement for implementation of quarantine protocols, awareness of carcass and placenta disposal, use of compartment-specific equipment, clothing, and boots, implementation of insect, bird, and rodent control measures, proper emphasis on calf management, and complying with biosecurity protocols for workers and visitors.

Study limitations: The study utilized a convenient sampling strategy by targeting volunteer farmers. However, it is crucial to note that this approach may introduce a potential bias known as 'participant bias'. This is due to the possibility of including interested farmers with a higher level of knowledge about the targeted topics, leading to a skewed representation of the broader population. Despite this limitation, the study's findings hold significance as there is limited knowledge or quantified data regarding dairy farm biosecurity in India. The results can help generate interest and awareness among the dairy chain stakeholders about the concept and implementation of farm biosecurity, laying a valuable foundation for future research in this field.

5. Conclusions

This study highlights that the mean biosecurity scores for both external and internal factors on the selected dairy farms were below the global average. Key areas requiring improvement include 'purchase and reproduction' and 'health management'. A positive correlation between herd size and overall biosecurity scores suggests that larger farms tend to have better biosecurity practices. The study clearly shows the need for carrying out a more quantitative assessment of the biosecurity practices in the region so that an evidence-based package of practices can be formulated.

Author Contributions: Conceptualization, P.D., D.V., M.S., J.S.B. and J.D.; data curation, P.D., D.V., N.C. and J.D.; formal analysis, P.D., I.C., D.V., M.S., J.S.B., N.C. and J.D.; funding acquisition, P.D., J.S.B. and J.D.; investigation, P.D., D.V. and M.S.; methodology, P.D., I.C., D.V., M.S., J.S.B. and J.D.; project administration, P.D.; resources, P.D., J.S.B., N.C. and J.D.; software, P.D., D.V. and N.C.; supervision, I.C., J.S.B. and J.D.; writing—original draft, P.D., D.V., M.S. and N.C.; writing—review and editing, P.D., I.C., J.S.B. and J.D. All authors have read and agreed to the published version of the manuscript.

Funding: The financial support received from the project 'Rashtriya Krishi Vikas Yojana (RKVY), India' is thankfully acknowledged. The author P.D. thankfully acknowledges the financial assistance provided by the National Agricultural Higher Education Project (IDP-NAHEP) of the Indian Council of Agricultural Research (ICAR) at Guru Angad Dev Veterinary and Animal Sciences University, India, to carry out the International Faculty Training program at Ghent University, Belgium.

Institutional Review Board Statement: The study was conducted according to the guidelines of the Declaration of Helsinki and approved by the Institutional Advisory Committee of Guru Angad Dev Veterinary and Animal Sciences University, Ludhiana (PGS/A-I/20/3307-08, approved on 26 October 2020). The study was survey-based, and there was no involvement of any invasive procedures or experimental protocols on the study subjects.

Informed Consent Statement: The study was completely voluntary and proper consent of the participants were obtained before enrollment in the study.

Data Availability Statement: The data presented in this study are available on request from the corresponding author. The data are not publicly available due to privacy and confidentiality agreements with the participants.

Acknowledgments: The authors thank Ghent University (Belgium) and Guru Angad Dev Veterinary and Animal Sciences University (India) for providing the necessary facilities for undertaking the research.

Conflicts of Interest: The authors declare no conflict of interest.

References

1. Miller, G.D.; Auestad, N. Towards a sustainable dairy sector: Leadership in sustainable nutrition. *Int. J. Dairy Technol.* **2013**, *66*, 307–316. [CrossRef]
2. Hernández-Castellano, L.E.; Nally, J.E.; Lindahl, J.; Wanapat, M.; Alhidary, I.A.; Fangueiro, D.; Grace, D.; Ratto, M.; Bambou, J.C.; de Almeida, A.M. Dairy science and health in the tropics: Challenges and opportunities for the next decades. *Trop. Anim. Health Prod.* **2019**, *51*, 1009–1017. [CrossRef] [PubMed]
3. McDaniel, C.J.; Cardwell, D.M.; Moeller, R.B.; Gray, G.C. Humans and Cattle: A Review of Bovine Zoonoses. *Vector-Borne Zoonotic Dis.* **2014**, *14*, 1–19. [CrossRef] [PubMed]
4. Shortall, O.; Ruston, A.; Green, M.; Brennan, M.; Wapenaar, W.; Kaler, J. Broken biosecurity? Veterinarians' framing of biosecurity on dairy farms in England. *Prev. Vet. Med.* **2016**, *132*, 20–31. [CrossRef] [PubMed]
5. Huber, N.; Andraud, M.; Sassu, E.L.; Prigge, C.; Zoche-Golob, V.; Käsbohrer, A.; D'Angelantonio, D.; Viltrop, A.; Żmudzki, J.; Jones, H.; et al. What is a biosecurity measure? A definition proposal for animal production and linked processing operations. *One Health* **2022**, *15*, 100433. [CrossRef] [PubMed]
6. Maye, D.; Chan, K.W. On-farm biosecurity in livestock production: Farmer behaviour, cultural identities and practices of care. *Emerg. Top. Life Sci.* **2020**, *4*, 521–530. [CrossRef]
7. Dhaka, P.; Chantziaras, I.; Vijay, D.; Bedi, J.S.; Makovska, I.; Biebaut, E.; Dewulf, J. Can Improved Farm Biosecurity Reduce the Need for Antimicrobials in Food Animals? A Scoping Review. *Antibiotics* **2023**, *12*, 893. [CrossRef] [PubMed]
8. Harun, M.H.; Shimelis, S.; Andargie, B.; Abdi, M.Y.; Bekere, H.Y. Assessment of biosecurity status in dairy cow farms. *Bull. Natl. Res. Cent.* **2022**, *46*, 152. [CrossRef]
9. Burkitbayeva, S.; Janssen, E.; Swinnen, J. Technology Adoption, Vertical Coordination in Value Chains, and FDI in Developing Countries: Panel Evidence from the Dairy Sector in India (Punjab). *Rev. Ind. Organ.* **2020**, *57*, 433–479. [CrossRef]
10. Pingale, V.; Singh, N. Potential of Dairy Industry with Special Reference to Generating Employment in India. *Ind. J. Labour Econ.* **2022**, *65*, 519–532. [CrossRef]
11. Annual Report 2022–2023, Department of Animal Husbandry & Dairying. Available online: https://dahd.nic.in/document/annual_report (accessed on 25 October 2023).
12. Singh, A.; Empowerment of Dairy Farmers for Enhancing their Income. Directorate of Extension Education Guru Angad Dev Veterinary and Animal Sciences University, Ludhiana and ICAR Agricultural Technology Application Research Institute Zone-1, Ludhiana. 17 January 2023. Available online: Chrome-extension://efaidnbmnnnibpcajpcglclefindmkaj/https://www.gadvasu.in/assests/uploads/images/Bulletin%20No.%203.pdf (accessed on 22 October 2023).
13. Kumar, A.; Mishra, A.K.; Saroj, S.; Joshi, P.K. Impact of traditional versus modern dairy value chains on food security: Evidence from India's dairy sector. *Food Policy* **2019**, *83*, 260–270. [CrossRef]
14. Chatterjee, P.; Bhaumik, S.; Chauhan, A.S.; Kakkar, M. Protocol for developing a Database of Zoonotic disease Research in India (DoZooRI). *BMJ Open* **2017**, *7*, e017825. [CrossRef]

15. Mangtani, P.; Berry, I.; Beauvais, W.; Holt, H.R.; Kulashri, A.; Bharti, S.; Sagar, V.; Nguipdop-Djomo, P.; Bedi, J.; Kaur, M.; et al. The prevalence and risk factors for human Brucella species infection in a cross-sectional survey of a rural population in Punjab, India. *Trans. R. Soc. Trop. Med. Hyg.* **2020**, *114*, 255–263. [CrossRef]
16. Nayak, P.; Sodha, S.V.; Laserson, K.F.; Padhi, A.K.; Swain, B.K.; Hossain, S.S.; Shrivastava, A.; Khasnobis, P.; Venkatesh, S.R.; Patnaik, B.; et al. A cutaneous Anthrax outbreak in Koraput District of Odisha-India 2015. *BMC Public Health* **2019**, *19*, 470. [CrossRef]
17. Sudan, V.; Tewari, A.K.; Singh, H. Serodiagnosis of Toxoplasma gondii infection in bovines from Kerala, India using a recombinant surface antigen 1 ELISA. *Biologicals* **2015**, *43*, 250–255. [CrossRef]
18. Keshavamurthy, R.; Singh, B.B.; Kalambhe, D.G.; Aulakh, R.S.; Dhand, N.K. Prevalence of *Coxiella burnetii* in cattle and buffalo populations in Punjab, India. *Prev. Vet. Med.* **2019**, *166*, 16–20. [CrossRef] [PubMed]
19. Mourya, D.T.; Yadav, P.D.; Shete, A.M.; Sathe, P.S.; Sarkale, P.C.; Pattnaik, B.; Sharma, G.; Upadhyay, K.J.; Gosavi, S.; Patil, D.Y.; et al. Cross-sectional Serosurvey of Crimean-Congo Hemorrhagic Fever Virus IgG in Livestock, India, 2013–2014. *Emerg. Infect Dis.* **2015**, *21*, 1837–1839. [CrossRef]
20. Hundal, J.S.; Sodhi, S.S.; Gupta, A.; Singh, J.; Chahal, U.S. Awareness, knowledge, and risks of zoonotic diseases among livestock farmers in Punjab. *Vet. World* **2016**, *9*, 186–191. [CrossRef]
21. Singh, B.B.; Kaur, R.; Gill, G.S.; Gill, J.P.S.; Soni, R.K.; Aulakh, R.S. Knowledge, attitude and practices relating to zoonotic diseases among livestock farmers in Punjab, India. *Acta Trop.* **2019**, *189*, 15–21. [CrossRef]
22. Singh, J.; Singh, B.B.; Tiwari, H.K.; Josan, H.S.; Jaswal, N.; Kaur, M.; Kostoulas, P.; Khatkar, M.S.; Aulakh, R.S.; Gill, J.P.S.; et al. Using Dairy Value Chains to Identify Production Constraints and Biosecurity Risks. *Animals* **2020**, *10*, 2332. [CrossRef]
23. Chatterjee, P.; Kakkar, M.; Chaturvedi, S. Integrating one health in national health policies of developing countries: India's lost opportunities. *Infect. Dis. Poverty* **2016**, *5*, 1–5. [CrossRef]
24. Kumar, N.; Tripathi, B.N. A serious skin virus epidemic sweeping through the Indian subcontinent is a threat to the livelihood of farmers. *Virulence* **2022**, *13*, 1943–1944. [CrossRef] [PubMed]
25. Postma, M.; Backhans, A.; Collineau, L.; Loesken, S.; Sjölund, M.; Belloc, C.; Emanuelson, U.; Grosse Beilage, E.; Stärk, K.D.C.; Dewulf, J. The biosecurity status and its associations with production and management characteristics in farrow-to-finish pig herds. *Animal* **2016**, *10*, 478–489. [CrossRef] [PubMed]
26. Damiaans, B.; Renault, V.; Sarrazin, S.; Berge, A.C.; Pardon, B.; Saegerman, C.; Dewulf, J. A risk-based scoring system to quantify biosecurity in cattle production. *Prev. Vet. Med.* **2020**, *179*, 104992. [CrossRef] [PubMed]
27. Cohen, J. *Statistical Power Analysis for the Behavioral Sciences*, 2nd ed.; Routledge: New York, NY, USA, 1988; ISBN 978-0-203-77158-7. [CrossRef]
28. Postma, M.; Backhans, A.; Collineau, L.; Loesken, S.; Sjölund, M.; Belloc, C.; Emanuelson, U.; Grosse Beilage, E.; Nielsen, E.O.; Stärk, K.D.C.; et al. Evaluation of the relationship between the biosecurity status, production parameters, herd characteristics and antimicrobial usage in farrow-to-finish pig production in four EU countries. *Porc. Health Manag* **2016**, *2*, 9. [CrossRef] [PubMed]
29. Bokma, J.; Dewulf, J.; Deprez, P.; Pardon, B. Risk factors for antimicrobial use in food-producing animals: Disease prevention and socio-economic factors as the main drivers? *Vlaams Diergeneeskd. Tijdschr.* **2018**, *87*, 188–200. [CrossRef]
30. Paquette, C.C.; Schemann, K.A.; Ward, M.P. Knowledge and attitudes of Australian livestock producers concerning biosecurity practices. *Aust. Vet. J.* **2020**, *98*, 533–545. [CrossRef] [PubMed]
31. Denis-Robichaud, J.; Kelton, D.F.; Bauman, C.A.; Barkema, H.W.; Keefe, G.P.; Dubuc, J. Biosecurity and herd health management practices on Canadian dairy farms. *J. Dairy Sci.* **2019**, *102*, 9536–9547. [CrossRef] [PubMed]
32. Menegon, F.; Capello, K.; Tarakdjian, J.; Pasqualin, D.; Cunial, G.; Andreatta, S.; Dellamaria, D.; Manca, G.; Farina, G.; Di Martino, G. Antibiotic Use in Alpine Dairy Farms and Its Relation to Biosecurity and Animal Welfare. *Antibiotics* **2022**, *11*, 231. [CrossRef] [PubMed]
33. Moya, S.; Navea, J.; Casal, J.; Ciaravino, G.; Yus, E.; Diéguez, F.J.; Benavides, B.; Tirado, F.; Allepuz, A. Government veterinarians' perceptions of routine biosecurity focused on dairy cattle farms in north-western and north-eastern Spain. *Front. Vet. Sci.* **2023**, *10*, 1043966. Available online: https://www.frontiersin.org/articles/10.3389/fvets.2023.1043966 (accessed on 23 March 2023). [CrossRef] [PubMed]
34. Renault, V.; Humblet, M.-F.; Pham, P.N.; Saegerman, C. Biosecurity at Cattle Farms: Strengths, Weaknesses, Opportunities and Threats. *Pathogens* **2021**, *10*, 1315. [CrossRef] [PubMed]
35. Vijay, D.; Bedi, J.S.; Dhaka, P.; Singh, R.; Singh, J.; Arora, A.K.; Gill, J.P.S. Qualitative Study on Antimicrobial Usage and Resistance in the Dairy Chain: A Situation Analysis and Solutions by Stakeholders from Punjab, India. *Antibiotics* **2022**, *11*, 1229. [CrossRef] [PubMed]
36. Patnaik, N.M.; Gupta, J.; Meena, B.S. Field level study to understand dimensions of antimicrobial use in dairy farms of Punjab. *Indian J. Dairy Sci.* **2020**, *73*, 457–463. [CrossRef]
37. Conan, A.; Goutard, F.L.; Sorn, S.; Vong, S. Biosecurity measures for backyard poultry in developing countries: A systematic review. *BMC Vet. Res.* **2012**, *8*, 240. [CrossRef] [PubMed]
38. Delsart, M.; Pol, F.; Dufour, B.; Rose, N.; Fablet, C. Pig Farming in Alternative Systems: Strengths and Challenges in Terms of Animal Welfare, Biosecurity, Animal Health and Pork Safety. *Agriculture* **2020**, *10*, 261. [CrossRef]
39. Sahlström, L.; Virtanen, T.; Kyyrö, J.; Lyytikäinen, T. Biosecurity on Finnish cattle, pig and sheep farms–Results from a questionnaire. *Prev. Vet. Med.* **2014**, *117*, 59–67. [CrossRef] [PubMed]

40. Can, M.F.; Altuğ, N. Socioeconomic implications of biosecurity practices in small-scale dairy farms. *Vet. Q.* **2014**, *34*, 67–73. [CrossRef]
41. Van Limbergen, T.; Dewulf, J.; Klinkenberg, M.; Ducatelle, R.; Gelaude, P.; Méndez, J.; Heinola, K.; Papasolomontos, S.; Szeleszczuk, P.; Maes, D. Scoring biosecurity in European conventional broiler production. *Poult. Sci.* **2018**, *97*, 74–83. [CrossRef] [PubMed]
42. Hidano, A.; Gates, M.C.; Enticott, G. Farmers' Decision Making on Livestock Trading Practices: Cowshed Culture and Behavioral Triggers Amongst New Zealand Dairy Farmers. *Front. Vet. Sci.* **2019**, *6*, 320. Available online: https://www.frontiersin.org/articles/10.3389/fvets.2019.00320 (accessed on 24 March 2023). [CrossRef] [PubMed]
43. Preite, L.; Barroso, P.; Romero, B.; Balseiro, A.; Gortázar, C. Struggling to improve farm biosecurity: Do free advice and subsidies hit the target? *Prev. Vet. Med.* **2023**, *212*, 105839. [CrossRef] [PubMed]

Disclaimer/Publisher's Note: The statements, opinions and data contained in all publications are solely those of the individual author(s) and contributor(s) and not of MDPI and/or the editor(s). MDPI and/or the editor(s) disclaim responsibility for any injury to people or property resulting from any ideas, methods, instructions or products referred to in the content.

animals

MDPI

Communication

Seroepidemiological Survey of Hepatitis E Virus in Intensive Pig Farming in Vojvodina Province, Serbia

Diana Lupulović [1,*], Marija Gnjatović [1], Jasna Prodanov-Radulović [2], Danica Ćujić [1], Vladimir Gajdov [2], Milena Samojlović [2] and Tamaš Petrović [2]

[1] Institute for the Application of Nuclear Energy INEP, Banatska 31b, 11080 Zemun, Serbia; marijad@inep.co.rs (M.G.); danicac@inep.co.rs (D.Ć.)

[2] Scientific Veterinary Institute "Novi Sad", Rumenački put 20, 21000 Novi Sad, Serbia; jasna@niv.ns.ac.rs (J.P.-R.); vladimir.g@niv.ns.ac.rs (V.G.); milena.s@niv.ns.ac.rs (M.S.); tomy@niv.ns.ac.rs (T.P.)

* Correspondence: diana.lupulovic@inep.co.rs; Tel.: +381-612124133

Simple Summary: Hepatitis E virus (HEV) is widespread in humans and pigs worldwide. HEV-1 and HEV-2 genotypes are restricted only to humans, while HEV-3 and HEV-4 have zoonotic potential and are common in humans and animals. The aim of our study was to determine the presence of anti-HEV antibodies in three hundred naturally infected domestic pigs from three different commercial farrow-to-finish farms in South Bačka on the territory of Vojvodina Province, Serbia. The presence of specific anti-HEV IgG was examined by an in-house ELISA, and doubtful results were retested by western blotting (WB). The overall detected seroprevalence was 40.66%, indicating that the Hepatitis E virus is circulating in the intensive farming pig population in Serbian farms.

Abstract: Hepatitis E virus (HEV) is the causative agent of acute hepatitis E infection in humans. Two epidemiological patterns of the disease exist—endemic and sporadic. Genotypes 1 (HEV-1) and 2 (HEV-2) are transmitted through contaminated water and are responsible for the outbreaks of many large-scale epidemics in developing countries of Asia and Africa. Genotypes 3 (HEV-3) and 4 (HEV-4) have zoonotic potential and cause sporadic cases and small outbreaks in high-income countries. The first case of HEV infection in swine was confirmed in 1997 and later detected in other animal species. The aim of this study was to investigate the seroprevalence of HEV infection in pig farms in Vojvodina province, Serbia. Three hundred blood samples were collected from five different categories of pigs from 3 different farms on the territory of the South Bačka district in Vojvodina (Serbia). The analyses were conducted by in-house ELISA, while the western blot method was used as a confirmatory test for doubtful results. The presence of HEV IgG was detected on all three examined farms. The established seroprevalence in Farm A was 37%, 31% in Farm B, and 51% in Farm C. The mean seroprevalence for all farms was 40.66%. A higher seroprevalence was found in fatteners compared to younger categories of pigs. We concluded that HEV is widespread on pig farms with intensive management. Further analyses should be conducted with the aim of implementing a surveillance program to prevent possible human infection.

Keywords: Hepatitis E virus; zoonosis; pigs; serology; ELISA; western blot

1. Introduction

Hepatitis E virus (HEV) is the causative agent of Hepatitis E infection, which poses a serious threat to public health. Two epidemiological patterns of the disease exist—endemic and sporadic types. Genotypes 1 (HEV-1) and 2 (HEV-2) infect only humans and are transmitted via the fecal-oral route from contaminated water during large epidemics. The disease is mainly spread in developing countries within Asia and Africa. The mortality rate with HEV-1 and HEV-2 could be between 0.2 and 4%, but, in pregnant women, it can be up to 30% in the third trimester [1,2].

Genotypes 3 (HEV-3) and 4 (HEV-4) have zoonotic potential, with swine being the main reservoir of infection. The first identification of swine HEV was in 1997 in the USA [3]. Since then, pigs have been considered a possible source of infection for human beings. Different strains of HEV have been detected in a wide variety of animal species, such as wild boar, deer, rabbit, camel, etc. [4].

Autochthonous, sporadic cases of HEV-3 have been recorded in many industrialized European countries, where it is increasingly recognized as a foodborne infection through the consumption of undercooked liver, meat, and meat products [5–7]. The infection is usually self-limiting, and no specific treatment is necessary. In some cases, the chronicity in immunocompromised patients is reported for the zoonotic HEV-3. Other modes of transmission of HEV-3 and HEV-4 have also been described, including blood transfusion and liver transplantation [8]. Hepatitis E virus can be transmitted via direct contact with swine and represents an occupational hazard. People who are exposed to close contact with pigs, slaughterhouse workers, veterinarians, butchers, and farmers have higher HEV seroprevalence in comparison to the general population [9,10].

Hepatitis E virus is a single-stranded positive-sense RNA particle [11]. According to the newest taxonomy, HEV belongs to the Hepeviridae family, the Orthepevirinae subfamily, and the genus *Paslahepevirus*. Genus *Paslahepevirus* consists of two species, named *Paslahepevirus alci* and *Paslahepevirus balayani* (previously known as Orhoherpesvirus A). Species *Paslahepevirus balayani* encompasses eight genotypes, HEV 1–8 [12]. HEV-1 and HEV-2 are restricted only to humans, while HEV-3 and HEV-4 are common in humans and animals. HEV-5 and HEV-6 were detected in wild boars in Japan. Genotype 7 (HEV-7) was found in a dromedary camel and was linked to chronic viral hepatitis in a patient after liver transplantation [13]. HEV-8 has been discovered in Bactrian camels in China. Although several HEV genotypes have been confirmed so far, only one serotype has been identified [11].

Pigs are the most important reservoir of the hepatitis E virus in nature. They are susceptible to infection but don't develop clinical signs of the disease. In general, the Hepatitis E virus has little impact on swine health status [3].

Pigs become infected at the age of 2 to 3 months. First, transient viremia occurs, lasting 1 to 2 weeks, and the virus is excreted in the feces for about 3 to 7 weeks [14]. Seroconversion in piglets starts between the 13th and 17th weeks of age when the effect of maternal antibodies ceases, and they develop their own immunity. Anti-HEV IgM and IgA antibodies can be detected first, followed by the appearance of IgG antibodies in the circulation [3]. Throughout their lives, pigs can be reinfected, and seroprevalence increases with age [15].

Determination of HEV seroprevalence in the pig population is of crucial importance for the assessment of epidemiological characteristics on pig farms and, consequently, the risk to human health. The results of seroprevalence vary between countries and farms and depend on the age of tested animals, the number of samples tested, and the methods of analysis. Salines et al. [16] analyzed the seroprevalence on pig farms from 43 European countries and found that the level of anti-HEV antibodies varied between 30 and 98%. In another research, based on 84 studies, the reported HEV seroprevalence on the country level was from 9.90% to 84.02% [17]. In addition, many studies have shown that seropositivity in slaughtering pigs is high and can even reach 100% [18,19]. These data point to a clear risk of an increase in human exposure to hepatitis E infection through contaminated meat.

Pig breeding represents one of the most important branches of agriculture in Serbia. According to data from 2022, 2,667,000 pigs in Serbia are reared yearly. Pig production is mostly concentrated in large farms, with intensive production in Vojvodina Province representing 41.43% of the total livestock production in the country. Also, pork meat production accounted for 58% of the total meat production [20,21]. Several studies were conducted in the last decade on the presence of the Hepatitis E virus in the swine population in Serbia. The established HEV seroprevalence in small backyard farms was 34.6% [22] and 52.25% in wild boar [23]. Furthermore, HEV RNA was identified in 29% of the tested pig livers at a slaughterhouse [24]. In humans, the presence of anti-HEV antibodies was detected in the population of blood donors, with the prevalence rate varying from 15% to 16.9% [25,26]. However, data on HEV infection in pigs in intensive farming in Serbia are scarce. Only one study reported the presence of antibodies against HEV in the restricted area of Belgrade City, with an overall seroprevalence of 55.33% [23].

The aim of our study was to determine the presence of anti-HEV antibodies in three hundred naturally infected domestic pigs from three large farrow-to-finish farms in South Bačka on the territory of Vojvodina Province, Serbia. This cross-sectional study encompasses a comparison of seroprevalence between the different categories of pigs. Our goal was also to characterize the circulation of Hepatitis E infection in the pig population on large farms with intensive farming and to assess the possible threat to human health.

2. Materials and Methods

2.1. Sampling Strategy and Study Region

In total, three hundred swine blood samples were taken from 3 different farrow-to-finish large farms (A, B, and C, with 1100, 1700, and 700 sows, respectively). One hundred blood samples were collected per farm from 20 animals belonging to 5 different production categories: suckling piglets, weaners, gilts, sows, and fatteners. The selected animals were purebred Landrace and crossbreeds of Landrace and Large White. At the moment of sampling, the animals were apparently healthy, with no evidence of clinical signs of disease. The selected farms were monitored regularly, according to the official veterinary health control programs by the Serbian Veterinary Directorate. All three premises were situated on the territory of Vojvodina Province in the northern part of Serbia (Figure 1).

All pigs included in the survey were properly ear-tagged. Blood sampling was performed by aseptic puncture of the jugular vein. A minimal amount of 10 mL of blood per animal was taken from the pigs in vacutainers without anticoagulant reagents. The samples were delivered to the virology laboratory in a hand fridge with ice, respecting the cold chain of transportation. In the laboratory, the sera were kept at room temperature (20 °C) until the spontaneous coagulation and clot retraction. The separated serum was centrifuged for 10 min at 1500 rpm, collected in microtubes, and then stored at −20 °C until testing.

2.2. Serological Examination

The presence of specific IgG antibodies in swine sera against the Hepatitis E virus was examined by in-house ELISA. Doubtful results near the cut-off values were retested by western blotting (WB), which is considered the "gold standard" for confirmation of results.

2.2.1. In-House ELISA

Anti-HEV IgG antibodies have been detected in pig blood sera by a previously described and validated in-house ELISA [27]. Recombinant HEV genotype 3 baculovirus protein Bac1-Δ-ORF2r, obtained by infecting *Trichoplusia ni* insect larvae, was used as an antigen in this test. Briefly, microtiter plates were coated with 15ng/well of antigen, blocked, washed, and incubated with swine sera (diluted 1:10 in blocking solution). Next, an HRP-conjugated goat anti-swine secondary antibody was added to each well. After the addition of the substrate, the reaction was stopped, and the optical density (OD) of the wells was read at a wavelength of 495 nm on an ELISA reader (Asys Expert 96, Biochrom, Cambridge, UK). Positive and negative swine sera pools were used as controls. The cut-off of the assay was established as 2.5 times the OD_{495} value of the negative sera pool. Examined samples with an optical density equal to or greater than the cut-off value was considered positive. In cases where OD values were doubtful, samples were subjected to western blot testing.

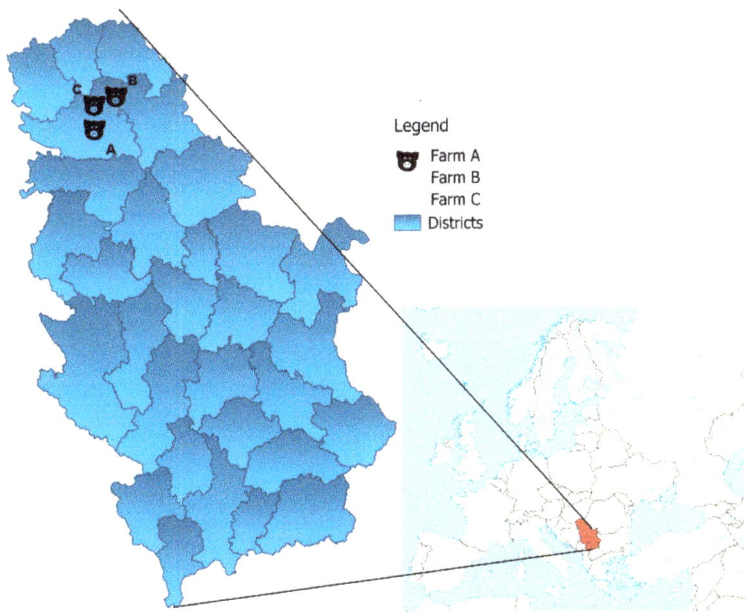

Figure 1. Mapping of locations of swine farms where blood sampling was conducted.

2.2.2. Western Blot

Antibodies from blood sera with doubtful in-house ELISA results were analyzed with western blot analysis using the recombinant ORF2 HEV gt3 protein, as described earlier by Jimenez de Oya et al. [27,28].

2.3. Statistical Analyses

Statistical analyses and data display were performed with Microsoft Excel 365 and ESRI ArcGIS Pro version 3.03. Pearson's chi-squared test was used to determine if there was an association between the seroprevalence among pig categories, as well as to compare the seropositivity rate between the farms. A p-value set as $p < 0.05$ was considered statistically significant.

2.4. Ethical Statement

Blood sampling was conducted in accordance with the recommendation of the Ethical Committee for the Protection of Animal Welfare of the University of Novi Sad. The owners of the farms agreed to have the pigs' blood sampled for research.

3. Results

Table 1 displays the number of HEV-seropositive animals on farms A (37%), B (31%), and C (54%), with the p-value indicating the presence or absence of a statistically significant difference in the number of positive samples between the production categories.

Table 1. Seroprevalence of anti-HEV IgG on farms A, B, and C by in-house ELISA.

Category	Farm A Pos/Tested (%)	Farm B Pos/Tested (%)	Farm C Pos/Tested (%)	p-Value
suckling piglets	5/20 (25%)	0/20 (0%)	15/20 (75%)	0.0000
weaners	2/20 (10%)	5/20 (25%)	16/20 (80%)	0.0000
gilts	12/20 (60%)	11/20 (55%)	3/20 (15%)	0.0070
sows	9/20 (45%)	4/20 (20%)	3/20 (15%)	0.0712
fatteners	9/20 (45%)	11/20 (55%)	17/20 (85%)	0.0255
Total:	37/100 (37%)	31/100 (31%)	54/100 (54%)	0.00274

The percentage of positivity among the different categories varied significantly, from 0% in suckling piglets on Farm B to 85% in fatteners on Farm C. Seropositivity rate differed significantly between all three examined farms and between categories of suckling piglets, weaners, gilts, and fatteners. Interestingly, a statistically significant difference was not recorded only in the category of sows ($p = 0.071$) (Table 1).

Western blot analysis was performed for the 11 pig sera with doubtful results in the ELISA: six had a positive result, as they reacted with the Bac1-Δ-ORF2r protein, as did the swine positive control serum (See Figure S1 of Supplementary Materials).

The mean value of anti-HEV IgG for the pig categories for all three farms is presented in Table 2.

Table 2. Seroprevalence of anti-HEV IgG in the different categories of pigs.

Category	HEV IgG (+) Pos/Tested (%)	HEV IgG (−) Neg/Tested (%)	p-Value
suckling piglets	20/60 (33.33%)	40/60 (66.67%)	0.00155
weaners	23/60 (38.33%)	37/60 (61.67%)	
gilts	26/60 (43.33%)	34/60 (56.67%)	
sows	16/60 (26.66%)	44/60 (73.34%)	
fatteners	37/60 (61.66%)	23/60 (38.34%)	
Total:	122/300 (40.66%)	178/300 (59.34%)	

An average of 40.66% (122/300) positivity was recorded for the 3 farms. The seroprevalence rate varied from 33.33% (20/60) in sucking piglets to 61.66% (37/60) in fatteners. The lowest seroprevalence (26.66%,16/60) was detected in the category of sows. The statistical analysis showed a significant difference between all categories of pigs ($p = 0.00155$).

4. Discussion

The presence of specific antibodies against HEV was proven by in-house ELISA in all categories of pigs (suckling piglets, weaners, gilts, sows, and fatteners) on tested industrial-type pig farms in the territory of Vojvodina Province, Serbia. The results of our study confirmed that HEV is ubiquitous on examined farms and that the virus circulates intensively in the pig population.

The average rate of anti-HEV antibodies for the three hundred blood samples tested was 40.66%. These data are similar to those previously described in other European countries. For instance, anti-HEV IgG seroprevalence of 48.4% in Spain, 49.8% in Germany, 40% in Bulgaria, 38.14% in France, 43.75% in Lithuania, and 44.06% in Poland has been described on pig farms [17,29–31]. However, significantly higher seroprevalences have been reported in China (60%), Greece (80%), and Argentina (80.1%) [32–34]. The unequal distribution of HEV seroprevalence can be explained by the different number of analyzed animals, the testing method, sampling plan, farming model, or study design [23,32].

In line with previous studies, our findings also suggest a notable presence of HEV among domestic pigs in Serbia A seroprevalence of 34.6% for anti-HEV antibodies was reported in pigs aged 3 to 4 months from small backyard farms with up to 20 sows [22]. Later on, 55.33% of HEV seropositive animals were detected on farms in the region of Belgrade City [23]. Based on these findings, we can conclude that HEV has been present in a high percentage of domestic pigs in Serbia for at least a decade. However, these data should be considered with caution, as they are limited only to the northern part of the country, while there is no available data for the southern part of Serbia.

A significantly higher seroprevalence was found on Farm C in relation to Farms A and B (Table 1). The HEV positivity rate on Farm C was 54%, while on Farms A and B, it was 37% and 31%, respectively. Farm C has the smallest number of breeding sows (700) compared to farms A and B, with 1100 and 1700 sows, respectively. There are reports that medium-size farms are at greater risk than large-sized farms [15,35,36]. The causes of different percentages of anti-HEV IgG on farms can also be explained by the location of the farm, the different breeding conditions, the number of animals in the facilities, the management on the farm, poor hygiene, and biosecurity measures [19,37]. The HEV seroprevalence in suckling piglets on Farm A was 25% (Table 1), which represents the passive immunity acquired through colostrum intake. This positivity gradually decreased, as only 10% of weaning pigs resulted positive, and increased again to 45% in sows. On Farm B, no seropositive animals were detected among newborn animals. In this case, seroconversion started earlier, which resulted in 25% of HEV-seropositive weaning pigs. Since the presence of HEV IgG on Farm B was found in only 20% of tested sows, we assumed that the passive immunity was weaker and seroconversion in piglets occurred earlier. Our results were in line with other studies. Meng et al. [3] found that piglets from highly seropositive sows had higher antibody levels in the first month of life, which began to decline significantly a few weeks after birth and almost disappeared by 8 to 9 weeks of age. Subsequently, HEV infection and seroconversion occurred at 13 to 14 weeks of age when most piglets developed their own immunity. Piglets originating from the sows with low titers of anti–HEV antibodies were seronegative at birth and seroconversion began earlier, i.e., they were shortly after birth susceptible to HEV infection. De Deus et al. [38] reported that IgG antibodies in piglets originating from highly positive mothers could be detected until the age of 9 weeks, while piglets born from weakly positive mothers had colostral antibodies only in the first 2–3 weeks of life and HEV infection began earlier. At the age of 15 weeks, seroconversion was detected in most animals. Casas et al. [39] described that the duration of maternal immunity in piglets is directly correlated with the level of anti-HEV antibodies in sows, i.e., the higher the level of antibodies in mothers, the longer the duration of maternal immunity.

The results on Farm C for the categories of suckling piglets, weaners, and sows differ from the results on Farms A and B. The presence of antibodies against HEV was detected in 75% of piglets and 85% of weaners, while only 15% of sows and gilts were seropositive. Possible reasons for these results could be the early weaning of piglets, overcrowding of facilities, and mixing of animals that were recorded on this farm, which are considered the main risk factors for infection [19]. In densely populated farms, piglets come into contact with the virus earlier, characteristic of infections transmitted by the enteric route [40,41].

It is interesting to note that HEV seropositivity increased from the category of suckling piglets (33.33%) to the fatteners (61.66%), except for sows, which presented a lower seroprevalence (26.66%) (Table 2). This data can be explained by the fact that gilts and sows are separated into compartments, where the mixing of animals is reduced. Although most authors published data on high seroprevalence in sows [23,39,42,43], there are also results in accordance with our findings. Meng et al. [44] examined many pigs and noted that there was considerable variation among sow categories, Clemente-Casares et al. [45] detected anti-HEV antibodies in 33.3% of primiparous sows and 42.8% of multiparous sows on farms in Spain, and Choi et al. [46] reported that only 8.8% of sows were seropositive on farms in Korea.

In our investigation, the highest level of anti-HEV antibodies (61.66%) was found in fattening pigs, with a range of 45% to 85% between farms. These results are consistent with previously published findings in the literature. For example, Vitral et al. [40] detected 97.3% of seropositive fatteners older than 25 weeks, and Pavia et al. [47] found the highest HEV antibodies rate in pigs older than 4.5 months. Likewise, a Spanish study showed that the proportion of anti-HEV IgG was higher in pigs older than 6 months than in younger ones [48] and Li et al. [17] reported that the seroprevalence rate was 66.20% in 5–8-month-old pigs.

5. Conclusions

In conclusion, this study shows that the Hepatitis E virus is widespread on farrow-to-finish farms with intensive production in Vojvodina Province, in the northern part of Serbia. A high circulation of HEV has been detected in all production categories and the overall seroprevalence on the tested farms was 40.66%. These data suggest that pigs are the reservoir of hepatitis E infection in the region and the HEV seroprevalence serves as an indicator of virus circulation within the population.

A higher seroprevalence was found in fatteners than in younger categories of pigs, indicating the possibility that HEV-seropositive animals can enter the slaughterhouses and food chain. Our results confirmed that ELISA is a suitable method for the detection of HEV infection in the pig population, since HEV viremia is short-lived in circulation, which makes diagnosis difficult.

In the future, the mitigation strategy should be focused on more effective biosecurity measures and farming practices to reduce HEV spread within both pig categories and farms. Further analyses should be conducted with the aim of implementing a surveillance program to prevent possible human infection.

Supplementary Materials: The following supporting information can be downloaded at: https://www.mdpi.com/article/10.3390/ani15020151/s1, Figure S1: Results of Western blot for pig sera: marker is placed in the first and last line (1st and 15th column); from 2 to 12 columns are added eleven examined swine sera; negative and positive HEV controls are placed on the position 13 and 14, respectively. Above the test sera are the marks "+" or "−" which means that the result is positive or negative in the Western blot technique (appearance of a molecular mass band of 65–68 kDa). Based on the results obtained, positive or negative test results for the presence of antibodies against HEV with in-house ELISA were defined.

Author Contributions: Conceptualization, D.L. and T.P.; methodology, D.Ć., J.P.-R. and M.G.; investigation, D.L. and M.S.; data curation, J.P.-R. and V.G.; writing—original draft preparation D.L., V.G. and T.P.; writing—review and editing, M.G., D.Ć. and M.S.; supervision D.L. and T.P. All authors have read and agreed to the published version of the manuscript.

Funding: This research was funded by the Ministry of Science, Technological Development and Innovation of Republic of Serbia by the Contract of implementation and funding of research work of NIV-NS in 2024, Contract No: 451-03-66/2024-03/200031, the World Bank and EU by the project: Serbia Accelerating Innovation and Entrepreneurship Project—SAIGE, 2022.

Institutional Review Board Statement: Ethical review and approval for this study are not required since serum samples that are used in this research were collected for the annual monitoring of swine for leptospirosis, brucellosis, Aujeszky disease, classical swine fever, and African swine fever according to the Rulebook of Measures for Health Protection of Animals in the Republic of Serbia for 2022 (Ministry of Agriculture, Forestry and Water Management of the Republic of Serbia, Veterinary Directorate, Official Gazette, 2022).

Informed Consent Statement: Informed consent was obtained from all subjects involved in the study.

Data Availability Statement: The data presented in this study are available on request from the corresponding author.

Acknowledgments: We greatly thank Juan Carlos Saiz for constructive criticism and valuable feedback for this research paper. We also would like to acknowledge laboratory technician Božica Ćulibrk for support in performing the analyses.

Conflicts of Interest: The authors declare no conflicts of interest.

References

1. Aggarwal, R.; Jameel, S. Hepatitis E. *Hepatology* **2011**, *54*, 2218–2226. [CrossRef]
2. Kumar, S.; Subhadra, S.; Singh, B.; Panda, B.K. Hepatitis E virus: The current scenario. *Int. J. Infect. Dis.* **2013**, *17*, e228–e233. [CrossRef] [PubMed]
3. Meng, X.J.; Purcell, R.H.; Halbur, P.G.; Lehman, J.R.; Webb, D.M.; Tsareva, T.S.; Haynes, J.S.; Thacker, B.J.; Emerson, S.U. A novel virus in swine is closely related to the human hepatitis E virus. *Proc. Natl. Acad. Sci. USA* **1997**, *94*, 9860–9865. [CrossRef] [PubMed]
4. Song, Y.J.; Park, W.J.; Park, B.J.; Lee, J.B.; Park, S.Y.; Song, C.S.; Lee, N.H.; Seo, K.H.; Kang, Y.S.; Choi, I.S. Hepatitis E virus infections in humans and animals. *Clin. Exp. Vaccine Res.* **2014**, *3*, 29–36. [CrossRef] [PubMed]
5. Tei, S.; Kitajima, N.; Takahashi, K.; Mishiro, S. Zoonotic transmission of Hepatitis E virus from deer to human beings. *Lancet* **2003**, *362*, 371–373. [CrossRef]
6. Colson, P.; Borentain, P.; Queyriaux, B.; Kaba, M.; Moal, V.; Gallian, P.; Heyries, L.; Raoult, D.; Gerolami, R. Pig liver sausage as a source of Hepatitis E virus transmission to humans. *J. Infect. Dis.* **2010**, *202*, 825–834. [CrossRef] [PubMed]
7. Garbuglia, A.R.; Scognamiglio, P.; Petrosillo, N.; Mastroianni, C.M.; Sordillo, P.; Gentile, D.; La Scala, P.; Girardi, E.; Capobianchi, M.R. Hepatitis E virus genotype 4 outbreak, Italy, 2011. *Emerg. Infect. Dis.* **2013**, *19*, 110–114. [CrossRef] [PubMed]
8. Hoad, V.C.; Gibbs, T.; Ravikumara, M.; Nash, M.; Levy, A.; Tracy, S.L.; Mews, C.; Perkowska-Guse, Z.; Faddy, H.M.; Bowden, S. First confirmed case of transfusion-transmitted Hepatitis E in Australia. *Med. J. Aust.* **2017**, *206*, 289–290. [CrossRef]
9. Lange, H.; Øverbø, J.; Borgen, K.; Dudman, S.; Hoddevik, G.; Urdahl, A.M.; Vold, L.; Sjurseth, S.K. Hepatitis E in Norway: Seroprevalence in humans and swine. *Epidemiol. Infect.* **2017**, *145*, 181–186. [CrossRef] [PubMed]
10. Wu, J.Y.; Lau, E.H.Y.; Lu, M.L.; Guo, C.; Guo, Z.M.; Yuanm, J.; Lu, J.H. An occupational risk of hepatitis E virus infection in the workers along the meat supply chains in Guangzhou, China. *One Health* **2022**, *14*, 100376. [CrossRef] [PubMed]
11. Kamar, N.; Dalton, H.R.; Abravanel, F.; Izopet, J. Hepatitis E virus infection. *Clin. Microbiol. Rev.* **2014**, *27*, 116–138. [CrossRef] [PubMed]
12. International Committee on Taxonomy of Viruses. Available online: https://ictv.global/report/chapter/hepeviridae/hepeviridae (accessed on 25 September 2024).

13. Letafati, A.; Taghiabadi, Z.; Roushanzamir, M.; Memarpour, B.; Seyedi, S.; Farahani, A.V.; Norouzi, M.; Karamian, S.; Zebardast, A.; Mehrabinia, M.; et al. Correction: From discovery to treatment: Tracing the path of hepatitis E virus. *Virol. J.* **2024**, *21*, 214. [CrossRef] [PubMed]
14. Meng, X.J. Hepatitis E virus: Animal reservoirs and zoonotic risk. *Vet. Microbiol.* **2010**, *140*, 256–265. [CrossRef] [PubMed]
15. Salines, M.; Andraud, M.; Rose, N. From the epidemiology of hepatitis E virus (HEV) within the swine reservoir to public health risk mitigation strategies: A comprehensive review. *Vet. Res.* **2017**, *48*, 31. [CrossRef]
16. Salines, M.; Barnaud, E.; Andraud, M.; Eono, F.; Renson, P.; Bourry, O.; Pavio, N.; Rose, N. Hepatitis E virus chronic infection of swine co-infected with Porcine Reproductive and Respiratory Syndrome Virus. *Vet. Res.* **2015**, *46*, 55. [CrossRef] [PubMed]
17. Li, P.; Ji, Y.; Li, Y.; Ma, Z.; Pan, Q. Estimating the global prevalence of hepatitis E virus in swine and pork products. *One Health* **2021**, *14*, 100362. [CrossRef] [PubMed]
18. Sooryanarain, H.; Meng, X.J. Swine hepatitis E virus: Cross-species infection, pork safety and chronic infection. *Virus Res.* **2020**, *284*, 197985. [CrossRef]
19. Meester, M.; Tobias, T.J.; Bouwknegt, M.; Kusters, N.E.; Stegeman, J.A.; van der Poel, W.H.M. Infection dynamics and persistence of hepatitis E virus on pig farms—A review. *Porcine Health Manag.* **2021**, *7*, 16. [CrossRef]
20. Statistical Yearbook of the Republic of Serbia. 2023. Available online: https://publikacije.stat.gov.rs/G2023/PdfE/G20232056.pdf (accessed on 6 December 2024).
21. Jeremić, M.; Lovre, K.; Matkovski, B. Serbian pork market analysis. *Econ. Agric.* **2018**, *65*, 1449–1460. [CrossRef]
22. Lupulovic, D.; Lazic, S.; Prodanov-Radulovic, J.; Oya, N.J.; Escribano-Romero, E.; Saiz, J.; Petrovic, T. First serological study of hepatitis E virus infection in backyard pigs from Serbia. *Food Environ. Virol.* **2010**, *2*, 110–113. [CrossRef]
23. Kureljušić, B.; Savić, B.; Jezdimirović, N.; Kureljušić, J.; Milićević, V.; Karabasil, N.; Vesković, M.S.; Žutić, J. Seroprevalence of hepatitis E in pigs and wild boars in the region of the city Belgrade. *J. Infect. Dev. Ctries.* **2020**, *14*, 669–673. [CrossRef]
24. Milojević, L.; Velebit, B.; Janković, V.; Mitrović, R.; Betić, N.; Simunović, S.; Dimitrijević, M. Prevalence, Genetic Diversity, and Quantification of the RNA Genome of the Hepatitis E Virus in Slaughtered Pigs in Serbia. *Animals* **2024**, *14*, 586. [CrossRef] [PubMed]
25. Petrović, T.; Lupulović, D.; Jiménez de Oya, N.; Vojvodić, S.; Blázquez, A.B.; Escribano-Romero, E.; Martín-Acebes, M.A.; Potkonjak, A.; Milošević, V.; Lazić, S.; et al. Prevalence of hepatitis E virus (HEV) antibodies in Serbian blood donors. *J. Infect. Dev. Ctries.* **2014**, *8*, 1322–1327. [CrossRef]
26. Delić, D.; Nešić, Z.I.; Žerjav, S.B.; Pešić, I.; Popović, N.; Simonović, J. Hepatitis E virus infection in Serbia: Epidemiology and clinical features. *Arch. Gastroenterohepatol.* **2003**, *22*, 53–56.
27. Jiménez de Oya, N.; Galindo, I.; Gironés, O.; Duizer, E.; Escribano, J.M.; Saiz, J.C. Serological Immunoassay for Detection of Hepatitis E Virus on theBasis of Genotype 3 Open Reading Frame 2 Recombinant Proteins Produced in *Trichoplusia ni* Larvae. *J. Clin. Microbiol.* **2009**, *47*, 3276–3282. [CrossRef] [PubMed]
28. Jiménez de Oya, N.; Galindo, I.; Romero, E.E.; Blázquez, A.B.; Padilla, J.A.; Halaihel, N.; Escribano, J.M.; Saiz, J.C. Expression and Immunoreactivities of Hepatitis E Virus Genotype 3 Open Reading Frame-2 (ORF-2) Recombinant Proteins Expressed in Insect Cells. *Food Environ. Virol.* **2009**, *1*, 77–84. [CrossRef]
29. Casas, M.; Pujols, J.; Rosell, R.; de Deus, N.; Peralta, B.; Pina, S.; Casal, J.; Martín, M. Retrospective serological study on hepatitis E infection in pigs from 1985 to 1997 in Spain. *Vet. Microbiol.* **2009**, *135*, 248–252. [CrossRef]
30. Baechlein, C.; Schielke, A.; Johne, R.; Ulrich, R.G.; Baumgaertner, W.; Grummer, B. Prevalence of Hepatitis E virus-specific antibodies in sera of German domestic pigs estimated by using different assays. *Vet. Microbiol.* **2010**, *144*, 187–191. [CrossRef] [PubMed]
31. Pismisheva, M.; Baymakova, M.; Golkocheva-Markova, E.; Kundurzhiev, T.; Pepovich, R.; Popov, G.T.; Tsachev, I. First serological study of hepatitis E virus infection in pigs in Bulgaria. *Comptes Rendus Acad. Bulg. Des Sci.* **2018**, *71*, 1001–1008.
32. Liu, D.; Liu, B.; He, Z.; Qiao, C.; Luo, Q.; Chen, X.; Wang, X.; Xiang, H.; Chen, J.; Zhang, P.; et al. Seroprevalence survey of Hepatitis E Virus in Domestic Pigs in Guangdong, China. *Animals* **2024**, *14*, 1861. [CrossRef] [PubMed]
33. Asimoula, S.; Tzika, E.; Alexopoulos, C.; Kyriakis, S.C.; Froesner, G. First report of serological evidence of hepatitis E virus infection in swine in northern Greece. *Acta Vet.* **2009**, *59*, 205–211. [CrossRef]
34. Di Cola, G.; Di Cola, G.; Fantilli, A.; Mamani, V.; Tamiozzo, P.; Martínez Wassaf, M.; Nates, S.V.; Ré, V.E.; Pisano, M.B. High circulation of hepatitis E virus (HEV) in pigs from the central region of Argentina without evidence of virus occurrence in pork meat and derived products. *Res. Vet. Sci.* **2023**, *164*, 105000. [CrossRef] [PubMed]
35. Bartolo, I.D.; Martelli, F.; Inglese, N.; Pourshaban, M.; Caprioli, A.; Ostanello, F.; Rugger, F.M. Widespread diffusion of genotype 3 hepatitis E virus among farming swine in Northern Italy. *Vet. Microbiol.* **2008**, *132*, 47–55. [CrossRef]

36. Hinjoy, S.; Nelson, K.E.; Gibbons, R.V.; Jarman, R.G.; Chinnawirotpisan, P.; Fernandez, S.; Tablerk, P.; Labrique, A.B.; Patchanee, P. A cross-sectional study of hepatitis E virus infection in pigs in different-sized farms in northern Thailand. *Foodborne Pathog. Dis.* **2013**, *10*, 698–704. [CrossRef] [PubMed]
37. Pavio, N.; Meng, X.J.; Renou, C. Zoonotic hepatitis E: Animal reservoirs and emerging risks. *Vet. Res.* **2010**, *41*, 46. [CrossRef]
38. de Deus, N.; Seminati, C.; Pina, S.; Mateu, E.; Martín, M.; Segalés, J. Detection of hepatitis E virus in liver, mesenteric lymph node, serum, bile and faeces of naturally infected pigs affected by different pathological conditions. *Vet. Microbiol.* **2007**, *119*, 105–114. [CrossRef]
39. Casas, M.; Cortés, R.; Pina, S.; Peralta, B.; Allepuz, A.; Cortey, M.; Casal, J.; Martín, M. Longitudinal study of hepatitis E virus infection in Spanish farrow-to-finish swine herds. *Vet. Microbiol.* **2011**, *148*, 27–34. [CrossRef] [PubMed]
40. Vitral, C.L.; Pinto, M.A.; Lewis-Ximenez, L.L.; Khudyakov, Y.E.; dos Santos, D.R.; Gaspar, A.M. Serological evidence of hepatitis E virus infection in different animal species from the Southeast of Brazil. *Mem. Inst. Oswaldo Cruz.* **2005**, *100*, 117–122. [CrossRef]
41. Guimarães, F.R.; Saddi, T.M.; Vitral, C.L.; Pinto, M.A.; Gaspar, A.M.C.; Souto, F.J.D. Hepatitis E Virus antibodies in swine herds of Mato Grosso state, central Brazil. *Braz. J. Microbiol.* **2005**, *36*, 223–226. [CrossRef]
42. Martinelli, N.; Pavoni, E.; Filogari, D.; Ferrari, N.; Chiari, M.; Canelli, E.; Lombardi, G. Hepatitis E virus in wild boar in the central Northern part of Italy. *Transbound. Emerg. Dis.* **2015**, *62*, 217–222. [CrossRef]
43. Seminati, C.; Mateu, E.; Peralta, B.; de Deus, N.; Martin, M. Distribution of hepatitis E virus infection and its prevalence in pigs on commercial farms in Spain. *Vet. J.* **2008**, *175*, 130–132. [CrossRef]
44. Meng, X.J.; Dea, S.; Engle, R.E.; Friendship, R.; Lyoo, Y.S.; Sirinarumitr, T.; Urairong, K.; Wang, D.; Wong, D.; Yoo, D.; et al. Prevalence of antibodies to the hepatitis E virus in pigs from countries where hepatitis E is common or is rare in the human population. *J. Med. Virol.* **1999**, *59*, 297–302. [CrossRef]
45. Clemente-Casares, P.; Pina, S.; Buti, M.; Jardi, R.; Martin, M.; Bofill-Mas, S.; Girones, R. Hepatitis E virus epidemiology in industrialized countries. *Emerg. Infect. Dis.* **2003**, *9*, 448–454. [CrossRef] [PubMed]
46. Choi, I.S.; Kwon, H.J.; Shin, N.R.; Yoo, H.S. Identification of Swine Hepatitis E Virus (HEV) and Prevalence of Anti-HEV Antibodies in Swine and Human Populations in Korea. *J. Clin. Microbiol.* **2003**, *41*, 3602–3608. [CrossRef] [PubMed]
47. Pavia, G.; Gioffrè, A.; Pirolo, M.; Visaggio, D.; Clausi, M.T.; Gherardi, M.; Samele, P.; Ciambrone, L.; Di Natale, R.; Spatari, G.; et al. Seroprevalence and phylogenetic characterization of hepatitis E virus in pig farms in Southern Italy. *Prev. Vet. Med.* **2021**, *194*, 105448. [CrossRef]
48. Jiménez de Oya, N.; de Blas, I.; Blázquez, A.B.; Martín-Acebes, M.A.; Halaihel, N.; Gironés, O.; Saiz, J.C.; Escribano-Romero, E. Widespread distribution of hepatitis E virus in Spanish pig herds. *BMC Res. Notes.* **2011**, *14*, 412. [CrossRef]

Disclaimer/Publisher's Note: The statements, opinions and data contained in all publications are solely those of the individual author(s) and contributor(s) and not of MDPI and/or the editor(s). MDPI and/or the editor(s) disclaim responsibility for any injury to people or property resulting from any ideas, methods, instructions or products referred to in the content.

animals

MDPI

Article

Bovine Coronavirus Prevalence and Risk Factors in Calves on Dairy Farms in Europe

Anna Catharina Berge [1,*] and Geert Vertenten [2,*]

[1] Veterinary Epidemiology Unit, Department of Internal Medicine, Reproduction and Population Medicine, Faculty of Veterinary Medicine, Ghent University, Salisburrylaan 133, 9820 Merelbeke, Belgium
[2] MSD Animal Health, Wim de Körverstraat 35, 5831 AN Boxmeer, The Netherlands
* Correspondence: anna.berge@ugent.be or cat@bergevetconsulting.com (A.C.B.); geert.vertenten@msd.com (G.V.); Tel.: +32-499-70-31-12 (A.C.B.); +32-2-776-60-16 (G.V.)

Simple Summary: This study evaluated the prevalence and risk factors in health, management, and biosecurity of bovine coronavirus (BCoV) in neonatal and weaned dairy calves on 125 dairy farms in 17 countries in Europe. All farms had BCoV antibodies in the bulk tank milk. BCoV was detected by RT-PCR in 80% of dairy herds, and 24% of neonatal calves, 23% of weaned calves, and 4% of fresh cows sampled were shedding BCoV in nasal and/or fecal samples. The biosecurity on 109 dairies scored with Biocheck was, on average, 60% (external score 71%, internal score 47%), indicating that many farms are poorly protected from the introduction of BCoV onto a farm and spread within a farm. Dry cow vaccination against BCoV reduced shedding in neonatal calves, whereas it was linked to increased shedding in weaned calves in these farms. Several calf husbandry factors, such as transition milk feeding, milk feeding levels, group housing, and weaning age, were associated with BCoV shedding in calves.

Abstract: This study evaluated prevalence and risk factors in health, management, and biosecurity of bovine coronavirus (BCoV) in neonatal and weaned dairy calves on 125 dairy farms in Europe. Nasal and fecal swabs from neonatal calves, weaned calves, and fresh cows were analyzed for BCoV using RT-PCR, and blood and bulk milk samples were collected for BCoV antibody levels using ELISA. Multiple logistic regression models with random effects of herds were used to evaluate the herd health status, husbandry, management, and biosecurity associated with BCoV shedding (nasal and/or fecal PCR positive samples) in neonatal and weaned calves. BCoV was detected in 80% of herds and in 24% of neonatal calves, 23% of weaned calves, and 5% of fresh cows. The biosecurity scored on 109 dairies with Biocheck.Ugent was, on average, 60% (external score 71%, internal score 47%), and there was no clear association between various biosecurity measures on BCoV shedding in calves. Dry cow vaccination against BCoV reduced shedding in neonatal calves, whereas it was linked to increased shedding in weaned calves in these farms. Several husbandry factors, including nutrition (transition milk feeding and milk feeding levels) and management (group housing and weaning age), were associated with BCoV shedding in calves.

Keywords: bovine coronavirus; preweaned calves; weaned calves; calf management; biosecurity

1. Introduction

Bovine coronavirus (BCoV) is an enveloped positive sense RNA virus that is widespread around the world and causes bovine enteric or respiratory infections, leading to important economic losses. BCoV can be associated with calf diarrhea, bovine respiratory disease (BRD), winter dysentery, and combinations thereof [1,2]. Infection is primarily via fecal–oral and, to a lesser extent, respiratory (aerosol) routes, and the same strains can cause both respiratory and enteric symptoms [1]. Enteric BCoV infections are generally most severe in calves, with yellow to blood-stained mucus-containing diarrhea, which progresses to a profuse watery diarrhea [3]. Winter dysentery is a sporadic acute, contagious hemorrhagic enterocolitis of cattle that occurs in an epidemic fashion in a herd in cattle of various ages. BCoV is part of the BRD complex, and its role and impact remain to be further investigated [4–6].

Within a herd, reservoirs for BCoV infection may be virus cycling in clinically or subclinical infected animals. The transmission of enteric BCoV occurs from carrier dam to offspring postpartum or from infected calves housed in proximity to unexposed ones. Respiratory BCoV may be transmitted via the fecal–oral route and respiratory (aerosol) routes [7]. The evaluation of the geographic spreading pattern indicates two clusters: a European cluster and an American-Asian cluster. Epidemiological studies indicate a spread between herds associated with cattle movements but also through other vectors. Sequencing of isolates from Danish and Swedish dairies has indicated a varied reinfection within herds and transmission between herds [8].

There are a few studies of BCoV antibodies in individual milk samples and bulk tank milk samples. Bulk milk samples from 2236 Swedish dairy herds indicated 89% of herds were antibody-positive and 52% had high levels of antibodies to BCoV [9]. Swedish studies indicate that milk samples and bulk tank milk samples could be used for monitoring programs for antibodies to BCoV and indicated that milk samples from fresh primiparous cows could be screened [10,11]. Another Swedish study of primiparous dairy cows found BCoV milk antibodies in 77% of organic herds and in 85% of the sampled conventional herds [12]. In Norway, BCoV antibodies were found in bulk tank milk in 72% of 1347 dairy herds [13]. Bulk tank milk samples from 341 dairy herds in England and Wales indicated that 100% of herds had BCoV antibodies [14]. There are limited and not comparable studies of enteric and respiratory BCoV virus shedding in Europe [15–22].

There are studies indicating the importance of BCoV in both enteric and respiratory diseases in cattle and evidence that the virus is widespread in Europe. However, there is insufficient data to determine the spread of the virus in European dairy herds; the importance of biosecurity in BCoV prevention; and insufficient studies that evaluate herd management, health, and productivity factors for BCoV infection in calves. The aim of this study was to evaluate the prevalence of BCoV in neonatal and weaned calves on a representative set of European dairy farms and to determine the major biosecurity and management risk factors associated with enteric and nasal shedding of the virus in these calves.

2. Materials and Methods

Farm enrollment: The aim of this cross-sectional field study was to have more than 130 dairy herds enrolled in European countries. The herd enrollment per country was approximately proportional to the national milk production. The herds were selected based on a convenience sample of 5–10 herds per country, identified by MSD Animal Health national technical teams and/or collaborating veterinary practitioners. The target minimum lactating herd size was 100 cows to have sufficient calves for sampling on one occasion. A derogation was made in Austria, where herds are, in general, small, and the herds were visited twice to obtain the targeted number of samples from calves.

Sampling: On each dairy farm, 5 to 20 calves under one month of age (neonatal calves: NC), 5 to 20 calves within one month of weaning (weaned calves: WC), and 5–10 fresh cows within one month of calving (fresh cows: FC) were sampled. From each sampled animal, a rectal fecal and a nasal swab were collected with a cotton-tipped swab, which were immersed in tubes with a viral transport media (Sigma virocult MW950SENT, MWE, Wilts, UK), and a venous blood sample was collected from the jugular vein in calves and from the jugular or tail vein from cows. A bulk tank milk sample was collected in a sterile 50 mL container for BCoV antibody detection.

Sample analysis: The samples were accompanied with a sampling record stating animal id; animal category (NC, WC, or FC); birth date; weaning date for WC and calving date for FC; and sampling date. Also, the animal health history within the previous 3 months was collected. The samples were sent chilled to the contracted laboratory in Boxmeer, the Netherlands. The presence of BCoV virus was analyzed with a duplex Real-Time RT-PCR (Kylt® Bovine Coronavirus In-Vitro Diagnostica, AniCon Labor GmbH, Hoeltinghausen, Germany). Ct-values < 35 were classified as positive samples, and Ct-values $\geq$ 35 were classified as negative. Bulk tank milk BCoV antibodies and serum BCoV antibodies were analyzed using a competitive monoscreen ELISA test (Bio-X 392/2, Bio-X Diagnostics S.A., Rochefort, Belgium). The outcomes of the serum and bulk tank milk samples were percentage inhibition (%inh) of the optical density readings. The samples' degree of positivity was categorized according to test manuals from 0 to 4 (%inh < 20 = 0, 20 $\leq$ %inh < 40 = 1 (+), 40 $\leq$ %inh < 60 = 2 (++), 60 $\leq$ %inh < 80 = 3 (+++), and %inh $\geq$ 80 = 4 (++++)).

Farm information collection: An extensive questionnaire was performed to determine the husbandry and management factors (Table S1). Management and herd health history was collected with a focus on young stock diarrhea and respiratory disease prevalence, vaccines, treatments, and preweaning calf nutrition with focus on colostrum and milk or milk replacer usage. The investigator also asked over 120 questions on biosecurity according to the Biocheck.Ugent dairy scoring system (https://biocheckgent.com/, accessed on 19 September 2024) (Table S2). All questionnaires and sampling records were transmitted to the principal study investigators through e-mail and hand-written records were transmitted with the samples to the contracted lab where all documents were scanned and transmitted electronically to the study investigator (G.V), who transmitted these further to the investigator (A.C.B) who evaluated documentation and submitted requests regarding missing or unclear data.

Data analysis: All data were entered in Excel, and all statistical analysis was performed using SAS version 9.4. Data were categorized and/or made into binary answers, according to their distributions to facilitate statistical analysis. The sampled animals where BCoV was detected in either in feces and/or nasal swabs were categorized as 'shedding BCoV' or a 'BCoV PCR-positive' animal. Prevalence estimates for positive nasal and fecal samples and animals within categories, farms, and countries were calculated. Average antibody levels within animal categories, farms, and countries were calculated and correlated with the bulk tank antibody levels with Pearson correlation analysis.

Biosecurity data analysis: The biosecurity questions were entered into the Biocheck.Ugent software for dairy and scored and categorized according to Biocheck.Ugent from 0 to 100% [23]. External biosecurity was composed of questions in 5 categories, including purchase and reproduction (BC-A), transports and carcass removal (BC-B), feed and water (BC-C), farm workers and visitors (BC-D), and control of vermin and pets (BC-E). The internal biosecurity was composed of 6 categories, including management of health (BC-F), calving (BC-G), calf (BC-H), milking parlor (BC-I), adult cattle (BC-J), and working and overall organization and equipment (BC-K). All categories have a subcategory score from 0 to 100% based upon weighted scores of questions, and these subcategories were further weighted and merged for scores for external, internal, and overall biosecurity according to the Biocheck algorithms.

Statistical analysis: Univariable and multivariable analysis were used to determine the risks for BCoV in neonatal calves and weaned calves in a separate analysis. Calf category stratified univariable and multinomial logistic models were performed for neonatal and weaned calves, as they may have different risk factors and due to interaction effects between calf category and other predictive variables. All univariable analyses included a random effect of the farm (variables tested and direction of association and level of significance can be found in Tables S1 and S2). Four logistic regression models with random effects of herds (population-averaged marginal models) were used to determine the odds of neonatal and weaned calves that were BCoV-positive on PCR conditional on predictive variables (Proc Glimmix). In all these models, variables were selected for inclusion in the models using a stepwise procedure, where the *p*-value for entry was 0.3 and a *p*-value for retention was 0.1. Interaction effects were tested for all variables retained in the final models. Statistical significance was declared at *p*-value $\leq$ 0.05 and a non-significant trend at 0.05 < *p*-

value ≤ 0.10. Confounding was evaluated by inspecting if the % difference is 10% or greater in the model with or without a confounding covariate. Two models evaluated biosecurity, management, and housing in relation to BCoV PCR-positive neonatal and weaned calves, including calf age as a covariate. In these models, Biocheck scores for the 11 individual biosecurity categories, as well as relevant individual questions, were statistically tested. Similarly, two multilevel logistic models evaluated the odds of BCoV PCR-positive neonatal and weaned calves on herd health factors, productivity, and antibody levels.

3. Results

A total of 126 farms were included from April 2021 to December 2022. This study included dairy farms in Austria, Belgium, Czech Republic, Denmark, France, Greece, Hungary, Ireland, Italy, the Netherlands, Poland, Portugal, Romania, Sweden, Slovakia, and Scotland (United Kingdom) (Table S3). The average lactating herd size was 370 cows, and 68% of herds were Holstein Friesian herds, and the remaining herds were either mixed breeds or other dairy breeds. There were, on average, 69 preweaned calves and 217 weaned non-pregnant calves. Sixty percent of the dairies reported using dry cow vaccinations against BCoV.

A map describing the sampled farms (Figure 1, created in Google maps) shows clustered sampling in some countries. For seven farms, the biosecurity information was not completed. From an additional seven farms, the individual calf sample information was missing essential data, such as age at the time of sampling, thus excluding their use in the study. The data presented here were for neonatal calves from 113 farms and for weaned calves from 116 farms. The results from multilevel models may have a varying number of calves included due to missing explanatory variables.

Figure 1. Map of dairy farms included in this BCoV prevalence and risk factor study. The coloring of farms indicates levels of BCoV shedding in calves; green—0%, 0% < orange < 10%, red $\geq$ 10%, and gray—excluded farms.

3.1. *BCoV Fecal and Nasal PCR-Positive*

BCoV was detected in 27% of neonatal calves, in 24% of nasal samples, and 17% of fecal swabs (Table 1A). BCoV was detected in 32% of weaned calves, 21% in nasal and 22% in fecal swabs (Table 1B). Four percent of fresh cows were shedding BCoV in four percent of nasal and five percent of fecal swabs. BCoV was detected in 48% of 113 herds in neonatal calves (Figure 2) and in 71% of 116 herds in weaned calves (Figure 3). There was a relatively even distribution of percentage of neonatal and weaned calves per farm shedding BCoV (Figures 2 and 3).

Table 1. BCoV detection in nasal and fecal swabs from neonatal calves (A) and weaned calves (B) on European dairies.

	Neonatal Calves BCoV Recovery *				Weaned Calves BCoV Recovery *		
A	**Fecal+**	**Fecal−**	**Sum (%)**	**B**	**Fecal+**	**Fecal−**	**Sum (%)**
Nasal+	196	124	320 (24%)	Nasal+	167	145	312 (21%)
Nasal−	40	994	1034 (76%)	Nasal−	160	984	1144 (79%)
Sum (%)	236 (17%)	1118 (83%)	1354	Sum (%)	327 (22%)	1129 (78%)	1456

* −/+ indicate detection of BCoV with RT-PCR.

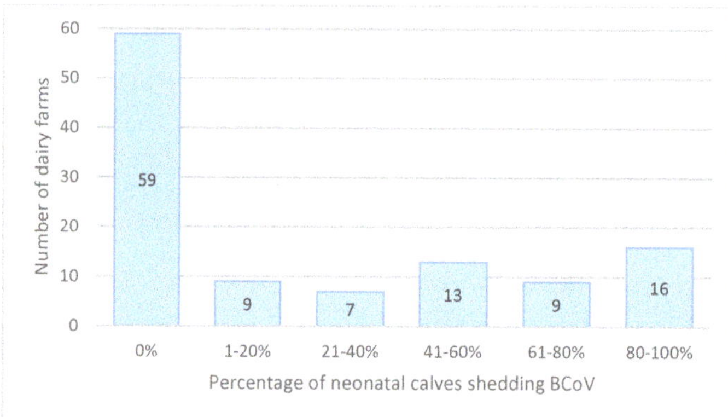

Figure 2. Number of herds (of 113) stratified by percentage of neonatal calves shedding BCoV.

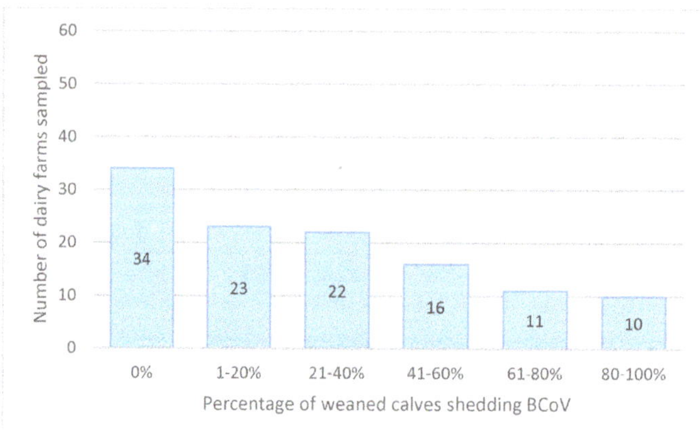

Figure 3. Number of herds (of 116) stratified by percentage of weaned calves shedding BCoV.

3.2. Bulk Tank Milk and Cattle BCoV Antibodies

Serum BCoV antibodies were detected in 92% of neonatal calves and 81% of weaned calves (at levels above 0 %inh in the ELISA). BCoV antibodies were detected in all bulk tank milk samples, the bulk tank milk (BTM) mean was 89 %inh (++++), and BTM antibodies in all farms were above 54 %inh (++) (Figure 4). There was a low correlation between bulk tank milk antibody levels and the animal antibody levels in the herds (R = 0.16). The shedding of BCoV in calves occurred in calves with all levels of antibodies (Figure 5).

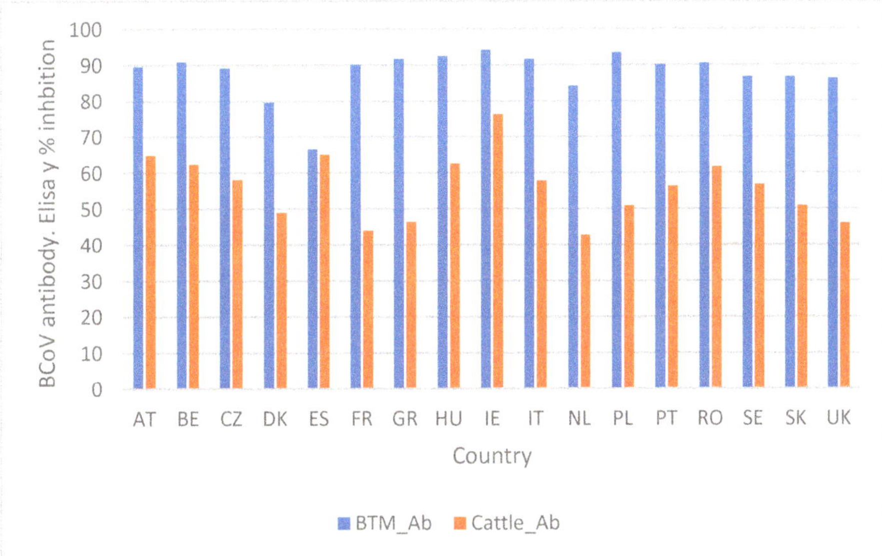

Figure 4. BCoV mean antibody levels (% inhibition in Elisa) in bulk tank milk (BTM_Ab) and the mean antibody levels of neonatal calves, weaned calves, and fresh cows (Cattle_Ab) in the enrolled European dairy farms stratified by country.

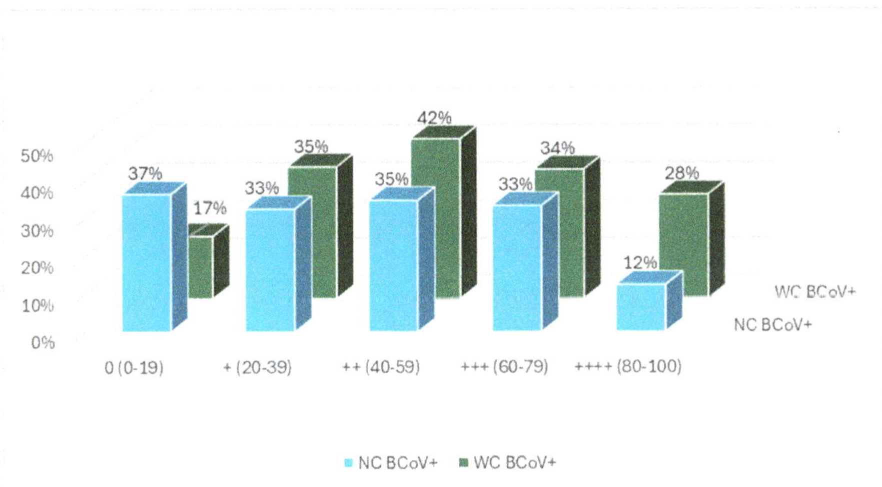

Figure 5. Percentage of calves shedding BCoV stratified by BCoV antibody levels in neonatal (NC BCoV+) and weaned (WC BCoV+) calves.

3.3. Biosecurity Associated with BCoV in Calves (NC and WC)

The biosecurity was scored with Biocheck. The scoring tool was used for 118 farms, and the average Biocheck score was 60% (external score 71%, internal score 48%). The distribution of the Biocheck scores for the categories A–K are described in Figure 6.

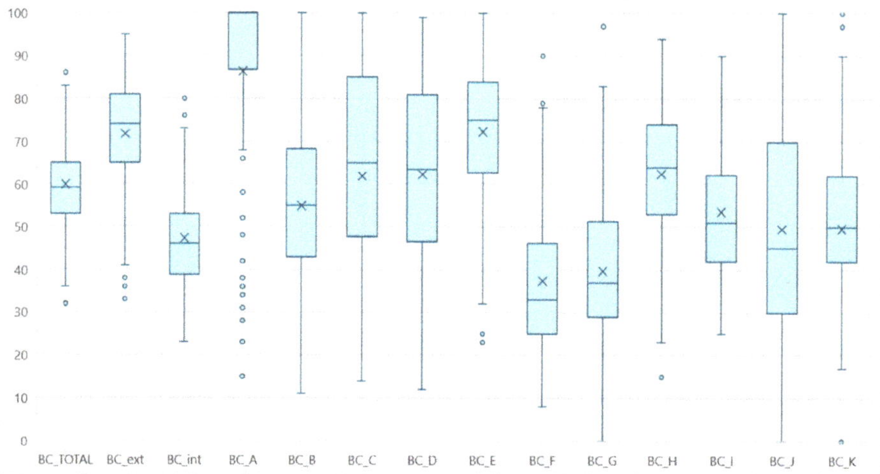

Figure 6. Boxplot distributions of Biocheck scores for 118 dairies. BC-ext (External biosecurity) includes BC-A (purchase and reproduction), BC-B (transports and carcass removal), BC-C (feed and water), BC-D (farm workers and visitors), and BC-E (control of vermin and pets). The BC-int (internal biosecurity) includes BC-F (management of health), BC-G (calving management), BC-H (calf management), BC-I (milking parlor), BC-J (adult cattle management), and BC-K (working and overall organization and equipment).

3.4. Management and Biosecurity Associated with BCoV in NC and WC

The logistic regression evaluating herd management and biosecurity factors associated with BCoV in neonatal calves included 1332 calves, of which 355 calves were shedding BCoV in 110 dairies (Table 2). This model indicates that the odds of BCoV shedding in neonatal calves increased 2.3 times for every subsequent week the first month of life. The odds of BCoV shedding in calves decreased by 55% in dairy herds that used dry cow BCoV vaccination. The odds of BCoV shedding in neonatal calves was 2.5 times higher with group housing starting the first week and 1.9 times higher the second week compared to the third week or later. There was a 41% reduction in odds of BCoV shedding in calves that received transition milk feeding the first week. The odds of BCoV shedding were 2.6 times higher in herds where the milk feeding volume the first month was less than 6 L/day compared to farms that fed more than 6 L/day. In this model, an increasing Biocheck B score was associated with higher odds of BCoV shedding in calves, whereas, with every increasing percentage of replacement heifers purchased in the previous year, there was a 25% increase in the odds of neonatal BCoV shedding.

The logistic regression evaluating herd management and biosecurity factors associated with BCoV in weaned calves included 1459 calves, of which 472 calves were shedding BCoV in 116 dairies (Table 3). This model indicates that the odds of BCoV shedding in weaned calves decreased 15% for every increasing month of life. The odds of BCoV shedding in calves was 1.81 times higher in dairy herds that used dry cow BCoV vaccination and was 2.24 times higher in herds using bovine respiratory disease (BRD) vaccines in calves compared to non-vaccinating herds. The odds of BCoV shedding in weaned calves was reduced by 30% in herds where weaning was, on average, prior to 10 weeks of age and reduced by 45% in herds where group housing was practiced prior to 8 weeks of age. None of the biosecurity measures implemented were significantly associated with BCoV shedding in weaned calves.

Table 2. Logistic regression model describing management and biosecurity factors associated with odds of shedding BCoV in neonatal calves.

					95% Conf. Int.		
Variable	**Param.**	**Est.**	**S.E**	**OR**	**Lower CI**	**Higher CI**	***p*-Value**
Intercept	cont.	−3.79	0.36	0.02	0.01	0.05	<0.01
Week of life (1–4)	cont.	0.83	0.08	2.29	1.96	2.68	<0.01
Dam BcoV–vaccination	Yes	−0.80	0.16	0.45	0.33	0.61	<0.01
	No	0.00	0.00	1.00	1.00	1.00	ref
Biocheck score B (%)	cont.	0.02	0.00	1.02	1.02	1.03	<0.01
% Heifers purchased	cont.	0.23	0.08	1.25	1.07	1.47	<0.01
Group housing started	week 1	0.92	0.19	2.51	1.73	3.65	<0.01
	week 2	0.64	0.21	1.90	1.26	2.86	<0.01
	week 3+	0.00	0.00	1.00	1.00	1.00	ref
Transition milk feeding	Yes	−0.52	0.21	0.59	0.39	0.90	<0.01
	No	0.00	0.00	1.00	1.00	1.00	ref
Milk/MR daily quantity	<6 L	0.97	0.18	2.63	1.85	3.74	<0.01
	≥6 L	0.00	0.00	1.00	1.00	1.00	ref

Table 3. Logistic regression model describing management and biosecurity factors associated with the odds of shedding BCoV in weaned calves.

					95% Conf. Interval		
Variable	**Param.**	**Est.**	**S.E**	**OR**	**Lower CI**	**Higher CI**	***p*-Value**
Intercept	cont.	−0.69	0.28	0.50	0.29	0.86	0.01
Month of life	cont.	−0.16	0.07	0.85	0.74	0.98	0.03
Dam BCoV vaccination	Yes	0.59	0.14	1.81	1.37	2.38	<0.01
	No	0.00	0.00	1.00	1.00	1.00	ref
BRD calf vaccination	Yes	0.80	0.13	2.24	1.73	2.89	<0.01
	No	0.00	0.00	1.00	1.00	1.00	ref
Calf weaning age	<70 days	−0.35	0.13	0.70	0.55	0.91	<0.01
	≥70 days	0.00	0.00	1.00	1.00	1.00	ref
Group housing started	<56 days	−0.60	0.13	0.55	0.43	0.71	<0.01
	≥56 days	0.00	0.00	1.00	1.00	1.00	ref

3.5. Herd Health Factors Associated with BCoV in NC and WC

The logistic regression evaluating calf and herd health, and the productivity factors associated with BCoV in neonatal calves included 1346 calves, of which 342 calves were shedding BCoV in 112 dairies (Table 4). This model indicates that the odds of BCoV shedding in neonatal calves increased 90% for every increasing week of life (weeks 1 to 4) and with increasing the BCoV antibody levels (0–4), the odds of shedding decreased 24%. The odds of BCoV shedding were 54% lower in herds with cow average yearly (305 day) milk production above the median of the herds and 68% higher in herds with BTM somatic cell counts above the median of the herds. The odds of BCoV shedding increased 1% for every % increase in the reported herd average % of calves that had diarrhea in the preweaning period. The odds of BCoV shedding were highest in herds reporting that diarrhea started predominantly in the first week of life and 66% lower in herds reporting diarrhea starting in week two. The odds of BCoV shedding in calves was 2.5 times higher in herds reporting a BRD outbreak in the previous year. In herds where BCoV shedding was detected in fresh cows, the calves were 4.1 times more likely to shed BCoV.

Table 4. Logistic regression model describing calf and herd-level health, and productivity factors associated with the odds of shedding BCoV in neonatal calves.

					95% Conf. Interval		
Variable	**Param.**	**Est.**	**S.E**	**OR**	**Lower CI**	**Higher CI**	***p*-Value**
Intercept	cont.	−2.30	0.29	0.10	0.06	0.18	<0.01
Week of life (1–4)	cont.	0.62	0.08	1.87	1.61	2.17	<0.01
Antibody level (0–5)	cont.	−0.27	0.05	0.76	0.69	0.85	<0.01
Ave 305d cow milk yield	≥median	−0.78	0.15	0.46	0.34	0.62	<0.01
	<median	0.00	0.00	1.00	1.00	1.00	ref
BTM SCC	≥median	0.52	0.15	1.68	1.25	2.27	<0.01
	<median	0.00	0.00	1.00	1.00	1.00	ref
% Calf diarrhea preweaning	cont.	0.01	0.00	1.01	1.01	1.02	<0.01
Primary week of diarrhea	week 1	0.00	0.00	1.00	1.00	1.00	ref
	week 2	−0.83	0.16	0.44	0.32	0.60	<0.01
	week 3	−0.26	0.22	0.77	0.50	1.18	0.17
BRD outbreak previous year	yes	0.90	0.18	2.47	1.73	3.51	<0.01
	no	0.00	0.00	1.00	1.00	1.00	ref
Cow BCoV shedding	yes	1.42	0.15	4.13	3.05	5.58	<0.01
	no	0.00	0.00	1.00	1.00	1.00	ref

The logistic regression evaluating calf and herd health and productivity factors associated with BCoV in weaned calves included 1410 calves, of which 459 calves were shedding BCoV in 113 dairies (Table 5). This model indicates that the odds of BCoV shedding in weaned calves decreased 15% for every increasing month of life (months 2–5), and with increasing the BCoV antibody levels, the odds of shedding increased 16%. The odds of BCoV shedding were 39% higher in herds with milk production above the median of the herds. The odds of BCoV shedding increased 1% for every % increase in the reported herd average % of calves that have diarrhea in the preweaning period. The odds of BCoV shedding in calves was 34% higher in herds reporting a BRD outbreak in the previous year. In herds where BCoV shedding was detected in the fresh cows, the calves were 2.5 times more likely to shed BCoV. There was also a significant but small increase in the odds (1.001) of BCoV shedding in herds, with an increasing number of weaned non-pregnant heifers.

Table 5. Logistic regression model describing calf and herd-level health, and productivity factors associated with the odds of shedding BCoV in weaned calves.

					95% Conf. Interval		
Variable	**Param.**	**Est.**	**S.E**	**OR**	**Lower CI**	**Higher CI**	***p*-Value**
Intercept	cont.	−1.30	0.26	0.27	0.16	0.45	<0.01
Months of life	cont.	−0.16	0.08	0.85	0.73	0.98	0.03
Antibody level (0–5)	cont.	0.15	0.05	1.16	1.05	1.27	<0.01
Ave 305d cow milk yield	≥median	0.33	0.13	1.39	1.09	1.79	0.01
	<median	0.00	0.00	1.00	1.00	1.00	ref
% Calf diarrhea preweaning	cont.	0.01	0.00	1.01	1.00	1.01	<0.01
BRD outbreak in previous year	yes	0.30	0.16	1.34	0.98	1.85	0.07
	no	0.00	0.00	1.00	1.00	1.00	ref

Table 5. *Cont.*

Variable	Param.	Est.	S.E	OR	95% Conf. Interval Lower CI	Higher CI	*p*-Value
Cow BCoV shedding in previous year	yes	0.93	0.13	2.54	1.98	3.28	<0.01
	no	0.00	0.00	1.00	1.00	1.00	ref
Number of weaned non-pregnant heifers	cont.	0.00	0.00	1.001	1.00	1.00	0.02

4. Discussion

This study has shown that BCoV is commonly present in both the respiratory and enteric pathway in neonatal and weaned calves in European dairies, with all study herds being seropositive to the virus, and the virus was present in 80% of the study herds. Within the herds, 27% of preweaning calves, 32% of weaned calves, and 4% of fresh cows were shedding BCoV (BCoV-positive fecal and/or nasal samples).

The study could not evaluate country-level prevalences due to clustered sampling in some countries, which is a weakness due to the convenience enrollment of farms. We also intended to sample large countries such as Germany and England, however, the national regulatory authorities had enrollment criteria and costs that were not compatible with this study. There were no visible changing trends in BCoV prevalence in calves from north to south or west to east in Europe. This comes not as a surprise, as numerous studies have shown the global widespread nature of BCoV in dairy cattle [24]. Studies in Asia (China and Thailand) indicates similar endemic situations as described in our study [25,26].

There are previously limited and not comparable studies of enteric and respiratory BCoV in Europe. A study of French dairy calves in 1983–1984 found coronavirus in 42% of the samples and strangely found it more prevalent in healthy calves compared to diarrhea calves [15]. A study of 65 dairy herds in Spain found BCoV in 20% of samples from calves shedding rotavirus [16]. BCoV was detected in 26% of 230 fecal samples from preweaned calves in 100 Austrian farms [17]. In Switzerland, BCoV was detected in 8% of samples from diarrheic dairy calves on 71 dairies [18]. In Sweden, BCoV was found in 3% samples from calves, and limited BCoV virus was found in less than 200 normal or diarrheic samples in Norway [19,20]. Published data on respiratory BCoV are sparse. A Belgian cross-sectional study of 128 BRD outbreaks (dairy, mixed dairy, and beef farms) in 2016 to 2018 indicated that BCoV was the most frequently isolated virus (38%) [5]. A study in Ireland of nasal swabs from 1364 BRD outbreaks found BCoV in 23% of the samples [21]. A Polish study of young stock indicated a seroprevalence of 73% and a PCR virus detection in 11% of the animals sampled [22]. A study of water buffalo and dairy cattle in the Campania region in Southern Italy similarly found high seroprevalence (31% of the animals sampled) [27].

Scandinavian studies have indicated that bulk tank milk serology can provide an estimate of the herd-level seroprevalence of BCoV [10,28]. However, our study has indicated that bulk tank milk BCoV antibody levels are of limited value to evaluate animal antibody levels or the level of BCoV infection in animals, and this could be due to the endemic situation in Europe and possibly also that serum levels in fresh cows and neonatal calves are impacted by dry cow vaccinations.

We found similar levels of nasal and fecal shedding in weaned calves, whereas nasal shedding was higher in preweaned calves. A clinical trial by Cho et al. indicated that, regardless of route of exposure, oral or nasal, BCoV was detected in nasal and fecal samples from all infected calves and that nasal shedding occurred prior to fecal shedding regardless of the inoculation route, and the fecal shedding persisted longer [29]. We noted this in our study too, where, in the second week, nasal shedding was twice as high as fecal shedding.

As it is important to know herd-level risk factors for BCoV to effectively prevent and control the virus, this study aimed to identify herd-level risk factors. Biosecurity is very important to reduce the spread of diseases between and within farms. A Swedish study found a clear association between higher herd biosecurity levels and a lower prevalence of BCoV herd infection [30]. However, in Sweden, they evaluated antibody levels, while, in our study, we focused on BCoV detected with PCR in nasal and fecal swabs. In this cross-sectional study, it is important to note that we can only evaluate associations; however, causal factors cannot be determined. It should be noted that most dairy farms do not have high levels of biosecurity and that partial measures may still lead to the dissemination of BCoV between and within farms. One of the major risks for the introduction of contagious pathogens is through live animal introduction [31]. In the neonatal calves, BCoV shedding increased with an increasing percentage of replacement heifers purchased. Higher biosecurity for farm workers and visitors (BC-B category) was found associated with increased shedding in neonatal calves. This may be confounding, since higher biosecurity for people may be practiced due to higher infectious disease pressure on the calves in farms. It has been shown that BCoV can persist on clothes, boots, and equipment up to 24 h after exposure to virus-shedding calves [32], and not providing boots for visitors was a risk factor for BCoV seropositivity in dairy herds in Sweden [33]. In Norway, there is a stakeholder voluntary preventive program against BRSV and BCoV in cattle herds, with antibody classification of herds and enhanced external biosecurity [34]. In the Norwegian BCoV control program, it was noted that BCoV introduction through professional visitors, such as AI personnel, could be prevented with good biosecurity measures. Unfortunately, the prevention of the introduction of BCoV through people is not practiced sufficiently in European dairies, and many of the study farms indicated that veterinarians and other professional visitors, such as inseminators and claw trimmers, that all come in direct contact with the animals, did not use farm-dedicated protective clothing and boots. With the high prevalence and an endemic situation of BCoV in dairy farms in Europe, it is likely that partial biosecurity is not effective in endemic situations with highly infectious viruses. Furthermore, there are studies indicating that neonatal calves infected with BCoV can become persistently infected [35].

For husbandry/management factors associated with BCoV shedding in neonatal calves, we noted that using transition milk feeding and higher milk or milk replacer feeding volumes could reduce BCoV shedding. There are now numerous studies showing the importance of transition milk or colostrum supplementation the first week of life on gut health [36]. The importance of a good nutritional status for optimal immunity [37] is important to emphasize in disease prevention programs. It was noted that group housing starting in the first and second weeks compared to after the third increased the shedding of BCoV in neonatal calves. This is likely due to increased contacts between the calves in the group housing systems. In weaned calves, there was a lower likelihood of BCoV shedding in systems where group housing started prior to 8 weeks of age and calves were weaned prior to 10 weeks of age. The initiation of group housing after or in combination with weaning is stressful for previously single-housed calves, and this grouping stress combined with increased contact with herd mates may have led to increasing the numbers or intensity of BCoV shedding. This study cannot determine why the delayed weaning was associated with a higher likelihood of BCoV shedding, as it was also noted that, with increasing months of age in the weaned calves, there was a reduced risk of BCoV shedding. In a study in Thailand, the only factor that was found to be protective against being seropositive was disinfecting diarrheal stools compared to doing nothing or simply washing away stools [25]. This practice is very rare in Europe. In a Dutch study, non-optimal hygienic housing was also identified as a risk factor for BCoV shedding in neonatal calves [38].

Our study showed a reduced level of shedding in neonatal calves with increasing antibody levels, and this is mostly colostral-derived BCoV antibody protection of the calves under 1 month of age. However, in weaned calves, shedding was higher in calves with higher antibody levels, and this is most likely indicating a mounting immunity due to active BCoV infections in the calves. In this study, 60% of farmers used a dry cow vaccine against calf enteric pathogens, including BCoV, to boost colostral immunity in the calves. This vaccination was shown to be effective in reducing BCoV shedding in neonatal calves. This was contrary to a clinical trial in Korea, where they concluded that "The effectiveness of maternal antibodies in preventing viral replication and shedding appeared limited" [6]. However, there was increased shedding in weaned calves in our study herds using BCoV dam vaccination. This could have been due to lower infection rates in the neonatal calves, reducing the calves' natural acquired immunity to infection post-weaning. At the time of the study, the nasal vaccine against BCoV that is currently on the market in the EU was not available for farmers. The involvement of BCoV in the BRD complex was and is still to a large degree underestimated by veterinary practitioners and calf owners. In our study, BCoV shedding was much higher in weaned calves in herds using BRD vaccination. There is increasing evidence of BCoV's role in BRD, and studies have shown that BCoV is also in the lower respiratory tract of cattle with respiratory disease [39]. It may thus be likely that BCoV was contributing to the BRD challenges, to which the farm was vaccinating against other respiratory pathogens. The use of vaccines for BCoV will likely continue to evolve, as it has been shown to be a valuable tool against various coronaviruses in multiple species [40].

Herd health and productivity factors were evaluated separate from management and nutrition factors, as these are associated, and including these factors together in one model may create confounding. In both neonatal and weaned calves, BCoV shedding was more common in herds where fresh cows were detected to be shedding BCoV. It could not be determined whether the positive results in the fresh cows where the result of reactivation of a persistently infected animal [35] or a recent infection in the fresh cows. BCoV shedding was also more common in neonatal and weaned calves in herds that reported a BRD outbreak in the previous year. There were some unexplainable associations with average cow yearly milk production in that it was associated with reduced shedding in neonatal calves and increased shedding in weaned calves. BCoV shedding in neonatal calves was also more common in dairy herds with BTM somatic cell counts above the median.

In our study, we noted that neonatal and weaned calves were more likely to shed BCoV in herds where diarrhea was more commonly seen in the preweaning period. This was also noted in a Swedish study where BCoV endemic and recently infected herds had more coughing and diarrhea in young stock and cows compared to BCoV free herds [41]. A study in the Netherlands evaluated 424 calves under 22 days of age from 108 dairy farms and found that *E. coli* K99 was more likely to be detected in feces from calves where BCoV has been detected [38]. A study in veal calves in the Netherlands similarly noted that calves shedding BCoV were more likely to have more severe diarrhea days, and this was coupled with a reduced weight gain in the preweaning period [42]. Since higher preweaning weight gain has been associated with higher first lactation milk yield, the negative production economic consequences of BCoV on dairies span beyond the treatment costs and immediate losses [43]. This study confirms that there are clear associations between BCoV being actively shed by calves on dairies and the herd-level calf health.

Our study adds to the body of evidence of the widespread distribution of BCoV in dairy calves. BCoV is associated with increased calf health challenges, such as severe neonatal diarrhea and calf respiratory disease. The economic losses due to the morbidity and mortality of calves, reduced growth performance, and decreased milk production of adult dairy cattle [2,41,42] indicate that BCoV cannot be disregarded in sustainable health dairy production.

5. Conclusions

This study has shown that BCoV is commonly present in both the respiratory and enteric pathway in neonatal and weaned calves in European dairies, with all study herds being seropositive to the virus, and the virus was present in 80% of the study herds. The overall biosecurity levels on dairies in Europe, in general, show weaknesses. It is likely that, in endemic situations, such as this one, the most important measures to reduce the disease impacts of BCoV on calves include optimizing immunity through management and vaccinations. BCoV dam vaccinations have a good protective effect on neonatal calves but cannot protect the weaned calves. Several health and management factors might reduce the risk of BCoV infections such as transition milk feeding, high daily milk volumes to calves, and grouping of calves after 3 weeks of age but prior to weaning.

Supplementary Materials: The following supporting information can be downloaded at https://www.mdpi.com/article/10.3390/ani14182744/s1: Table S1. Univariable farm clustered logistic regression analysis of variables of herd management and health factors associated with BCoV shedding in neonatal and weaned calves. Table S2. Univariable farm clustered logistic regression analysis of the variables of biosecurity (Biocheck) factors associated with BCoV shedding in neonatal and weaned calves. Table S3. Dairy herds enrolled in this study, including lactating herd size: neonatal calves, weaned calves, and fresh cows sampled; bulk tank milk antibody and mean serum antibody for all cattle sampled: neonatal calves, weaned calves, and fresh cows; the percentage of animals per category shedding; and the Biocheck scores (total, external, and internal).

Author Contributions: Conceptualization, A.C.B. and G.V.; methodology, A.C.B. and G.V.; validation, A.C.B. and G.V.; formal analysis, A.C.B.; investigation, A.C.B. and G.V.; resources, G.V.; data curation, A.C.B.; writing—original draft preparation, A.C.B.; writing—review and editing, A.C.B. and G.V.; visualization, A.C.B.; supervision, G.V.; project administration, G.V.; funding acquisition, G.V. All authors have read and agreed to the published version of the manuscript.

Funding: This study was funded by MSD Animal Health.

Institutional Review Board Statement: The study was conducted according to the guidelines of the Declaration of Helsinki and approved by the Animal Research Ethics Subcommittee (AREC) of the University College Dublin (AREC-21-16-McAloon).

Informed Consent Statement: Written informed consent was obtained from the owners of the animals.

Data Availability Statement: The data are not publicly available due to privacy restrictions.

Acknowledgments: We would like to express our appreciation and extend our gratitude to the MSD Animal Health national technical teams, collaborating veterinary practitioners, and all participating dairy farms.

Conflicts of Interest: The authors declare that this study received funding from MSD Animal Health. Author Geert Vertenten is employed by the company MSD Animal Health. Author Anna Catharina Berge declares that the research was conducted in the absence of any commercial or financial relationships that could be construed as potential conflicts of interest.

References

1. Vlasova, A.N.; Saif, L.J. Bovine Coronavirus and the Associated Diseases. *Front. Vet. Sci.* **2021**, *8*, 643220. [CrossRef] [PubMed]
2. Saif, L.J. Bovine respiratory coronavirus. *Vet. Clin. N. Am. Food Anim. Pract.* **2010**, *26*, 349–364. [CrossRef] [PubMed]
3. Boileau, M.J.; Kapil, S. Bovine coronavirus associated syndromes. *Vet. Clin. N. Am. Food Anim. Pract.* **2010**, *26*, 123–146. [CrossRef] [PubMed]
4. Ellis, J. What is the evidence that bovine coronavirus is a biologically significant respiratory pathogen in cattle? *Can. Vet. J.* **2019**, *60*, 147–152.
5. Pardon, B.; Callens, J.; Maris, J.; Allais, L.; Van Praet, W.; Deprez, P.; Ribbens, S. Pathogen-specific risk factors in acute outbreaks of respiratory disease in calves. *J. Dairy Sci.* **2020**, *103*, 2556–2566. [CrossRef]
6. Cho, K.O.; Hoet, A.E.; Loerch, S.C.; Wittum, T.E.; Saif, L.J. Evaluation of concurrent shedding of bovine coronavirus via the respiratory tract and enteric route in feedlot cattle. *Am. J. Vet. Res.* **2001**, *62*, 1436–1441. [CrossRef]
7. Park, S.J.; Kim, G.Y.; Choy, H.E.; Hong, Y.J.; Saif, L.J.; Jeong, J.H.; Park, S.I.; Kim, H.H.; Kim, S.K.; Shin, S.S.; et al. Dual enteric and respiratory tropisms of winter dysentery bovine coronavirus in calves. *Arch. Virol.* **2007**, *152*, 1885–1900. [CrossRef]
8. Liu, L.; Hägglund, S.; Hakhverdyan, M.; Alenius, S.; Larsen, S.E.; Belak, S. Molecular epidemiology of bovine coronavirus on the basis of comparative analyses of the S gene. *J. Clin. Microbiol.* **2006**, *44*, 957–960. [CrossRef]
9. Traven, M.; Björnerot, L.; Larsson, B. Nationwide survey of antibodies to bovine coronavirus in bulk milk from Swedish dairy herds. *Vet. Rec.* **1999**, *144*, 527–529. [CrossRef]
10. Ohlson, A.; Blanco-Penedo, I.; Fall, N. Comparison of Bovine coronavirus-specific and Bovine respiratory syncytial virus-specific antibodies in serum versus milk samples detected by enzyme-linked immunosorbent assay. *J. Vet. Diagn. Investig.* **2014**, *26*, 113–116. [CrossRef]
11. Ohlson, A.; Alenius, S.; Traven, M.; Emanuelson, U. A longitudinal study of the dynamics of bovine corona virus and respiratory syncytial virus infections in dairy herds. *Vet. J.* **2013**, *197*, 395–400. [CrossRef] [PubMed]
12. Wolff, C.; Emanuelson, U.; Ohlson, A.; Alenius, S.; Fall, N. Bovine respiratory syncytial virus and bovine coronavirus in Swedish organic and conventional dairy herds. *Acta Vet. Scand.* **2015**, *57*, 2. [CrossRef] [PubMed]

13. Toftaker, I.; Sanchez, J.; Stokstad, M.; Nødtvedt, A. Bovine respiratory syncytial virus and bovine coronavirus antibodies in bulk tank milk—Risk factors and spatial analysis. *Prev. Vet. Med.* **2016**, *133*, 73–83. [CrossRef]
14. Paton, D.J.; Christiansen, K.H.; Alenius, S.; Cranwell, M.P.; Pritchard, G.C.; Drew, T.W. Prevalence of antibodies to bovine virus diarrhoea virus and other viruses in bulk tank milk in England and Wales. *Vet. Rec.* **1998**, *142*, 385–391. [CrossRef] [PubMed]
15. De Rycke, J.; Bernard, S.; Laporte, J.; Naciri, M.; Popoff, M.R.; Rodolakis, A. Prevalence of various enteropathogens in the feces of diarrheic and healthy calves. *Ann. Rech. Vet.* **1986**, *17*, 159–168.
16. Garcia, A.; Ruiz-Santa-Quiteria, J.A.; Orden, J.A.; Cid, D.; Sanz, R.; Gomez-Bautista, M.; de la Fuente, R. Rotavirus and concurrent infections with other enteropathogens in neonatal diarrheic dairy calves in Spain. *Comp. Immunol. Microbiol. Infect. Dis.* **2000**, *23*, 175–183. [CrossRef]
17. Haschek, B.; Klein, D.; Benetka, V.; Herrera, C.; Sommerfeld-Stur, I. Detection of bovine torovirus in neonatal calf diarrhoea in Lower Austria and Styria (Austria). *J. Vet. Med. B Infect. Dis. Vet. Public Health* **2006**, *53*, 160–165. [CrossRef]
18. Uhde, F.L.; Kaufmann, T.; Sager, H.; Albini, S.; Zanoni, R.; Schelling, E.; Meylan, M. Prevalence of four enteropathogens in the faeces of young diarrhoeic dairy calves in Switzerland. *Vet. Rec.* **2008**, *163*, 362–366. [CrossRef]
19. Björkman, C.; Svensson, C.; Christensson, B.; de Verdier, K. *Cryptosporidium parvum* and *Giardia intestinalis* in calf diarrhoea in Sweden. *Acta Vet. Scand.* **2003**, *44*, 145–152. [CrossRef]
20. Gulliksen, S.M.; Jor, E.; Lie, K.I.; Hamnes, I.S.; Løken, T.; Åkerstedt, J.; Østerås, O. Enteropathogens and risk factors for diarrhea in Norwegian dairy calves. *J. Dairy Sci.* **2009**, *92*, 5057–5066. [CrossRef]
21. O'Neill, R.; Mooney, J.; Connaghan, E.; Furphy, C.; Graham, D.A. Patterns of detection of respiratory viruses in nasal swabs from calves in Ireland: A retrospective study. *Vet. Rec.* **2014**, *175*, 351. [CrossRef] [PubMed]
22. Socha, W.; Larska, M.; Rola, J.; Bednarek, D. Occurrence of Bovine Coronavirus and other Major Respiratory Viruses in Cattle in Poland. *J. Vet. Res.* **2022**, *66*, 479–486. [CrossRef] [PubMed]
23. Damiaans, B.; Renault, V.; Sarrazin, S.; Berge, A.C.; Pardon, B.; Saegerman, C.; Dewulf, J. A risk-based scoring system to quantify biosecurity in cattle production. *Prev. Vet. Med.* **2020**, *179*, 104992. [CrossRef] [PubMed]
24. Zhu, Q.; Li, B.; Sun, D. Advances in Bovine Coronavirus Epidemiology. *Viruses* **2022**, *14*, 1109. [CrossRef] [PubMed]
25. Ven, S.; Arunvipas, P.; Lertwatcharasarakul, P.; Ratanapob, N. Seroprevalence of bovine coronavirus and factors associated with the serological status in dairy cattle in the western region of Thailand. *Vet. World* **2021**, *14*, 2041–2047. [CrossRef]
26. Geng, H.L.; Meng, X.Z.; Yan, W.L.; Li, X.M.; Jiang, J.; Ni, H.B.; Liu, W.H. Prevalence of bovine coronavirus in cattle in China: A systematic review and meta-analysis. *Microb. Pathog.* **2023**, *176*, 106009. [CrossRef]
27. Ferrara, G.; Iovane, V.; Improda, E.; Iovane, G.; Pagnini, U.; Montagnaro, S. Seroprevalence and Risk Factors for Bovine Coronavirus Infection among Dairy Cattle and Water Buffalo in Campania Region, Southern Italy. *Animals* **2023**, *13*, 772. [CrossRef]
28. Toftaker, I.; Toft, N.; Stokstad, M.; Sølverød, L.; Harkiss, G.; Watt, N.; O'Brien, A.; Nødtvedt, A. Evaluation of a multiplex immunoassay for bovine respiratory syncytial virus and bovine coronavirus antibodies in bulk tank milk against two indirect ELISAs using latent class analysis. *Prev. Vet. Med.* **2018**, *154*, 1–8. [CrossRef]
29. Cho, H.C.; Kim, Y.; Cho, Y.I.; Park, J.; Choi, K.S. Evaluation of bovine coronavirus in Korean native calves challenged through different inoculation routes. *Vet. Res.* **2024**, *55*, 74. [CrossRef]
30. Ohlson, A.; Emanuelson, U.; Traven, M.; Alenius, S. The relationship between antibody status to bovine corona virus and bovine respiratory syncytial virus and disease incidence, reproduction and herd characteristics in dairy herds. *Acta Vet. Scand.* **2010**, *52*, 37. [CrossRef]
31. Toftaker, I.; Ågren, E.; Stokstad, M.; Nødtvedt, A.; Frössling, J. Herd level estimation of probability of disease freedom applied on the Norwegian control program for bovine respiratory syncytial virus and bovine coronavirus. *Prev. Vet. Med.* **2020**, *181*, 104494. [CrossRef] [PubMed]
32. Oma, V.S.; Klem, T.; Traven, M.; Alenius, S.; Gjerset, B.; Myrmel, M.; Stokstad, M. Temporary carriage of bovine coronavirus and bovine respiratory syncytial virus by fomites and human nasal mucosa after exposure to infected calves. *BMC Vet. Res.* **2018**, *14*, 22. [CrossRef] [PubMed]
33. Ohlson, A.; Heuer, C.; Lockhart, C.; Traven, M.; Emanuelson, U.; Alenius, S. Risk factors for seropositivity to bovine coronavirus and bovine respiratory syncytial virus in dairy herds. *Vet. Rec.* **2010**, *167*, 201–206. [CrossRef] [PubMed]
34. Stokstad, M.; Klem, T.B.; Myrmel, M.; Oma, V.S.; Toftaker, I.; Østerås, O.; Nødtvedt, A. Using Biosecurity Measures to Combat Respiratory Disease in Cattle: The Norwegian Control Program for Bovine Respiratory Syncytial Virus and Bovine Coronavirus. *Front. Vet. Sci.* **2020**, *7*, 167. [CrossRef]
35. Kanno, T.; Ishihara, R.; Hatama, S.; Uchida, I. A long-term animal experiment indicating persistent infection of bovine coronavirus in cattle. *J. Vet. Med. Sci.* **2018**, *80*, 1134–1137. [CrossRef]
36. Conneely, M.; Berry, D.P.; Murphy, J.P.; Lorenz, I.; Doherty, M.L.; Kennedy, E. Effect of feeding colostrum at different volumes and subsequent number of transition milk feeds on the serum immunoglobulin G concentration and health status of dairy calves. *J. Dairy Sci.* **2014**, *97*, 6991–7000. [CrossRef]
37. Ockenden, E.M.; Russo, V.M.; Leury, B.J.; Giri, K.; Wales, W.J. Preweaning Nutrition and Its Effects on the Growth, Immune Competence and Metabolic Characteristics of the Dairy Calf. *Animals* **2023**, *13*, 829. [CrossRef]
38. Bartels, C.J.; Holzhauer, M.; Jorristma, R.; Swart, W.A.; Lam, T.J. Prevalence, prediction and risk factors of enteropathogens in normal and non-normal faeces of young Dutch dairy calves. *Prev. Vet. Med.* **2010**, *93*, 162–169. [CrossRef]

39. Rahe, M.C.; Magstadt, D.R.; Groeltz-Thrush, J.; Gauger, P.C.; Zhang, J.; Schwartz, K.J.; Siepker, C.L. Bovine coronavirus in the lower respiratory tract of cattle with respiratory disease. *J. Vet. Diagn. Investig.* **2022**, *34*, 482–488. [CrossRef]
40. Tizard, I.R. Vaccination against coronaviruses in domestic animals. *Vaccine* **2020**, *38*, 5123–5130. [CrossRef]
41. Duse, A.; Ohlson, A.; Stengarde, L.; Traven, M.; Alenius, S.; Bengtsson, B. Associations between Bovine Coronavirus and Bovine Respiratory Syncytial Virus Infections and Productivity, Health Status and Occurrence of Antimicrobial Resistance in Swedish Dairy Herds. *Antibiotics* **2021**, *10*, 641. [CrossRef] [PubMed]
42. Renaud, D.L.; Rot, C.; Marshall, J.; Steele, M.A. The effect of *Cryptosporidium parvum*, rotavirus, and coronavirus infection on the health and performance of male dairy calves. *J. Dairy Sci.* **2021**, *104*, 2151–2163. [CrossRef] [PubMed]
43. Gelsinger, S.L.; Heinrichs, A.J.; Jones, C.M. A meta-analysis of the effects of preweaned calf nutrition and growth on first-lactation performance. *J. Dairy Sci.* **2016**, *99*, 6206–6214. [CrossRef] [PubMed]

Disclaimer/Publisher's Note: The statements, opinions and data contained in all publications are solely those of the individual author(s) and contributor(s) and not of MDPI and/or the editor(s). MDPI and/or the editor(s) disclaim responsibility for any injury to people or property resulting from any ideas, methods, instructions or products referred to in the content.

animals

Article

Pen-Based Swine Oral Fluid Samples Contain Both Environmental and Pig-Derived Targets

Grzegorz Tarasiuk [1], Marta D. Remmenga [2], Kathleen C. O'Hara [2], Marian K. Talbert [2], Marisa L. Rotolo [3], Pam Zaabel [4], Danyang Zhang [5], Luis G. Giménez-Lirola [1] and Jeffrey J. Zimmerman [1,*]

1 Department of Veterinary Diagnostic and Production Animal Medicine, College of Veterinary Medicine, Iowa State University, Ames, IA 50011, USA; tarasiuk@iastate.edu (G.T.)
2 USDA: VS: Strategy and Policy, Center for Epidemiology and Animal Health, Fort Collins, CO 80526, USA
3 PIC North America, Hendersonville, TN 37075, USA
4 National Pork Board, Des Moines, IA 50325, USA
5 Department of Statistics, College of Liberal Arts and Sciences, Iowa State University, Ames, IA 50010, USA
* Correspondence: jjzimm@iastate.edu; Tel.: +1-515-294-1073

Simple Summary: The transfer of a target directly from pigs or indirectly from the environment into pen-based oral fluid samples was documented using a fluorescent tracer (red food coloring). Pens of ~30, ~60, and ~125 14-week-old pigs (32 pens/size) on commercial swine farms received one of two treatments: (1) pig exposure, i.e., ~3.5 mL of tracer solution sprayed into the mouth of 10% of the pigs in the pen; (2) environmental exposure, i.e., 20 mL of tracer solution was poured on the floor in the center of the pen. Oral fluids collected one day prior to treatment (baseline fluorescence control) and immediately after treatment were tested for fluorescence. Data were evaluated by receiver operating characteristic (ROC) analysis, with Youden's J statistic used to set a threshold. Pretreatment oral fluid samples with fluorescence responses above the ROC threshold were removed from further analysis (7 of 96 samples). Analysis of post-treatment samples showed detectable levels of red food coloring in oral fluid samples from 78 of 89 pens (87.6%), including 43 of 47 (91.5%) pens in which pigs were directly exposed and 35 of 42 (83.3%) pens in which the tracer was placed in the environment. Thus, oral fluid samples contain both pig-derived and environmental targets.

Abstract: Laboratory methods for detecting specific pathogens in oral fluids are widely reported, but there is little research on the oral fluid sampling process itself. In this study, a fluorescent tracer (diluted red food coloring) was used to test the transfer of a target directly from pigs or indirectly from the environment to pen-based oral fluid samples. Pens of ~30, ~60, and ~125 14-week-old pigs (32 pens/size) on commercial swine farms received one of two treatments: (1) pig exposure, i.e., ~3.5 mL of tracer solution sprayed into the mouth of 10% of the pigs in the pen; (2) environmental exposure, i.e., 20 mL of tracer solution was poured on the floor in the center of the pen. Oral fluids collected one day prior to treatment (baseline fluorescence control) and immediately after treatment were tested for fluorescence. Data were evaluated by receiver operating characteristic (ROC) analysis, with Youden's J statistic used to set a threshold. Pretreatment oral fluid samples with fluorescence responses above the ROC threshold were removed from further analysis (7 of 96 samples). Based on the ROC analyses, oral fluid samples from 78 of 89 pens (87.6%), contained red food coloring, including 43 of 47 (91.5%) pens receiving pig exposure and 35 of 42 (83.3%) pens receiving environmental exposure. Thus, oral fluid samples contain both pig-derived and environmental targets. This methodology provides a safe and quantifiable method to evaluate oral fluid sampling vis-à-vis pen behavior, pen size, sampling protocol, and target distribution in the pen.

Keywords: surveillance; oral fluids; sampling; tracer; environmental sampling; swine

1. Introduction

Oral fluid specimens are widely used in the swine industry because they are easily collected, contain detectable concentrations of pathogen-specific nucleic acids and antibodies, and provide the means to efficiently establish the exposure status of the group under surveillance [1]. Reflecting their usefulness to pork producers, approximately 2 million polymerase chain reaction (PCR) assays were performed on swine oral fluid samples at six independent veterinary diagnostic laboratories in the U.S. between 2019 and 2022 for influenza A virus (IAV), *Mycoplasma hyopneumoniae*, porcine circovirus type 2 (PCV2), porcine deltacoronavirus (PDCoV), porcine epidemic diarrhea virus (PEDV), porcine reproductive and respiratory syndrome virus (PRRSV), and transmissible gastroenteritis virus (TGEV). Further, the use of oral fluids is widely reported in the refereed literature. Thus, a search in Google Scholar™ using the terms "oral fluid" and "swine" (accessed November 2023) produced 1820 results for the years 2015 to 2023. The majority of these publications addressed various aspects of swine infectious diseases, with oral fluids serving an ancillary role. That is, few publications focused on optimizing or understanding the process of collecting oral fluids. Among the few, White et al. [2] analyzed oral fluid sampling time in the context of pig participation, Kittawornrat and Zimmerman [3] reviewed pig behaviors related to oral fluid collection, and Tarasiuk et al. [4] researched the effect of group size in the context of pig participation in oral fluid sampling.

Laboratory detection of specific pathogens in oral fluids is widely researched but there is an incomplete understanding of the original source of the target (pig vs. environment). The objective of this study was to evaluate the transfer of a target directly from pigs or indirectly from the environment into pen-based oral fluid samples. In terms of experimental design, this question can be addressed using infectious agents when the work is carried out in research settings. However, in commercial swine herds, the intentional use of infectious agents poses ethical and legal risks to both the collaborating producer and neighboring farms. Under such circumstances, the research is best carried out using noninfectious markers.

Readily detectable noninfectious tracers have previously been used in animal studies. As reviewed by Algeo et al. [5], tetracycline has been used in oral vaccination studies in coyotes, foxes, raccoons, skunks, feral swine, and other species. Tetracycline binds to calcium ions and is then deposited in teeth and bone (calcified tissues) [6]. In vaccination field studies, animals that consume the vaccine can be identified by ultraviolet visualization of tetracycline in cross-sectioned teeth and/or bone [5]. Similarly, Rhodamine B was used to monitor the uptake of edible oral vaccines in jackrabbits [7], deer [8], and wild boar [9]. Rhodamine B enters the circulatory system and is excreted in urine and feces but it is also sequestered in tissues, hair, and whiskers [10]. Thereafter, the uptake of oral bait vaccines was determined by fluorescent bands of Rhodamine B in hair and whiskers [11].

Tetracycline and Rhodamine B are useful for studies in wildlife, but cannot be used in studies involving food animals. Therefore, red food coloring (Red Food Coloring, McCormick and Co., Inc., Hunt Valley, MD, USA) was used as a safe alternative in this research. The commercial product is composed of two food colors (Red No. 40 and Red No. 3) that meet the criteria of the US Federal Food, Drug, and Cosmetic Act for use in food, drugs, and cosmetics, as defined in 21 CFR §74.303 and §74.340. In addition, synthetic food colors are known to fluoresce at specific wavelengths that are measurable using standard laboratory equipment [12]. Thus, fluorescence provided the means to determine whether swine transferred the target (red food coloring) to the pen-based oral fluid sample.

2. Materials and Methods

2.1. Experimental Design

The study was conducted in commercial production sites using pens holding ~30, ~60, or ~125 pigs, with 32 pens in each size category. Pigs were placed into facilities at 3 weeks of age and were 10–14 weeks of age at the time of the study. In brief, pigs were provided red food coloring suspended in a sucrose solution, a pen-based oral fluid sample was collected, and the sample was tested for fluorescence to determine whether the food coloring was transferred to the sample. To account for naturally occurring fluorescent compounds in the environment, oral fluid samples were also collected one day prior to treatment. Two routes of exposure were evaluated: direct (pig) and indirect (pen floor), in 16 pens for each pen size category. Direct treatment consisted of spraying the solution (~3.5 mL) into the oral cavity of individual restrained pigs (10% of the pigs in the pen). Indirect administration consisted of indirectly exposing the pigs by placing the solution (20 mL) on the floor in the center of the pen. Oral fluid samples were tested for fluorescence using a SpectraMax® i3x Multi-Mode Microplate Reader (Molecular Devices LLC, San Jose, CA, USA).

2.2. Commercial Production Sites

The study was conducted in 3 commercial wean-to-finish production sites (Midwest USA) with curtain-sided, tunnel-ventilated barns built between 2010 and 2018. Pens were fully slatted and equipped with nipple drinkers and standard dry feeders. Barns used in the study were designed to house pigs in pens of ~30, ~60, or ~125 pigs. The study was conducted in 16 pens in each pen size group for each treatment, with pens selected randomly (random.org accessed on 15 May 2023) from all available pens, excluding hospital and recovery pens.

2.3. Oral Fluid Collection and Handling

Oral fluid samples were collected using 100% cotton rope (1.3 cm 3-strand twisted 100% cotton rope, Skydog Rigging Equipment, Lake in the Hills, IL, USA). The rope was placed on the walkway side of the pen ~100 cm from one corner of the pen with the end of the rope suspended at the pigs' shoulder height. Ropes were hung for 30 min in pens of 30 pigs, 45 min in pens of 60 pigs, and 60 min in pens of 125 pigs. At the end of the collection period, the oral fluid sample was collected by placing the rope in a plastic bag and hand-squeezing to release the fluid. The oral fluid that accumulated in the bottom of the plastic bag was then poured into a 50 mL tube (Southern Labware®, Cumming, GA, USA) and transported on dry ice to the laboratory. In the laboratory, oral fluid samples were centrifuged at 3300× g for 90 min (J6-HC, Beckman Coulter, Inc., 5350 Lakeview Parkway S drive Indianapolis, IN, USA) at 22 °C, after which the supernatant was transferred to 50 mL tubes (Southern Labware®, Cumming, GA, USA) and stored at −20 °C until processed for testing.

2.4. Treatments

2.4.1. Florescence Control Samples

A pretreatment oral fluid sample (background control) was collected from each pen to account for the presence of naturally fluorescing substances in the pen environment, e.g., urine, feces, or feed mycotoxins [13–15]. Before collecting the background control oral fluid sample, a mock indirect treatment solution (20 mL) consisting of 50 g of granulated white sugar in 100 mL of distilled water was placed in the approximate center of the pen.

2.4.2. Direct (Pig) and Indirect (Pen Floor) Treatments

Treatments were performed at the pen level, with each of the 2 treatments randomly assigned to 16 pens (random.org accessed on 15 May 2023) in each pen size category. For both direct and indirect treatment, a stock solution containing a fluorescent food coloring was prepared by mixing 50 g of granulated white sugar in 100 mL of 60 °C distilled deionized water and, after cooling to 20 °C, adding commercial red food coloring (McCormick & Co., Inc., Hunt Valley, MD, USA) in a proportion of 2 parts sugar solution to 1 part food coloring. For direct treatment, a convenience sample of individual pigs (10% of the pen) were restrained, marked for individual identification (Ideal® Prima® Spray-On, Neogen Corp., Lansing, MI, USA), and ~3.5 mL of the stock solution sprayed (Dual Action Spray Bottle, Uline 12575 Uline Drive, Pleasant Prairie, WI, USA) into the buccal cavity.

For indirect treatment, 20 mL of the stock solution was poured directly on the floor in the approximate center of the pen to allow pigs free access.

In pens receiving direct treatment, the entire sampling process was recorded using video cameras (GoPro® 7 cameras, GoPro Inc., San Mateo, CA, USA) strategically placed to clearly record pig interactions with the rope. "Pig–rope contact" was defined as an individually identified pig clearly taking the rope into its mouth. Subsequently, pig–rope contacts were counted at one-minute intervals over the sampling period, i.e., any pig–rope contact during a one-minute interval was counted as "one". That is, 3 pigs exposed in pens of 30 pigs had a maximum of 90 possible contacts (3 pigs × 30 min). Thus, in pens of 60 and 125 pigs, the maximum possible contacts were 270 (6 pigs × 45 min) and 720 (12 pigs × 60 min), respectively. Pig participation at the pen level was defined as the number of marked pigs that had one or more pig–rope contacts during the sampling time. Maximum values for pig participation were 3, 6, and 12 in the direct treatments for pen size categories of 30, 60, and 125, respectively.

2.4.3. Relative Fluorescence Units (RFUs)

Pen-based oral fluid samples were tested for fluorescence using the SpectraMax® i3x Multi-Mode Microplate Reader (Molecular Devices LLC) and control software (SoftMax® Pro 6.5 Molecular Devices LLC), with the response reported in relative fluorescence units (RFUs). Initially, the optimized excitation and emission wavelengths of the red food coloring were established as 530 nm and 570 nm, respectively, using the control software spectral optimization function. For testing, oral fluid samples were thawed overnight at 4 °C and centrifuged at 3300× *g* for 10 min (J6-HC, Beckman Coulter, Inc., Indianapolis, IN, USA) at 22 °C. Thereafter, the supernatant was transferred to a 50 mL tube, vortexed, and tested in duplicate by pipetting 100 μL into one well in each of two 96-well black-walled and clear bottom plates (Thermo Fisher Scientific™ Nunc, Waltham, MA, USA). Concurrently, a standard curve based on the stock solution (Section 2.4.2) was run in duplicate on each plate (blank, 1:100 and then 2-fold dilutions through 1:102,400). To account for background and for the effect of the oral fluid matrix on fluorescence readings, the diluent for the standard curve dilutions was created by pooling 0.5 mL of the pretreatment oral fluid samples collected from each pen (n = 32) in a pen size category. Thus, 3 separate and specific standard curves were created, i.e., one for each pen size category.

2.5. Data Analysis

All analyses were performed using R software v4.3.1 [16] and RStudio v2023.03.0. Prior to analyzing the data, the mean RFU of each pen-based oral fluid sample (Section 2.4) was calculated from technical duplicates and log transformed (lnRFU). Mean lnRFU and 95% confidence intervals were calculated for each treatment and pen size combination using the summarise() function of the R dplyr() package v1.1.3 [17]. Thereafter, the lnRFU threshold value that optimally differentiated between pre- and post-treatment oral fluid sample lnRFU responses (Youden's J statistic) was determined by receiver operating characteristic (ROC) analysis using the pROC package v1.17.0.1 [18]. Six ROC analyses were performed, i.e., one for each combination of pen size (n = 3) and treatment (n = 2).

To maximize the specificity of the estimates, pen samples (n = 7) with pretreatment background lnRFU responses above the lnRFU threshold (plus a 0.001 tolerance) were excluded from all subsequent analyses. For the remaining eligible samples (n = 89), post-treatment mean lnRFUs and 95% confidence intervals were recalculated for each combination of treatment and pen size. Pen samples (n = 43) with post-treatment lnRFU values greater than the threshold (plus a 0.001 threshold) were retained for the remainder of the analysis. Thereafter, lnRFU responses were calculated as a percent (%) of the exposure stock solution (Section 2.4.3) based on a linear calibration using the standard curve run on each plate and the inverse predict function of R chemCal() package v0.2.3 [19]. Pen size category mean lnRFU values for the direct and indirect treatment groups were compared by analysis of variance (ANOVA) followed by pairwise comparisons using R-base functions aov() and TukeyHSD() when the overall F-statistic was significant. This analysis was then repeated for the mean concentration of red food coloring (percent stock solution). Post hoc tests *p*-values were adjusted using Tukey's method to control a familywise error rate of 0.05. The relationship between the total number of pig–rope contacts (count data derived from the video recordings) and post-treatment oral fluid sample lnRFU responses was evaluated using a linear regression fit using the geom_smooth function with method = lm in ggplot2 [20]. The relationship between pen-level pig participation and post-treatment oral fluid sample lnRFU responses was also evaluated using a linear fit.

3. Results

The pre- and post-treatment lnRFU responses (mean, 95% CI) and ROC response thresholds by pen size and treatment route are presented in a stepwise fashion in Table 1 (Steps 1, 2).

Table 1. Detection of fluorescence in pen-based oral fluid samples reported as log-transformed relative fluorescence units (ln(RFU)) following exposure to food coloring by route (direct, indirect) and pen size (30, 60, or 125 pigs per pen).

		Step-Wise Analysis	Pen Size (Sampling Time) 30 Pigs per Pen (30 min)	60 Pigs per Pen (45 min)	125 Pigs per Pen (60 min)
Direct treatment [2]	1.	Pre-treatment lnRFU means based on 16 pens per pen size (95% CI)	11.86 [1] (11.78, 11.93)	11.87 [1] (11.77, 11.96)	11.78 [1] (11.73, 11.82)
	2.	Post-treatment lnRFU means based on 16 pens per pen size (95% CI)	12.17 (12.05, 12.30)	12.30 (12.21, 12.38)	13.36 (13.19, 13.52)
	3.	ROC threshold (lnRFU) [3]	12.06	12.10	12.45
	a.	Pre-treatment pen samples > ROC threshold [4]	1 of 16 pens	0 of 16 pens	0 of 16 pens
	b.	Post-treatment pen samples > ROC threshold (lnRFU mean, 95% CI) [5]	11 of 15 pens 12.28 [a] (12.16, 12.40)	16 of 16 pens 12.30 [a] (12.21, 12.38)	16 of 16 pens 13.36 [b] (13.19, 13.52)
Indirect treatment [2]	1.	Pre-treatment means based on 16 pens for each pen size (95% CI)	11.83 (11.76, 11.90)	11.78 (11.67, 11.89)	11.76 (11.67, 11.84)
	2.	Post-treatment means based on 16 pens for each pen size (95% CI)	12.13 (12.02, 12.24)	12.17 (12.08, 12.25)	12.04 (11.94, 12.14)
	3.	ROC threshold (lnRFU) [3]	11.96	11.99	11.88
	a.	Pre-treatment pen samples > ROC threshold [4]	2 of 16 pens	2 of 16 pens	2 of 16 pens
	b.	Post-treatment pen samples > ROC threshold [5] (lnRFU mean, 95% CI)	12 of 14 pens 12.16 [a] (12.06, 12.26)	13 of 14 pens 12.18 [a] (12.12, 12.25)	10 of 14 pens 12.11 [a] (12.04, 12.17)

[1] Fluorescence in pen-based oral fluid samples quantified using a multi-mode microplate reader (SpectraMax® i3x Multi-Mode Microplate Reader, Molecular Devices LLC, San Jose, CA, USA) and control software (SoftMax® Pro 6.5 Molecular Devices LLC), read as relative fluorescent units, and then log transformed (ln(RFU)) for analysis and reporting. [2] Direct treatment: 10% of pigs in pen individually restrained and stock solution (~3.5 mL) sprayed into buccal cavity. Indirect treatment: stock solution (20 mL) poured on floor in center of pen. [3] Threshold derived from receiver operating characteristics analysis (R package pROC v1.17.0.1 [18]) of pre- and post-treatment ln(RFU) results. A separate ROC was performed for each combination of (treatment × pen size). [4] Pretreatment oral fluid samples collected from each pen (16 per pen size category) one day prior to treatment. Pens with pretreatment samples > ROC threshold were disqualified from further analysis on the basis of excess background fluorescence. [5] Post-treatment oral fluid samples collected immediately after treatment. Estimates in Step 3b do not include pens eliminated in Step 3a. [a,b] Tukey's test $p < 0.001$.

The ROC analyses produced six independent lnRFU threshold values by pen size category and route of treatment (Table 1, Step 3). The pretreatment pen sample lnRFU exceeded the ROC threshold in 7 (7.35%) of the 96 pens in the study. Therefore, these pens were excluded from subsequent analyses due to excessive background fluorescence (Table 1, Step 3a).

Among the remaining 89 pens, ROC analysis of post-treatment lnRFU data showed that food coloring was detected in 43 of 47 (91.5%) post-treatment samples from pens receiving direct treatment and 35 of 42 (83.3%) pens receiving indirect treatment (Table 1, Step 3b). Among the 43 direct treatment pens in which food coloring was detected, ANOVA of the post-treatment lnRFU data showed a difference in response ($p < 0.001$) by pen size. Post hoc pairwise comparisons between direct treatment groups detected no difference in mean lnRFU in samples from pens of 30 vs. 60 pigs (Tukey's test, $p = 0.986$), but comparisons of 30 vs. 125 pigs (Tukey's test $p < 0.001$) and 60 vs. 125 pigs (Tukey's test $p < 0.001$) detected significant differences. In contrast, no difference in response was detected among pen sizes for pens receiving indirect treatment (ANOVA, $p = 0.171$).

Among the 78 pen samples reported in Table 1, Step 3b, the concentration of red food coloring was calculated as a percent (%) of the stock solution based on an analysis of the standard curves run on each plate. In pens receiving direct treatment, the mean concentrations (95% CI) were 0.0047% (0.0034, 0.0060), 0.0049% (0.0038, 0.0060), and 0.046% (0.031, 0.062) for pens sizes of 30, 60, and 125 pigs, respectively. In pens receiving indirect treatment, the mean concentrations (95% CI) were 0.0036% (0.0028, 0.0045), 0.0037% (0.0032, 0.0042), and 0.0033% (0.0029, 0.0038) in pen size categories of 30, 60, and 125 pigs, respectively. Among pens receiving direct treatment, a significant difference in the concentration of red food coloring was found among pen sizes (ANOVA, $p < 0.001$). Post hoc pairwise comparisons detected no difference in pens of 30 vs. 60 pigs (Tukey's test, $p = 0.999$), but a significant difference was found in pens of 30 vs. 125 pigs (Tukey's test, $p < 0.001$) and 60 vs. 125 pigs (Tukey's test, $p < 0.001$). No difference in the concentration of red food coloring was detected among pen sizes for pens receiving indirect treatment (ANOVA, $p = 0.628$).

Among the 43 direct treatment pens (Table 1, Step 3b), pen-level pig participation (number of marked pigs contacting the rope at least once in the pen) and pen total contacts (number of one-minute intervals in which any marked pig in the pen contacted with a rope) are presented in Figure 1A,B. By pen size, mean pen-level pig participation was 2.8 of 3 pigs (2.6, 3.0), 5.4 of 6 pigs (5.0, 5.9), and 8.8 of 12 (7.9, 9.7) for pens of 30, 60, and 125 pigs, respectively. The mean number of pig–rope contacts was 35.9 (95% CI 25.1, 46.7) out of a possible total of 90 for pens of 30, 60.8 (52.1, 69.6) out of a possible total of 270 for pens of 60, and 74.1 (62.4, 85.8) out of a possible total of 720 for pens of 125 pigs. Linear regression analysis (lm() function of base R) showed that the number of contact events had a significant linear relationship to a pen's RFU response for pens of 30 pigs, i.e., on average, each contact event produced an increase of 0.0096 in lnRFU. The slopes were not significant for pens of 60 ($p = 0.154$) or 125 ($p = 0.226$). Slopes were not significant for pen-level pig participation for any of the pen sizes ($p > 0.257$).

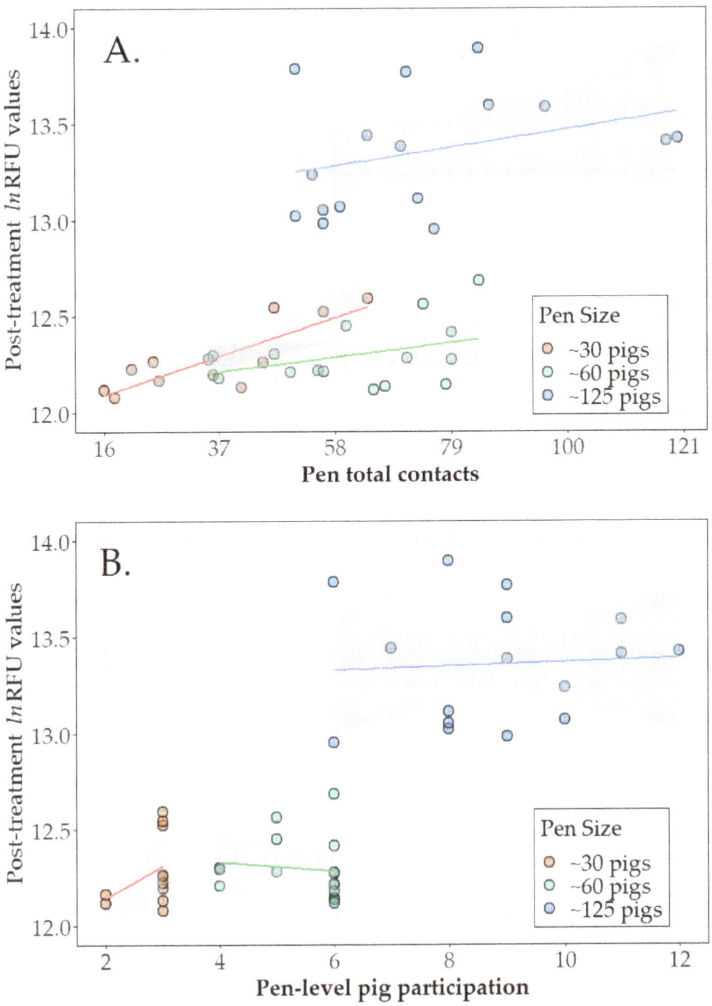

Figure 1. Direct treatment results. Scatterplots of post-treatment relative fluorescent units (ln(RFU)) in pen-based oral fluids versus (**A**) number of pig–rope contacts by individually marked pigs and (**B**) number of marked pigs that participated with fitted linear regression lines and 95% confidence intervals. Figure 1 is based on 43 pens receiving direct treatment and classified as positive in the ROC analyses (Table 1 Step 3). Note that 10% of pigs per pen were individually treated and marked, i.e., 3, 6, and 12 pigs in pens of ~30, ~60, and ~125 pigs, respectively. Oral fluid sampling time varied by pen size, i.e., 30 min in pens of ~30 pigs, 45 min in pens of ~60 pigs, and 60 min in pens of ~125 pigs.

4. Discussion

In humans and animals, hormones, proteins, antibodies, viruses, and other bioelements in the circulatory system enter the buccal cavity via the salivary glands, mucus membranes, and gingival crevices by ultrafiltration, transudation, passive diffusion, and/or active transport [21]. This occurs as a highly dynamic process. For example, Brill and Krasse [22] demonstrated the transfer of fluorescein dye from the circulatory system into gingival crevicular fluid within 30 s of intravenous injection in dogs. Similar results were observed in humans who consumed capsules containing fluorescein dye [23] and humans intravenously injected with albumin labeled with iodine-131 [24]. In the context of infectious diseases, Madonia et al. [25] demonstrated the transfer of Coxsackie B-1 virus from the circulatory system into oral fluids by injecting the virus into a marginal ear vein of rabbits and detecting the virus in oral fluids within two minutes. In human medicine, human immunodeficiency virus (HIV) [26] and HIV antibody [27] were reported in oral fluids collected from patients with immunodeficiency syndrome (AIDS) or AIDS-related complex (ARC). Moving quickly, in 1995, the FDA approved an HIV antibody testing system based on oral fluids rather than blood [28].

Ongoing research has produced a number of practical and widely used applications for oral fluids, e.g., assessment of drug use [29] and oral fluid-based testing of human populations for HIV [30], measles [31], hepatitis A [32], severe acute respiratory syndrome coronavirus 2 (SARS-CoV-2) testing [33], and others. In pigs, the first diagnostic application of oral fluids was reported by Prickett et al. in 2008 [34]. Specifically, PRRSV RNA and PRRSV antibodies were detected in oral fluids collected from pigs inoculated with the virus under experimental conditions. Since then, an additional 16 bacterial and 25 viral pathogens of swine have been detected in oral fluids under experimental and field conditions [35]. Because the approach is both convenient and effective, oral-fluid-based testing is widely used by swine producers and veterinarians for a variety of pathogens [36]. In 2022, 460,150 PCR assays were performed on oral fluids for seven major pathogens of swine (IAV, *Mycoplasma hyopneumoniae*, PCV2, PDCoV, PEDV, PRRSV, and TGEV) at five diagnostic laboratories (Iowa, Kansas, Ohio, Minnesota, and South Dakota) accredited by the American Association of Veterinary Laboratory Diagnosticians [37].

While the detection of nucleic acids in oral fluids may reflect a pathogen's systemic distribution, it may also represent local replication in oropharyngeal tissues. As reviewed by Lilja et al. [38], in humans, bacterial and viral pathogens that replicate in tonsils include Epstein–Barr virus, adenovirus, influenza virus, parainfluenza virus, coxsackie A virus, herpes virus, rhino virus, coronavirus, *Streptococcus pyogenes*, *Neisseria gonorrhoeae*, *Corynebacterium diphtheriae*, and *Borrelia vincentii*. Similarly, in pigs, a variety of bacterial and viral pathogens replicate in and/or colonize tonsils including *Salmonella typhimurium*, *Salmonella choleraesuis*, *Streptococcus suis*, African swine fever virus, classical swine fever virus, PCV2, PRRSV, foot-and-mouth disease virus, and others [39]. As reviewed by Munguía-Ramírez et al. [35], all of these pathogens have been detected by PCR in swine oral fluids.

Thus, diagnostic targets performed on oral fluid may represent systemic infection or replication in buccal tissues. In swine and other animals, diagnostic targets in oral fluid samples may also reflect their presence in the environment [40]. Normal swine behavior includes rooting, biting, smelling, and tasting the environment. Diagnostic targets encountered in the environment during this exploratory process may be retained in the oral cavity, deposited in the oral fluid sample, and detected by testing [3]. The diagnostic relevance of this process was first illustrated by Johnson et al. [41] who showed that the source of PRRSV antibody detected in oral fluid samples from PRRSV negative pigs was porcine dry-sprayed plasma in the feed provided to the animals. The consistency and repeatability of the transfer of targets from the environment into oral fluid samples is exemplified by the routine detection of enteric pathogens in oral fluids, e.g., *Brachyspira* ssp. [42], *Lawsonia intracellularis* [42], PDCoV [43], PEDV [44,45], and TGEV [46].

Despite the strength of the empirical evidence, the detection of environmental targets in oral fluid samples has not been proven under controlled conditions. To address this deficiency, we documented the transfer of a tracer directly from pigs or indirectly from the pigs' environment into pen-based oral fluid samples in active swine farms over a range of pen sizes. To address the ethical and legal risks posed by the use of infectious agents in commercial farms, and yet precisely define exposure conditions in the field, an inert tracer was used. Broadly speaking, the use of chemical tracers in field research is well-documented [47]. Examples in veterinary medicine include the use of tetracycline to document oral rabies vaccine uptake in racoons [48] and Rhodamine B to document classical swine fever virus oral vaccine uptake in wild boar [9,49]. In this study, the requirements for the tracer included safety, availability, low cost, ease of detection, and suitability for use in food animals. None of the previously described tracers meet these qualifications. In contrast, the food colorings used in this study (Red No. 3 and Red No. 40) are approved by the U.S. FDA for use in foods, drugs, and cosmetics and are widely available in food markets at a reasonable cost. Further, under proper storage conditions, i.e., cool, dry, and shielded from light, the manufacturer ascribes a shelf life of 1440 days [50].

This study made use of the fact that food colorings fluoresce at specific wavelengths that may be detected with standard laboratory equipment [12]. The optimum fluorescence for the red food coloring used in this study was achieved at 530 nm and 570 nm for excitation and emission, respectively. The possible presence of naturally fluorescent compounds in the livestock environment (background fluorescence), e.g., urine, mycotoxins, and feces [13–15], was accounted for by including pretreatment oral fluid sample RFU responses in the ROC analysis. In the experimental design, the transfer of a target from pigs into oral fluids was observed from direct exposure of 10% of pigs in the pen or from indirect exposure of 20 mL tracer solution on the floor of the pen. It is worth noting that when the amount of dye introduced into the environment was held constant, there was no significant difference in the mean lnRFU or concentration as percent of the stock solution for oral fluids samples collected across different pens sizes. This suggests that, as pen size and pig numbers increase, diagnostic targets in the environment are nevertheless consistently deposited in oral fluid samples. As expected, when the amount of dye introduced into pens was increased by direct exposure of a higher number of animals, the mean lnRFU in oral fluid samples increased. This supports the idea that higher burdens of biologic targets are likely to increase the concentration, and therefore likelihood of detection, of the target in oral fluid samples. Variations in the experimental design using food coloring (or other safe tracers) could be used to address other questions regarding behavioral and mechanical aspects of swine oral fluid sampling, e.g., holding sampling time and pen size constant and changing number of pigs exposed. Regardless, the present study demonstrated the repeatability of the transfer of targets into oral fluid samples and described a new tool for further epidemiological studies.

5. Conclusions

The objective of this study was to evaluate the repeatability of the transfer of a target directly from pigs or indirectly from the environment into pen-based oral fluid samples over a range of pen sizes in commercial swine farms. Red food coloring was used as the target because of its safety, availability, and detectability (fluorescence) profile. Direct exposure of ~10% of the pigs in a pen resulted in detection of the fluorescent marker in 73.3% (11 of 15) of the pen-based oral fluid samples from pens of ~30 pigs and 100% of samples from pens of ~60 and ~125 pigs (16 of 16 pens for each). For environmental targets, the marker was detected in 85.7% (12 of 14 pens), 92.9% (13 of 14 pens), and 71.4% (10 of 14 pens) of oral fluid samples from pens holding ~30, ~60, and ~125 pigs, respectively. Thus, targets to which the pigs have access are routinely transferred into oral fluid samples. Although the surrogate target used in this study was intentionally administered to pigs or placed in the environment, the results provided a rational explanation for the routine detection of swine enteric pathogens, i.e., pathogens not normally in the upper respiratory tract or oropharynx, in pen-based oral fluid samples. In conclusion, on commercial swine farms, the use of red food coloring provided a safe, inexpensive, specific, and analytically sensitive method for documenting the transfer of the target directly from pigs or indirectly from the environment into pen-based oral fluid samples.

Author Contributions: Conceptualization, G.T., L.G.G.-L. and J.J.Z.; investigation, G.T. and J.J.Z.; methodology, G.T., M.D.R., K.C.O., M.L.R., P.Z., L.G.G.-L. and J.J.Z.; analysis, M.D.R., M.K.T. and D.Z.; writing, G.T., M.L.R., K.C.O., D.Z., P.Z. and J.J.Z., visualization, G.T., M.D.R., K.C.O., D.Z., M.K.T. and J.J.Z.; funding acquisition, G.T., M.D.R., K.C.O., M.L.R., P.Z. and J.J.Z. All authors have read and agreed to the published version of the manuscript.

Funding: This research was funded by United States Department of Agriculture, Animal and Plant Health Inspection Service, award identification number AP22VSSP0000C047. This work was funded by a cooperative agreement with the USDA Center for Epidemiology and Animal Health. The findings and conclusions reported are those of the author(s) and should not be construed to represent any official USDA or U.S. Government determination or policy.

Institutional Review Board Statement: The animal study protocol was approved by the Institutional Animal Care and Use Committee of Iowa State University (IACUC-22-187, 28 September 2023).

Informed Consent Statement: Informed consent was obtained from commercial farm owner involved in the study.

Data Availability Statement: The data that support the findings of this study are available on request from the corresponding author.

Acknowledgments: The authors thank J. Connor, T. Bauman, and P. Thomas for providing access to production facilities.

Conflicts of Interest: The authors declare no conflict of interest.

References

1. Henao-Diaz, A.; Giménez-Lirola, L.; Baum, D.H.; Zimmerman, J. Guidelines for oral fluid-based surveillance of viral pathogens in swine. *Porc. Health Manag.* **2020**, *6*, 28. [CrossRef] [PubMed]
2. White, D.; Rotolo, M.; Olsen, C.; Wang, C.; Prickett, P.; Kittawornrat, A.; Panyasing, Y.; Main, R.; Rademacher, C.; Hoogland, M.; et al. Recommendations for pen-based oral-fluid collection in growing pigs. *J. Swine Health Prod.* **2014**, *22*, 138–141.
3. Kittawornrat, A.; Zimmerman, J.J. Toward a better understanding of pig behavior and pig welfare. *Anim. Health Res. Rev.* **2010**, *12*, 25–32. [CrossRef] [PubMed]
4. Tarasiuk, G.; Remmenga, M.; O'Hara, K.; Rotolo, M.; Zaabel, P.; Zimmerman, J. Effect of pen size and number of ropes on behaviors associated with oral fluid sampling. In Proceedings of the American Association of Swine Veterinarians Annual Meeting, Aurora, CO, USA, 4–7 March 2023. [CrossRef]
5. Algeo, T.P.; Norhenberg, G.; Hale, R.; Montoney, A.; Chipman, R.B.; Slate, D. Oral rabies vaccination variation in tetracycline biomarking among Ohio raccoons. *J. Wildl. Dis.* **2013**, *49*, 332–337. [CrossRef] [PubMed]
6. Skinner, H.C.; Nalbandian, J. Tetracyclines and mineralized tissues: Review and perspectives. *Yale J. Biol. Med.* **1975**, *48*, 377–397.
7. Evans, J.; Griffith, R.E., Jr. A fluorescent tracer and marker for animal studies. *J. Wildl. Manag.* **1973**, *37*, 73–81. [CrossRef]
8. Phillips, G.E.; Goldade, D.A.; VerCauteren, K.C.; Ott, T.L.; Wagner, D.C. Iophenoxic acid and Rhodamine B as biomarkers of bovine tuberculosis vaccine bait uptake by white-tailed deer. In Proceedings of the Vertebrate Pest Conference, Waikoloa, HI, USA, 5 March 2014. [CrossRef]
9. Baruzzi, C.; Coats, J.; Callaby, R.; Cowan, D.P.; Massei, G. Rhodamine B as a long-term semi-quantitative bait marker for wild boar. *Wildl. Soc. Bull.* **2017**, *41*, 271–277. [CrossRef]
10. Jacob, J.; Jones, D.A.; Singleton, G.R. Retention of the bait marker Rhodamine B in wild house mice. *Wildl. Res.* **2002**, *29*, 159–164. [CrossRef]
11. Fry, T.L.; Atwood, T.; Dunbar, M.R. Evaluation of rhodamine B as a biomarker for raccoons. *Hum. Wildl. Interact.* **2010**, *4*, 275–282. Available online: https://www.jstor.org/stable/24868847 (accessed on 12 September 2023).
12. Chen, G.Q.; Wu, Y.M.; Wang, J.; Zhu, T.; Gao, S.M. Fluorescence spectroscopy study of synthetic food colors. *Guang Pu Xue Yu Guang Pu Fen Xi* **2009**, *29*, 2518–2522. (In Chinese)
13. Leiner, M.J.; Hubmann, M.R.; Wolfbeis, O.S. The total fluorescence of human urine. *Anal. Chim. Acta* **1987**, *198*, 13–23. [CrossRef]
14. Kim, M.S.; Lefcourt, A.M.; Chen, Y.R. Optimal fluorescence excitation and emission bands for detection of fecal contamination. *J. Food Prot.* **2003**, *66*, 1198–1207. [CrossRef]
15. Hruska, Z.; Yao, H.; Kincaid, R.; Brown, R.; Cleveland, T.; Bhatnagar, D. Fluorescence excitation–emission features of aflatoxin and related secondary metabolites and their application for rapid detection of mycotoxins. *Food Bioprocess Techol.* **2014**, *7*, 1195–1201. [CrossRef]
16. R Core Team. *R: A Language and Environment for Statistical Computing*; R Foundation for Statistical Computing: Vienna, Austria, 2023.
17. Wickham, H.; François, R.; Henry, L.; Müller, K.; Vaughan, D. dplyr: A Grammar of Data Manipulation. R Package Version 1.1.3. 2023. Available online: https://CRAN.R-project.org/package=dplyr (accessed on 15 September 2023).
18. Robin, X.; Turck, N.; Hainard, A.; Tiberti, N.; Lisacek, F.; Sanchez, J.; Müller, M. pROC: An open-source package for R and S+ to analyze and compare ROC curves. *BMC Bioinform.* **2011**, *12*, 77. [CrossRef] [PubMed]
19. Ranke, J. chemCal: Calibration Functions for Analytical Chemistry. R Package Version 0.2.3. 2022. Available online: https://CRAN.R-project.org/package=chemCal (accessed on 20 November 2023).
20. Wickham, H. *ggplot2: Elegant Graphics for Data Analysis*; Springer: New York, NY, USA, 2016.

21. Bellagambi, F.G.; Lomonaco, T.; Salvo, P.; Vivaldi, F.; Hangouët, M.; Ghimenti, S.; Biagini, D.; Di Francesco, F.; Fuoco, R.; Errachid, A. Saliva sampling: Methods and devices. An overview. *TrAC Trends Anal. Chem.* **2020**, *124*, 115781. [CrossRef]
22. Brill, N.; Krasse, B.O. The passage of tissue fluid into the clinically healthy gingival pocket. *Acta Odontol. Scand.* **1958**, *16*, 233–245. [CrossRef]
23. Brill, N.; Björn, H. Passage of tissue fluid into human gingival pockets. *Acta Odontol. Scand.* **1959**, *17*, 11–21. [CrossRef]
24. Schein, A.H.; Tung, F. Appearance of parenterally administered proteins in saliva. *Nature* **1962**, *196*, 1092–1093. [CrossRef]
25. Madonia, J.V.; Bahn, A.N.; Calandra, J.C. Salivary excretion of Coxsackie b-1 virus in rabbits. *Appl. Microbiol.* **1966**, *14*, 394–396. [CrossRef] [PubMed]
26. Groopman, J.; Salahuddin, S.; Sarngadharan, M.; Markham, P.; Gonda, M.; Sliski, A.; Gallo, R. HTLV-III in saliva of people with AIDS-related complex and healthy homosexual men at risk for AIDS. *Science* **1984**, *226*, 447–449. [CrossRef] [PubMed]
27. Archibald, D.W.; Zon, L.; Groopman, J.E.; McLane, M.F.; Essex, M. Antibodies to human T-lymphotropic virus type III (HTLV-III) in saliva of acquired immunodeficiency syndrome (AIDS) patients and in persons at risk for AIDS. *Blood* **1986**, *67*, 831–834. [CrossRef]
28. Pugliese, G. FDA Approves Oral HIV Test. *Infect. Control Hosp. Epidemiol.* **1995**, *16*, 186. [CrossRef]
29. Yeh, C.K.; Christodoulides, N.J.; Floriano, P.N.; Miller, C.S.; Ebersole, J.L.; Weigum, S.E.; McDevitt, J.; Redding, S.W. Current development of saliva/oral fluid-based diagnostics. *Tex. Dent. J.* **2010**, *127*, 651–661.
30. Bird, A.G.; Gore, S.M.; Jolliffe, D.W.; Burns, S.M. Anonymous HIV surveillance in Saughton prison, Edinburgh. *Aids* **1992**, *6*, 725–734. [CrossRef]
31. Ramsay, M.; Brugha, R.; Brown, D. Surveillance of measles in England and Wales: Implications of a national saliva testing programme. *Bull. World Health Organ.* **1997**, *75*, 515–521. [PubMed]
32. Jacobson, S.K.; Buttery, R.; Parry, J.V.; Perry, K.R.; Wreghitt, T.G. Investigation of a hepatitis A outbreak in a primary school by sequential saliva sampling. *Clin. Diagn. Vir.* **1995**, *3*, 173–180. [CrossRef] [PubMed]
33. Tsang, N.N.Y.; So, H.C.; Ng, K.Y.; Cowling, B.J.; Leung, G.M.; Ip, D.K.M. Diagnostic performance of different sampling approaches for SARS-CoV-2 RT-PCR testing: A systematic review and meta-analysis. *Lancet Infect. Dis.* **2021**, *21*, 1233–1245. [CrossRef] [PubMed]
34. Prickett, J.; Simer, R.; Christopher-Hennings, J.; Yoon, K.J.; Evans, R.B.; Zimmerman, J. Detection of porcine reproductive and respiratory syndrome virus infection in porcine oral fluid samples: A longitudinal study under experimental conditions. *J. Vet. Diagn. Investig.* **2008**, *20*, 156–163. [CrossRef] [PubMed]
35. Munguía-Ramírez, B.; Armenta-Leyva, B.; Giménez-Lirola, L.; Wang, C.; Zimmerman, J.J. Surveillance on swine farms using antemortem specimens. In *Optimizing Pig Herd Health and Production*; Maes, D., Segalés, J., Eds.; Burleigh Dodds Science Publishing: Cambridge, UK, 2023; pp. 97–138. [CrossRef]
36. Bjustrom-Kraft, J.; Christopher-Hennings, J.; Daly, R.; Main, R.; Torrison, J.; Thurn, M.; Zimmerman, J. The use of oral fluid diagnostics in swine medicine. *J. Swine Health Prod.* **2018**, *26*, 262–269. [CrossRef]
37. Trevisan, G.; (Iowa State University, Ames, IA, USA). Personal communication, 2023.
38. Lilja, M.; Räisänen, S.; Stenfors, L.E. Initial events in the pathogenesis of acute tonsillitis caused by Streptococcus pyogenes. *Int. J. Pediatr. Otorhinolaryngol.* **1998**, *45*, 15–20. [CrossRef]
39. Horter, D.C.; Yoon, K.J.; Zimmerman, J.J. A review of porcine tonsils in immunity and disease. *Anim. Health Res. Rev.* **2003**, *4*, 143–155. [CrossRef] [PubMed]
40. Prickett, J.R.; Zimmerman, J.J. The development of oral fluid-based diagnostics and applications in veterinary medicine. *Anim. Health Res. Rev.* **2010**, *11*, 207–216. [CrossRef] [PubMed]
41. Johnson, J.; Main, R.; Zimmerman, J. Exogenous source of PRRSV antibody in positive oral-fluid ELISA results. *J. Swine Health Prod.* **2012**, *20*, 215.
42. Frana, T.; Warneke, H.; Stensland, W.; Kinyon, J.; Bower, L.; Burrough, E. Comparative detection of Lawsonia intracellularis, Salmonella, and Brachyspira from oral fluids and feces. In Proceedings of the American Association of Swine Veterinarians Annual Meeting, Dallas, TX, USA, 1–4 March 2014; pp. 67–69.
43. Zhang, J. Porcine deltacoronavirus: Overview of infection dynamics, diagnostic methods, prevalence and genetic evolution. *Virus Res.* **2016**, *226*, 71–84. [CrossRef] [PubMed]
44. Bower, L.; Madson, D.; Hoang, H.; Sun, D.; Giménez-Lirola, L.; Magstadst, D.; Arruda, P.; Wilberts, B.; Yoon, K. Utility of oral fluid sampling and testing for monitoring PEDV in herds. In Proceedings of the American Association of Swine Veterinarians Annual Meeting, Dallas, TX, USA, 1–4 March 2014; pp. 61–62.
45. Bjustrom-Kraft, J.; Woodard, K.; Giménez-Lirola, L.; Rotolo, M.; Wang, C.; Sun, Y.; Lasley, P.; Zhang, J.; Baum, D.; Gauger, P.; et al. Porcine epidemic diarrhea virus (PEDV) detection and antibody response in commercial growing pigs. *BMC Vet. Res.* **2016**, *12*, 99. [CrossRef]
46. Magtoto, R.; Poonsuk, K.; Baum, D.; Zhang, J.; Chen, Q.; Ji, J.; Piñeyro, P.; Zimmerman, J.; Giménez-Lirola, L.G. Evaluation of the serologic cross-reactivity between transmissible gastroenteritis coronavirus and porcine respiratory coronavirus using commercial blocking enzyme-linked immunosorbent assay kits. *mSphere* **2019**, *4*, 10–1128. [CrossRef]
47. Savarie, P.J.; Johns, B.E.; Gaddis, S.E. A review of chemical and particle marking agents used for studying vertebrate pests. In Proceedings of the Vertebrate Pest Conference, Newport Beach, CA, USA, 3–5 March 1992; Volume 15. Available online: https://escholarship.org/uc/item/90q7x14m (accessed on 10 October 2023).

48. Olson, C.A.; Mitchell, K.D.; Werner, P.A. Bait ingestion by free-ranging raccoons and nontarget species in an oral rabies vaccine field trial in Florida. *J. Wildl. Diseas.* **2000**, *36*, 734–743. [CrossRef]
49. Rossi, S.; Staubach, C.; Blome, S.; Guberti, V.; Thulke, H.H.; Vos, A.; Koenen, F.; Le Potier, M.F. Controlling of CSFV in European wild boar using oral vaccination: A review. *Front. Microbiol.* **2015**, *6*, 1141. [CrossRef]
50. McCormic for Chefs. Available online: https://www.mccormickforchefs.com/en-us/products/mccormick-culinary/red-food-color (accessed on 26 December 2023).

Disclaimer/Publisher's Note: The statements, opinions and data contained in all publications are solely those of the individual author(s) and contributor(s) and not of MDPI and/or the editor(s). MDPI and/or the editor(s) disclaim responsibility for any injury to people or property resulting from any ideas, methods, instructions or products referred to in the content.

animals

Article

Active Participatory Regional Surveillance for Notifiable Swine Pathogens

Giovani Trevisan [1,†], Paul Morris [2,†], Gustavo S. Silva [1], Pormate Nakkirt [2], Chong Wang [1,2], Rodger Main [1] and Jeffrey Zimmerman [2,*]

1 Department of Veterinary Diagnostic and Production Animal Medicine, College of Veterinary Medicine, Iowa State University, Patterson Hall, 1800 Christensen Drive, Ames, IA 50011-1134, USA; trevisan@iastate.edu (G.T.); gustavos@iastate.edu (G.S.S.); chwang@iastate.edu (C.W.); rmain@iastate.edu (R.M.)

2 Department of Statistics, College of Liberal Arts and Sciences, Iowa State University, Snedecor Hall, 2438 Osborn Drive, Ames, IA 50011-4009, USA; psmorris@iastate.edu (P.M.); pnakk@iastate.edu (P.N.)

* Correspondence: jjzimm@iastate.edu; Tel.: +1-5125-294-1073

† These authors contributed equally to this work.

Simple Summary: Effective, sustainable regional surveillance for the early detection of notifiable swine pathogens has been difficult to achieve. Regional surveillance based on clinical signs (syndromic surveillance) is not diagnostically sensitive and specific. Surveillance based on farm-by-farm testing is burdensome and costly. Borrowing the strengths of each approach, we evaluated an active participatory surveillance design in which regional status was determined by targeted sampling of 10 poor-doing pigs in each participating farm followed by screening in credentialed laboratories. The analysis showed that at 0.1% prevalence (18 infected farms among 17,521 farms) and a farm-level detection probability of 30%, active participatory surveillance would detect $\geq$ 1 positive farms with 67%, 90%, and 97% probability when producer participation was 20%, 40%, and 60%, respectively. Depending on the specimen collected (serum or swab sample) and test format (nucleic acid or antibody detection), the cost per round of sampling ranged from EUR 0.016 to EUR 0.032 (USD 0.017 to USD 0.034 USD) per pig in the region. The techniques and technologies required for active participatory surveillance are widely available and in common use. Implementation would require coordination among producers, industry groups, and animal health authorities.

Abstract: We evaluated an active participatory design for the regional surveillance of notifiable swine pathogens based on testing 10 samples collected by farm personnel in each participating farm. To evaluate the performance of the design, public domain software was used to simulate the introduction and spread of a pathogen among 17,521 farms in a geographic region of 1,615,246 km^2. Using the simulated pathogen spread data, the probability of detecting $\geq$ 1 positive farms in the region was estimated as a function of the percent of participating farms (20%, 40%, 60%, 80%, 100%), farm-level detection probability (10%, 20%, 30%, 40%, 50%), and regional farm-level prevalence. At 0.1% prevalence (18 positive farms among 17,521 farms) and a farm-level detection probability of 30%, the participatory surveillance design achieved 67%, 90%, and 97% probability of detecting $\geq$ 1 positive farms in the region when producer participation was 20%, 40%, and 60%, respectively. The cost analysis assumed that 10 individual pig samples per farm would be pooled into 2 samples (5 pigs each) for testing. Depending on the specimen collected (serum or swab sample) and test format (nucleic acid or antibody detection), the cost per round of sampling ranged from EUR 0.017 to EUR 0.032 (USD 0.017 to USD 0.034) per pig in the region. Thus, the analysis suggested that an active regional participatory surveillance design could achieve detection at low prevalence and at a sustainable cost.

Keywords: surveillance; participatory surveillance; swine; early detection; targeted sampling

1. Introduction

In this study, a swine farm is defined as a specific geographic location where a population of pigs under one management system is raised; a region is defined as a contiguous geographical area within which the farms under surveillance are located. Swine farms are diverse in size and structure, production type, housing, and management, but the trend over the last few decades has been toward fewer and larger farms. As an example, the number of U.S. farms with pigs declined from 168,450 in 1995 to 68,300 in 2012 [1] while the average farm inventory increased from 302 pigs to 1044 [2]. This period also saw the emergence of specialized swine farms and widespread adoption of the practice of moving young pigs from breeding-specific farms to feeding-specific farms. Thus, in 2019, Denmark, France, Germany, and Spain cumulatively imported 15.7 million and exported 20.5 million live pigs [3] and, in the U.S., 63.4 million live pigs were transported from one state to finishing farms in other states [4]. Overall, these changes have contributed to improved production efficiency but complicated disease control. Logically, the greater movement of animals, personnel, and material facilitates the spread of infectious agents. For example, porcine epidemic diarrhea virus spread to at least 12 states within 8 weeks of its initial detection in the U.S. [5,6].

Under these circumstances, the early detection of notifiable swine pathogens is essential but difficult. In Brazil (1978), an unrecognized outbreak of African swine fever virus (ASFV) in the index farm was followed by its spread to 11 states. Eradication took 8 years and cost ~USD 20 million [7]. In a 1997–1998 outbreak in the Netherlands, a retrospective analysis determined that classical swine fever virus (CSFV) was spreading in the country 5 to 7 weeks prior to its recognition [8]. Eradication was ultimately accomplished at a cost of ~USD 2.3 billion [9]. In the United Kingdom (2001), foot-and-mouth disease virus (FMDV) infections went unnoted and the virus was disseminated widely via the movement of infected animals. Eradication took 6 months, led to the euthanasia of 4 million animals, and cost ~USD 4.0 billion [10]. At present, the ASFV pandemic initiated in 2007 continues to expand despite the recognition that *"an early detection system for ASF could facilitate early reporting and response (and limit) the spread of the disease"* [11].

The need for effective, on-going regional surveillance is obvious, but a workable design is not [12]. Surveillance based on "down-the-road" testing to prove farms free from infection is often performed in government-supported eradication programs, e.g., Aujeszky's disease [13], but is costly and administratively burdensome. Syndromic surveillance [14], i.e., detection based on reports of clinical signs consistent with the pathogen(s) of interest, should meet the need, but Poppensiek and Budd, cited in [15], found that *"The greatest single difficulty in a disease-reporting program proved to be the failure of vets to file reports"*. Exploring this problem, Gates et al. [16] found that the reluctance to report arose from feelings of uncertainty, fear of the consequences of reporting, distrust of authorities, and unfamiliarity with the reporting process. Participatory surveillance, i.e., including members of the population at risk in the surveillance data collection process [17–20], has improved syndromic surveillance, but its effectiveness is limited by the participants' clinical experience and the inherent diagnostic ambiguity of clinical signs. Thus, Elbers et al. [21] estimated that a diagnosis of CSFV based solely on clinical signs achieved a diagnostic sensitivity of 73% and a diagnostic specificity of 53%.

The objective of this study was to characterize the performance of a surveillance design best described as "collecting and testing a few targeted samples from each of many farms in the region". In more formal terms, we analyzed the performance and cost of test-based, regional, active participatory surveillance based on the targeted sampling of 10 poor-doing pigs by farm personnel (producer, staff, and/or veterinarian) followed by screening for the pathogen of interest in credentialed laboratories. Because our objective was to explore the feasibility and performance of this general design, the analysis did not include a sampling and testing process for a specific pathogen. However, a key assumption in the cost analysis was that 10 pigs would be sampled on each participating farm and the individual pig samples combined into two pools (5 pigs per pool) for antibody or nucleic acid testing.

2. Study Design

The study was conducted in three phases. In Phase 1, the spread of a notifiable (but unspecified) pathogen was simulated in a population of 17,521 swine farms holding 51,515,699 pigs in a geographic region of 1,615,246 km^2 for a period of 70 days. Based on the simulated farm status (negative or positive), Phase 2 estimated the probability of detecting ≥ 1 positive farms in the region as a function of farm-level sensitivity (10%, 20%, 30%, 40%, 50%), percent of farms participating in the surveillance program (20%, 40%, 60%, 80%, 100%), and farm-level prevalence in the region. Since the objective was to broadly evaluate the performance of the surveillance design, Phase 1 (pathogen spread) and Phase 2 (detection) simulations were performed over a range of parameters. In Phase 3, active participatory surveillance was analyzed in terms of the cost per farm and the cost per pig in inventory per round of sampling and testing.

2.1. Phase 1: Simulating Pathogen Spread—Animal Disease Spread Model (ADSM)

The Animal Disease Spread Model (ADSM) is public-domain software designed to simulate the spread of infectious agents in livestock populations [22]. ADSM uses a static, fixed population and defines the population of animals at a single geographic location, i.e., a farm, as the epidemiological unit. For Phase 1 simulations, a population of swine farms was created from publicly available concentrated animal feeding operation (CAFO) permit data provided by the appropriate authorities in the states of Colorado, Iowa, Kansas, Minnesota, Missouri, Nebraska, Oklahoma, and South Dakota. State datasets were collated into a single ADSM-compatible file. Farms determined to be inactive or with data quality issues were removed, resulting in a final data set consisting of 17,521 farms (Table 1). Swine packing plant locations and slaughter capacities were included in the population file to account for their role in indirect pathogen spread [23], with latitude and longitude generated from their addresses [24] using Google Maps (www.google.com/maps accessed 1 April 2021).

Table 1. Population of farm sites used in Phase 1 (pathogen spread) and Phase 2 (probability of detection) simulations by production type and pig inventory [1].

Total Pig Inventory	Breeder Sites (No. Pigs)	Breeder–Feeder Sites (No. Pigs)	Feeder Sites (No. Pigs)	Total Sites (No. Pigs)
≤1000	476 (140,823)	650 (254,214)	3296 (980,356)	4422 (1,375,393)
1001 to 4999	474 (1,244,232)	294 (641,930)	10,493 (29,374,884)	11,261 (31,261,046)
≥5000	132 (1,556,655)	120 (1,595,964)	1586 (15,726,642)	1838 (18,879,261)
TOTAL	1082 (2,941,710)	1064 (2,492,108)	15,375 (46,081,882)	17,521 (51,515,700)

[1] Swine farm data based on publicly available animal feeding operation permit data provided by the appropriate authorities in the U.S. states of Colorado, Iowa, Kansas, Minnesota, Missouri, Nebraska, Oklahoma, and South Dakota.

ADSM (version 3.510.0) software required the identification of each farm site by production type (breeder, feeder, or breeder/feeder), inventory (number of animals), and geolocation (latitude and longitude). For the majority of farms, production type was provided in the state datasets or derived from the site name, e.g., "Smith Sow Farm". Using this approach, 13,041 of 17,521 farms in the population file were assigned to production type: 209 (1.6%) breeder/feeder, 702 (5.4%) breeder, and 12,130 (93.0%) feeder. The remaining 4481 farms were randomly assigned [25] to production type proportional to state-level production types or, if state data were not adequately reported, the overall proportions in the database. State-level permit data described the capacity (inventory) of each farm either as the number of pigs or as "animal units". In the latter case, animal units were converted to the number of pigs on the basis of one animal unit per 2.5 pigs weighing ≥ 24.9 kg (≥55 pounds) or 10 pigs weighing < 24.9 kg (<55 pounds) [26]. With one exception, all states reported farm location by latitude and longitude, by ZIP Code (i.e., postal code), or by county (i.e., an administrative subdivision of a state). For farms without precise geolocation, the spsample function in the sp R package (version 1.4-5) [27] was used to randomly generate a latitude and longitude within the geographic unit associated with the record (i.e., ZIP Code, county, or state).

2.1.1. ADSM Simulations

The ADSM software was designed to simulate the spread of a designated pathogen in a defined livestock population by setting parameter values representative of the pathogen's transmission characteristics and industry production practices. In this study, ADSM simulations were performed over a range of parameter values (Table 2) to provide spread estimates generalizable to a variety of notifiable pathogens. Although disease control options are available in ADSM, e.g., movement restrictions, vaccination, and farm depopulation, they were not implemented so as to allow the unrestricted spread of the hypothetical agent within the region.

Table 2. Phase 1 (pathogen spread): parameters used in simulating the spread of a notifiable swine pathogen in a defined region [1].

<table>
<tr><th>Spread Parameters</th><th>Parameter Definitions and/or Values</th></tr>
<tr><td>1. Index farm.
a. Location (pig density).
b. Production type.</td><td>1. First positive farm in each simulation.
a. County-level pig density: low (1.1–3.3 pigs per km²), medium (15.9–25.3 pigs per km²), high (106.8–214.5 pigs per km²).
b. Breeder, breeder–feeder, or feeder.</td></tr>
<tr><td>2. Direct contact.
a. Distance for direct contact.
b. Daily movement rate.
c. Probability of infecting a negative farm.</td><td>2. Transmission by moving infectious animals among sites.
a. BETAPert distribution, min 0.5 km, mode 100 km, max 1000 km.
b. Fixed rate as specified by farm type:
<table>
<tr><th></th><th></th><th>Destination</th><th></th></tr>
<tr><th>Source farm</th><th>Breeder [28]</th><th>Feeder [28]</th><th>Packing Plant [29]</th></tr>
<tr><td>Breeder–Feeder</td><td>NA</td><td>0.0204</td><td>0.0310</td></tr>
<tr><td>Breeder</td><td>0.0014</td><td>0.0687</td><td>0.0310</td></tr>
<tr><td>Feeder</td><td>NA</td><td>0.0348</td><td>0.0310</td></tr>
</table>
c. Probabilities tested 0.2, 0.4, and 0.6.</td></tr>
<tr><td>3. Indirect contact.
a. Distance for indirect contact.
b. Daily indirect contact rate.
c. Probability of infecting a negative farm.</td><td>3. Transmission by movement of people, fomites, etc.
a. BETAPert distribution, min 0.5 km, mode 100 km, max 1000 km.
b. Fixed rate as specified by farm type:
<table>
<tr><th></th><th></th><th>--------- Destination ---------</th><th></th></tr>
<tr><th>Source farm</th><th>Breeder</th><th>Feeder</th><th>Packing Plant</th></tr>
<tr><td>Breeder–Feeder</td><td>NA</td><td>0.0204</td><td>0.0310</td></tr>
<tr><td>Breeder</td><td>0.0014</td><td>0.0687</td><td>0.0310</td></tr>
<tr><td>Feeder</td><td>NA</td><td>0.0348</td><td>0.0310</td></tr>
</table>
c. Probabilities tested 0.05, 0.1, and 0.15.</td></tr>
<tr><td>4. Local area spread.
a. Probability of infecting a negative farm.</td><td>4. Daily probability of spread to farms ≤ 1 km from infected farm.
a. Exponential drop off. Probabilities tested 0.001, 0.01, and 0.1.</td></tr>
</table>

[1] Pathogen spread simulations performed using public domain software [22] and a population of 17,521 farms (51,515,700 pigs) in a contiguous geographic region (1,615,246 km²).

For simplicity, each spread scenario began with a single index farm (Table 2). A total of 30 index farms were identified by randomly selecting 10 farms from each of the 3 pig density categories (1.1–3.3 pigs per km^2, 15.9–25.3 pigs per km^2, 106.8–214.5 pigs per km^2) using features built into R (version 4.1.0) software [25]. Each of these 30 individual index farms was successively categorized as a breeder, breeder–feeder, or feeder in simulations. This process ensured that the spread scenarios covered the range of possible outcomes that could arise due to differences in pig density in the region surrounding the index farm and in the direct and indirect contact rates among production types. A total of 2430 pathogen spread scenarios were simulated based on all combinations of county-level pig density (n = 3), index farm within county density (10 farms in each of the 3 county-level pig densities), index farm production type (n = 3, i.e., breeder, feeder, breeder–feeder), spread by direct contact (n = 3, i.e., probability levels 0.2, 0.4, 0.6), spread by indirect contact (n = 3, i.e., probability levels 0.05, 0.10, 0.15), and area spread (n = 3, i.e., probability levels 0.001, 0.010, 0.100). Each scenario was replicated 100 times to account for the stochastic nature of the ADSM simulations.

2.1.2. ADSM Automation

After constructing the initial pathogen spread scenario using the ADSM Scenario Creator, the creation of the subsequent 2429 scenarios was automated. In brief, the Scenario Creator output a directory that contained an SQLite database ("ScenarioX.db") into the ADSM workspace that housed the information ADSM used to run the scenario. An R script [25] was written to copy the database file and update the unique parameter values for each scenario (Table 2), thereby creating additional SQLite databases suitable to be imported and run on ADSM.

The procedure was performed as follows:

1. A new file directory in the ADSM workspace was created using the base R function dir.create and the saved template scenario "ScenarioX.db" was copied from the initial simulation. This file was renamed using the base R function file.rename, e.g., "new_scenario.db".
2. A connection was created between R and the SQLite database using the dbConnect function from the RSQLite R package (version 2.2.4) [30] to access the "new_scenario.db" file for editing, (i.e., con = dbConnect(SQLite(), dbname = "new_scenario.db")).
3. Once the connection was opened, the dbGetQuery function was used to bring the SQLite table to be edited into the R environment as a data frame. For example, the R code created a data frame in the R environment named "Population" using the ScenarioCreator_unit table from the SQLite database. This table contained the entire population file input during the ADSM scenario creation process.
    a. Population <- dbGetQuery(con, "SELECT * FROM ScenarioCreator_unit").
    b. Other tables altered using this procedure included those containing the direct spread parameters (ScenarioCreator_directspread), indirect spread parameters (ScenarioCreator_indirectspread), and local area spread parameters (ScenarioCreator_airbornespread).
4. R functions were then used to update the data frame to fit the new desired scenario (i.e., production types, initial infection statuses, or transmission probabilities).
5. The dbWriteTable function with the overwrite option specified as TRUE was used to replace the SQLite table in the database file with the newly edited data frame. For example, dbWriteTable(con, name = "ScenarioCreator_unit", value = Population, overwrite = TRUE).
6. Rerunning the dbConnect line exactly as written in Step 2 saved the SQLite database file with the changes included (i.e., con = dbConnect(SQLite(), dbname = "new_scenarioX.db")).

At the conclusion of the scenario creation process, the spread scenarios were run using batch processing (see https://github.com/NAVADMC/ADSM/wiki/Batch-processing-of-scenarios-using-ADSM-Auto-Scenario-Runner accessed on 1 May 2021).

2.1.3. Phase 1: Spread Results

Phase 1 simulation results (100 iterations for each of the 2430 scenarios) are reported in Table 3 as the mean number of infected farms on simulation day 70 for all possible combinations of the spread parameter values listed in Table 2, i.e., index farm county pig density (n = 3 pig densities), index farm type (n = 3), probability of transmission by direct contact (n = 3 levels), indirect contact (n = 3 levels), and area spread (n = 3 levels). All parameters in the model affected the outcome, but holding all other parameters constant (ceteris paribus), it can be seen that the probability of transmission by direct contact, i.e., the movement of infectious animals among sites, was the most impactful in terms of the total number of infected farms on day 70.

Table 3. Phase 1 (pathogen spread): results of the simulated regional spread of a notifiable swine pathogen reported as the mean number of infected farms on simulation day 70 for specific spread scenarios [1].

Index Farm Location and Type [2]		Area Spread	Direct Contact 0.2 Indirect Contact			Direct Contact 0.4 Indirect Contact			Direct Contact 0.6 Indirect Contact		
			0.05	0.10	0.15	0.05	0.10	0.15	0.05	0.10	0.15
Low-density county 1.1–3.3 pigs per km^2	BF	0.001	4	4	5	16	16	21	63	67	81
	B		10	11	12	46	54	60	191	218	242
	F		6	7	8	25	31	36	99	113	138
	BF	0.010	5	5	5	19	22	25	86	96	107
	B		11	12	15	59	64	75	245	279	318
	F		7	8	10	31	39	45	136	160	182
	BF	0.100	13	14	15	85	97	119	430	427	534
	B		34	42	48	247	298	311	1127	1223	1366
	F		19	25	33	135	164	190	637	754	864
Medium-density county 15.9–25.3 pigs per km^2	BF	0.001	4	4	4	16	17	19	64	76	79
	B		10	11	12	47	55	62	195	218	250
	F		6	7	9	27	31	37	99	126	150
	BF	0.010	5	5	6	21	23	29	88	102	112
	B		12	13	15	64	67	80	260	304	336
	F		7	10	10	35	41	49	141	164	187
	BF	0.100	16	18	18	119	120	146	456	510	595
	B		42	49	58	292	317	393	1325	1396	1539
	F		23	30	40	155	200	209	742	859	956
High-density county 106.8–214.5 pigs per km^2	BF	0.001	5	5	5	19	20	21	66	74	85
	B		11	12	13	51	58	65	200	233	271
	F		6	8	9	26	34	41	108	132	151
	BF	0.010	6	7	8	26	31	33	109	123	141
	B		14	16	18	71	80	94	285	322	358
	F		10	11	13	42	51	55	164	182	210
	BF	0.100	47	56	67	244	286	330	982	1015	1288
	B		81	89	102	446	500	575	1661	1936	2093
	F		57	72	85	310	368	432	1239	1345	1553

[1] Pathogen spread simulations (100 per scenario × 10 index farms) were performed using public domain software [22] in a population of 17,521 farms (51,515,699 pigs) in a contiguous geographic region (1,615,246 km^2). [2] BF (breeder–feeder), B (breeder), F (feeder).

2.2. Phase 2. Simulating Pathogen Detection

Among the 2430 spread scenarios simulated in Phase 1, the 360 scenarios indicated in Table 4 were selected for use in Phase 2. Each of the 12 groups of spread scenarios shown in

Table 4 consisted of 30 scenarios, i.e., 10 index farms in each pig density category, with each index farm successively classified as one of 3 production types (breeder, breeder–feeder, and feeder). For each of these 360 spread scenarios, the detection of ≥ 1 positive farms in the region was simulated under 25 pathogen detection settings based on farm-level sensitivity (10%, 20%, 30%, 40%, or 50%) and farm participation in the surveillance program (20%, 40%, 60%, 80%, or 100% of farms in the region). For the 20%, 40%, 60%, and 80% participation levels, farm participation was allocated uniformly across the 3 pig inventory size categories in Table 1 through simple random sampling without replacement. For example, simulations at 20% participation included 20% of the farms in the ≤1000 category, 20% in the 1001–4999 category, and 20% in the ≥5000 category. For each participation level, 1000 farm groupings were randomly selected using R software [25] to match the 1000 surveillance simulations. By definition, 100% participation did not require participant selection.

Table 4. Spread parameter values from Phase 1 (pathogen spread) selected for use in Phase 2 (probability of detection) simulations.

Index Farm Location [1]	Area Spread	Spread Probabilities								
		Direct Contact 0.2 Indirect Contact			Direct Contact 0.4 Indirect Contact			Direct Contact 0.6 Indirect Contact		
		0.05	0.10	0.15	0.05	0.10	0.15	0.05	0.10	0.15
Low-density county	0.001	-	-	-	-	-	-	-	-	-
	0.010	-	✓	-	-	✓	-	-	✓	-
	0.100	-	-	-	-	-	-	-	-	-
Medium-density county	0.001	-	-	-	-	-	-	-	-	-
	0.010	-	✓	-	-	✓	-	-	✓	-
	0.100	-	-	-	-	-	-	-	-	-
High-density county	0.001	-	-	-	-	-	-	-	-	-
	0.010	-	✓	-	-	✓	-	-	✓	-
	0.100	-	✓	-	-	✓	-	-	✓	-

[1] Index farm located in low-density county (1.1–3.3 pigs per km^2), medium-density county (15.9–25.3 pigs per km^2), or high-density county (106.8–214.5 pigs per km^2).

An R function [25] was written to perform the surveillance simulations for each combination of spread scenario replicate and detection setting, as described below. Because each spread scenario was replicated 100 times, there were a total of 100,000 iterations (1000 iterations for each of the 100 replicates) for each combination of spread scenario and detection setting.

For iteration i, where i = 1, . . ., 1000:

1. Assign as participants the ith set of participating farms from the list of pre-selected sets corresponding to the current setting's participation level. For 100% participation, the entire population of farms were participants.
2. For each participating farm, the farm infection status (negative, positive) was identified for days 7, 14, 21, 28, 35, 42, 49, 63, and 70 of the ADSM spread simulation.
3. For each of the days listed in Step 2, participating farms classified as positive were "tested" (simulated) independently in R using the rbinom function with the probability of detection equaling the assigned farm-level sensitivity. Thus, for each positive farm and where p was the assigned farm-level sensitivity, rbinom (n = 1, size = 1, prob = p) randomly generated a 0 or 1, where 1 indicated that the infection was detected.

2.2.1. Phase 2: Detection Results

For each of the 25 detection settings, results were reported as the probability of detection by regional farm-level prevalence, with the probability of detection calculated as the percentage of iterations in which ≥ 1 true positive farms "tested" positive in the simulations. Results are provided in Table 5 and Figures 1 and 2 by regional prevalence, farm-level sensitivity, and producer participation. The analysis showed that detection was dependent on the interactions between producer participation and farm-level sensitivity, but high probabilities of detection were achieved at low prevalence over a wide range of participation and sensitivity values. For example, at 0.1% prevalence (18 positive farms among 17,521 farms) and a farm-level detection probability of 30%, the participatory surveillance design achieved 67%, 90%, and 97% probabilities of detecting ≥ 1 positive farms in the region when producer participation was 20%, 40%, and 60%, respectively.

Table 5. Probability of detecting ≥ 1 positive farms as a function of regional prevalence, farm-level sensitivity (%), and producer participation (%).

Regional Prevalence [1]	Farm-Level Sensitivity (%) [2]	Producer Participation 20%	40%	60%	80%	100%
0.1% (18 farms)	10	0.304	0.520	0.672	0.777	0.850
	20	0.519	0.777	0.900	0.957	0.982
	30	0.671	0.900	0.972	0.993	0.998
	40	0.776	0.956	0.993	0.999	1.000
	50	0.849	0.982	0.998	1.000	1.000
0.2% (35 farms)	10	0.506	0.760	0.886	0.946	0.975
	20	0.760	0.945	0.989	0.998	1.000
	30	0.885	0.989	0.999	1.000	1.000
	40	0.945	0.998	1.000	1.000	1.000
	50	0.975	1.000	1.000	1.000	1.000
0.3% (53 farms)	10	0.657	0.885	0.963	0.988	0.996
	20	0.885	0.988	0.999	1.000	1.000
	30	0.962	0.999	1.000	1.000	1.000
	40	0.988	1.000	1.000	1.000	1.000
	50	0.996	1.000	1.000	1.000	1.000

[1] Farm-level prevalence in a population of 17,521 farms in a defined region (1,615,246 km^2). [2] Farm-level sensitivity is the probability of a positive test on samples from an infected farm.

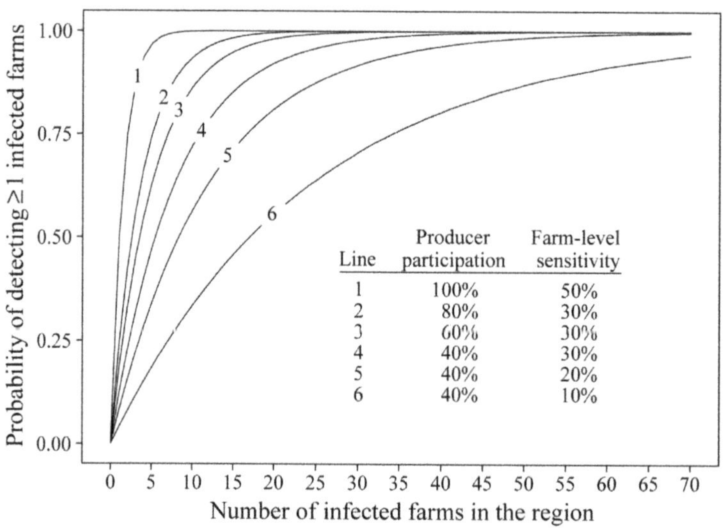

Figure 1. Illustration of the interaction between producer participation, farm-level detection sensitivity, and number of positive farms on the probability of detecting ≥ 1 positive farms in the region.

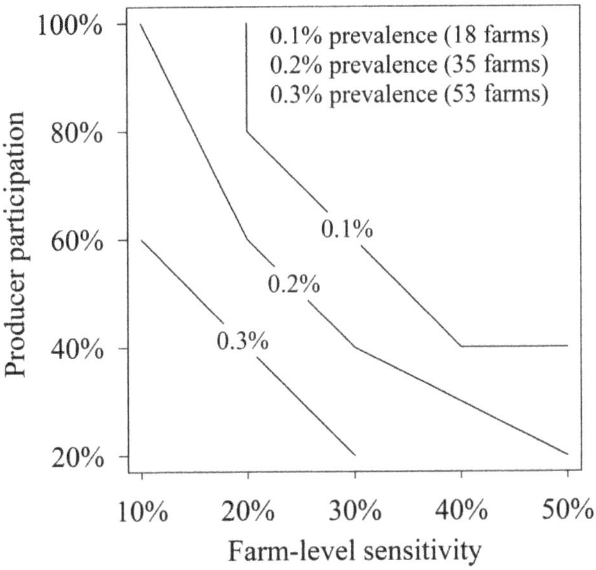

Figure 2. As shown for 3 prevalence levels, various combinations of producer participation and farm-level sensitivity produced $\geq$ 95% probability of detecting $\geq$ 1 positive herds in the region. Farm-level sensitivity is the probability of a positive test on samples from an infected farm.

2.3. Phase 3: Cost of Sampling and Testing

The cost analysis assumed that ante mortem specimens (blood, blood swabs, nasal swabs, oral swabs, or fecal swabs) would be collected from 10 poor-doing pigs in each participating farm, combined into 2 pooled samples (5 pigs per pool), shipped to a credentialed laboratory in an insulated shipping container with coolant, and tested by polymerase chain reaction (PCR) or antibody ELISA. Since the general surveillance design did not define a specific testing protocol, costs were estimated for 3 cases: serum samples tested by PCR, swab samples tested by PCR, and serum samples tested by ELISA. To further evaluate the impact of testing costs on overall program costs, 3 price levels for PCR (EUR 18.65, EUR 23.32, EUR 27.98/USD 20.00, USD 25.00, USD 30.00) and antibody ELISA (EUR 4.66, EUR 7.00, EUR 9.33/USD 5.00, USD 7.50, USD 10.00) were used in the estimates.

The estimated cost of a single round of sampling used the inputs and costs listed in Table 6 and assumed 100% producer participation. Costs listed in Table 6 are the mean of prices quoted by 3 companies for the distribution of products in the U.S. Supplies for collecting serum samples included single-use blood collection tubes and needles (n = 10), tubes in which to pool samples (n = 2), and disposable gloves (2 pairs). Supplies for swab samples included swabs (n = 10), transport medium, tubes in which to pool samples (n = 2), and disposable gloves (2 pairs). Package shipment costs reflect rates paid by clients of the Iowa State University Veterinary Diagnostic Laboratory (Ames, IA, USA) and may vary.

The analysis assumed that the labor and materials required to collect, process, and package samples for shipment would be provided by the farm and, therefore, were not included in the cost analysis. Likewise, it was assumed that blood samples would be centrifuged and the serum pooled (5 pigs per pool) prior to shipment in order to avoid processing charges at the laboratory. Some costs that would be expected in the normal course of sampling and testing were also not included. For example, no attempt was made to account for the added cost of duplicate sampling or testing, e.g., the cost of an additional tube and needle for a second attempt at blood collection from a pig or the cost of retesting a non-negative sample in the laboratory. Likewise, the cost analysis did not include the costs required to administer and coordinate the program.

Table 6. Cost per sampling per farm to collect samples and ship to laboratory [1].

Category	Cost per Item [2]		No. Items	Cost per Sampling	
A. Sample collection. Assumes 10 pigs/farm/sampling					
Option 1. Serum samples					
Blood collection tubes (single-use)	0.525	0.563	10	5.26	5.64
Blood collection needles (single-use)	0.572	0.613	10	5.72	6.13
Plastic tube for pooling 5 samples [3]	0.171	0.183	2	0.35	0.37
Disposable gloves	€0.064	$0.069	4 (2 pairs)	0.26	0.28
				€11.58	$12.42
Option 2. Swab samples (blood, nasal, oral, or fecal)					
Sample collection swabs	0.532	0.570	10	5.32	5.70
Transport medium, e.g., phosphate-buffered saline	0.036	0.039	5 mL	0.36	0.39
Plastic tube for pooling 5 samples	0.171	0.183	2	0.35	0.37
Disposable gloves	€0.064	$0.069	4 (2 pairs)	0.26	0.28
				€6.29	$6.74
B. Shipment of samples to the laboratory					
Insulated shipping container	5.167	5.540	1	5.17	5.54
Cold packs to ship with samples	6.529	7.000	1	6.53	7.00
Parcel shipping charge	€13.991	$15.000		13.99	15.00
				€25.69	$27.54

[1] EUR (€) 1.00 = USD ($) 1.0721 U.S. (https://www.federalreserve.gov/releases/h10/current/ accessed on 19 June 2023). [2] Means of prices provided by three distributors in the U.S. [3] Cost analysis assumed blood samples would be centrifuged and serum pooled (5 pigs per pool) prior to shipment to avoid sample processing charges at the laboratory.

2.3.1. Phase 3: Estimated Cost of Sampling and Testing

The results of the cost analysis are listed in Table 7 for 3 "specimen by test" combinations with 3 costs for each test. The estimates are given in terms of the average cost per farm in the region, the average cost per pig in the region, and the average cost per pig in inventory for the farm size categories given in Table 1, i.e., farms with ≤1000 pigs, 1001–4999 pigs, and ≥5000 pigs. Using a PCR cost of EUR 23.32 (USD 25.00) or ELISA cost of EUR 7.00 (USD 7.50) per sample, the cost of sampling and testing would be approximately EUR 0.03 (USD 0.03) or EUR 0.02 (USD 0.02) per pig in the region, respectively. On a farm basis, given that sample size and test costs are the same for all farms, the cost per pig increases as the farm pig inventory decreases, as shown in Table 7.

Table 7. Cost of sampling and testing by specimen and test based on 3 test cost options.

	Serum Tested by PCR			Swabs Tested by PCR			Serum Tested by ELISA		
Cost per test / Denominator	EUR 18.65 USD 20.00	EUR 23.32 USD 25.00	EUR 27.98 USD 30.00	EUR 18.65 USD 20.00	EUR 23.32 USD 25.00	EUR 27.98 USD 30.00	EUR 4.66 USD 5.00	EUR 7.00 USD 7.50	EUR 9.33 USD 10.00
Per farm in region [1]	EUR 74.56 USD 79.94	EUR 83.89 USD 89.94	EUR 93.22 USD 99.94	EUR 69.28 USD 74.27	EUR 78.60 USD 84.27	EUR 87.93 USD 94.27	EUR 46.58 USD 49.94	EUR 51.25 USD 54.94	EUR 55.91 USD 59.94
Per pig in region [1]	EUR 0.025 USD 0.027	EUR 0.029 USD 0.031	EUR 0.032 USD 0.034	EUR 0.023 USD 0.025	EUR 0.027 USD 0.029	EUR 0.030 USD 0.032	EUR 0.016 USD 0.017	EUR 0.018 USD 0.019	EUR 0.019 USD 0.020
Per pig in inventory Farms of ≤ 1000 pigs [2]	EUR 0.240 USD 0.257	EUR 0.270 USD 0.289	EUR 0.299 USD 0.321	EUR 0.223 USD 0.239	EUR 0.253 USD 0.271	EUR 0.283 USD 0.303	EUR 0.150 USD 0.161	EUR 0.165 USD 0.177	EUR 0.180 USD 0.193
Farms of 1001–4999 pigs [3]	EUR 0.027 USD 0.029	EUR 0.030 USD 0.032	EUR 0.034 USD 0.036	EUR 0.025 USD 0.027	EUR 0.028 USD 0.030	EUR 0.032 USD 0.034	EUR 0.017 USD 0.018	EUR 0.019 USD 0.020	EUR 0.021 USD 0.022
Farms of ≥ 5000 pigs [4]	EUR 0.007 USD 0.008	EUR 0.008 USD 0.009	EUR 0.009 USD 0.010	EUR 0.007 USD 0.007	EUR 0.007 USD 0.008	EUR 0.008 USD 0.009	EUR 0.005 USD 0.005	EUR 0.006 USD 0.005	EUR 0.006 USD 0.006

[1] Estimates based on 17,521 farms in the region, holding 51,515,700 pigs. [2] Estimates based on 4422 farms with a mean inventory of 311 pigs. [3] Estimates based on 11,261 farms with a mean inventory of 2776 pigs. [4] Estimates based on 1838 farms with a mean inventory of 10,272 pigs.

3. Discussion

Concerning surveillance systems, Thacker et al. [31] advised that "*Simplicity should be a guiding principle .... Simple systems are easy to understand and implement, cost less than complex systems, and provide flexibility*". Consistent with the theme of "simplicity", the active participatory regional surveillance design was based on targeted sampling of 10 live but poor-doing pigs on participating farms by farm personnel, followed by testing in credentialed laboratories. The design differed most from traditional surveillance in that it focused on the status of the region rather than the status of individual farms. The result was fewer samples per farm yet sensitive regional surveillance at a manageable cost (Tables 5 and 7).

Targeted sampling, already recommended for the surveillance of CSFV [32] and ASFV [33], addressed the problem of detection in populations characterized by heterogeneity and low prevalence [34]. That is, commercial swine farms separate animals into barns and pens by age, stage, and function, i.e., conditions that are inconsistent with the independence and homogeneity assumptions underlying the traditional power formula based on simple random sampling. Thus, Crauwels et al. [35] reported that random sampling would be unlikely to include a CSFV-positive pig for several weeks following its introduction into a naïve farm; that is, until it had spread sufficiently and infected a sufficient proportion of the population.

The surveillance design called for sampling live, poor-doing pigs because notifiable pathogens may not produce remarkable clinical signs and early mortalities, including CSFV [36], ASFV [37], and FMDV [38]. Kirkland et al. [36] cautioned that CSFV strains of low or moderate virulence could circulate without notable clinical signs for 4–8 weeks. Schulz et al. [39] concluded that, depending on the virulence of the isolate, it could take up to a month for ASFV-related mortalities to be noted. Thus, sampling poor-doing live pigs would facilitate early detection by eliminating the expectation of telltale clinical signs and/or conspicuous mortalities.

Sampling live pigs also anticipates the need to quickly resolve ambiguous ("non-negative") test results. The typical response to a non-negative surveillance sample result is retesting the original sample using the original test or a confirmatory assay. If the retest result is conclusive, the question is resolved. If not, the fact that the samples originated from live pigs means that it is likely possible to re-sample and retest the original pigs and/or their penmates to quickly resolve the issue. On the other hand, if the samples originated from dead pigs or pigs no longer on the farm, the resolution will require extensive sampling of animals on the farm of origin and on other epidemiologically relevant farms.

In this "generic" surveillance scenario it was not necessary to designate specific specimens to be collected or tests to be performed. While blood and serum are traditional surveillance specimens, a variety of more easily collected antemortem specimens are increasingly used in diagnostics and surveillance, e.g., blood swabs, nasal swabs, oropharyngeal swabs, and rectal swabs [40]. Trevisan et al. [41] documented this trend for the period 2007 to 2018 in a study of 547,873 diagnostic cases submitted to 4 Midwestern U.S. veterinary diagnostic laboratories for porcine reproductive and respiratory syndrome virus (PRRSV) testing. In 2007, 51% of the diagnostic cases included serum samples; in 2018, 21% of cases included serum samples, 35% included oral fluid samples, and 11% included processing fluid samples.

Sample collection by farm personnel working under the supervision of the farm veterinarian is common practice in many parts of the world. The use of easily collected antemortem samples will facilitate producer participation and is consistent with sampling by lay personnel. Regardless of the specimen(s) selected for use, the ability of lay participants to collect diagnostic samples is supported by the literature. In human medicine, Branson [42] reported that 156,121 (94.5%) of 165,194 self-collected dried blood spot specimens were acceptable for human immunodeficiency virus testing. The remaining 5.5% were disqualified for insufficient quantity, contamination, or excessive time between sampling and submission. Similarly, Tsang et al. [43] found no loss in diagnostic accuracy with self-collected oronasal swabs or oral fluid samples in a systematic review of 23 refereed studies involving severe acute respiratory syndrome coronavirus 2 (SARS-CoV-2) testing. In the veterinary literature, formal examples of the use of producer-collected samples include field studies on CSFV [44], PRRSV [45], and antibiotic resistance in *Escherichia coli* [46].

The use of farm personnel in sample collection acknowledges the fact that those who work with the pigs are also those most aware of recent changes in pig health and are best qualified to identify the appropriate animals to sample. Relevant to program sustainability, the use of farm personnel reduces sampling costs by eliminating the need to employ program samplers and integrates both scalability and responsiveness into the design. That is, because farm personnel are already on the farms, sample size and/or frequency can be quickly adjusted in response to changing circumstances, e.g., increased after the initial detection of the target to improve case finding and decreased after the threat is contained to reduce costs.

The final point in the surveillance design is the testing of samples in credentialed diagnostic laboratories. "Credentialed", in this case, refers to laboratories operating under national or international standards, e.g., ISO/IEC 17025 [47,48]. Such laboratories have operational quality management systems, proper equipment, and the technical expertise to reliably perform testing. Further, many of these laboratories are equipped with laboratory information management systems and the capacity to report test results electronically, thereby facilitating timely reporting to participants and, if needed, animal health authorities. Alternatively, testing in the field using point-of-care test devices is sometimes suggested as a means to expedite the discovery of notifiable pathogens. This may be possible in the future, but not at present. Hobbs et al. [49] reported the most fundamental problem: "*Inadequate regulatory guidance and poor industry oversight has led to a proliferation of point-of-care tests of varying quality and fitness for purpose* ...". A number of issues would need to be addressed if point-of-care tests are to be used for notifiable pathogens, but at a minimum, an accounting system for tracking kits and test results will need to be in place to avoid misuse.

The performance analyses of the active participatory regional surveillance design were based on simulations of the spread (Phase 1) and detection (Phase 2) of an unspecified pathogen in a population of naïve swine farms representative of the Midwest U.S. The swine farm dataset was created using concentrated animal feeding operation (CAFO) permit records from eight U.S. states. CAFO permitting requirements and data quality were not standardized across states, but the data required for the pathogen spread simulations (farm geolocation, inventory, and production type) were provided in most cases. After resolving inconsistencies and missing data (Section 2.1), the dataset consisted of 17,521 swine farms holding 51,515,699 pigs in a region of ~1,582,000 km^2 [50]. As a point of reference, the geographic area of Belgium, France, Germany, Luxembourg, the Netherlands, Portugal, and Spain is ~1,584,000 km^2 [51].

The purpose of Phase 1 was to create datasets of swine farms of known infection status (negative, positive) for use in the detection simulations (Phase 2). The Animal Disease Spread Model (ADSM) [15] software used in Phase 1 provided substantial modeling flexibility and has previously been used to simulate the spread of ASFV in Vietnam [49], Aujeszky's disease virus in Thailand [52], CSFV in the Republic of Serbia [53], and PRRSV in both Uganda [54] and Canada [28]. A total of 2430 pathogen spread scenarios were simulated (Section 2.1.1) based on combinations of county-level pig densities, index herd production types, and the probabilities of transmission by direct contact, indirect contact, and area spread. Among these scenarios, 360 spread scenarios representing the range of outcomes were selected for use in the Phase 2 detection simulations.

The objective of Phase 2 was to estimate the probability of detecting ≥ 1 positive farms in the region as a function of farm-level sensitivity, percent of farms participating in the surveillance program, and regional farm-level prevalence. The results were expressed in terms of the probability of detection by regional herd prevalence (from Phase 1 simulation results) rather than time-to-detection because of the diversity of spread rates simulated in the ADSM software. A difficulty when applying targeted sampling to surveillance is the absence of agreed-upon methods for calculating sample size and associated farm-level sensitivities. Using a modeling approach, Nielsen et al. [32,33] reported that targeted sampling of 5 sick or dead pigs in a population of 1000 pigs would detect CSFV 4 to 37 days and ASFV 13 days post-introduction with 95% probability. In the present study, a sample size of 10 pigs was considered a practical number for on-farm collection. The present study was not pathogen-specific and, in the absence of citable estimates, a conservative range of farm-level detection sensitivities (10%, 20%, 30%, 40%, and 50%) was used in the Phase 2 detection analysis. Similarly, data to inform the level of farm participation in this voluntary regional surveillance program were lacking. Consequently, adopting an approach that would inform administrators if such a program were to be initiated, a range of participation levels (20%, 40%, 60%, 80%, 100%) were evaluated for their effect on detection by prevalence.

In Phase 3, the regional surveillance design was analyzed for the cost per farm (17,521 farms) and per pig in the region (51,515,700 pigs) for one round of sampling. Options evaluated in the analysis included specimen (serum vs. swab samples), assay format (PCR or ELISA), and 3 assay cost options. The analysis was based on the present costs of materials for collecting, shipping, and testing in credentialed laboratories (Table 6). The cost analysis assumed that the labor required to collect and package samples would be provided by participant swine producers and that administrative costs would be borne by existent animal health agencies. Using test costs of EUR 23.32 (USD 25.00) per PCR and EUR 7.00 (USD 7.50) per ELISA, the cost per farm in the region ranged from EUR 51.25 (USD 54.94) for serum tested by ELISA to EUR 83.89 (USD 89.94) for serum tested by PCR. The cost per pig in the region ranged from EUR 0.018 (USD 0.019) to EUR 0.029 (USD 0.031) for the same scenarios.

Lee et al. [55] reported that U.S. swine producers would be willing to pay USD 0.581 (EUR 0.542) per pig per year to reduce the risk of losses from notifiable pathogens. While Lee et al. [55] focused on biosecurity, the regional surveillance design described herein would facilitate early detection and elimination, provide evidence of freedom from disease, and support access to international markets. Thus, the net effect is the amelioration of the major economic losses expected after the introduction of a notifiable pathogen at a price close to the producers' cost constraints.

The dataset of 17,521 farms used in this study was assembled from CAFO permits and, therefore, may be considered representative of the region. From Table 1, it can be seen that the 4422 farms (25.2%) in the smallest farm category (≤1000 pigs) held 2.67% of the pigs in the region. The 11,261 farms (71.8%) in the mid-size category (1001 to 4999 pigs) held an additional 64.3% of the pigs in the region. By definition, smaller farms have fewer pigs and, therefore, surveillance cost per pig is higher (Table 7). To be successful, participatory surveillance requires broad engagement. While it may be possible to further reduce costs associated with sampling, transport, and testing, it would be prudent to explore the means to incentivize small producer participation.

4. Conclusions

The regional active participatory surveillance design evaluated in this study is simple and adaptable to the surveillance of a variety of pathogens, farm animal species, or regions. Simplicity and clarity in sampling, testing, and reporting are central to the success of a participatory program because voluntary programs depend on the full confidence of the participants. In truth, there is little innovation in the proposed surveillance framework; the personnel, testing, and reporting systems are largely in place. The only possible novelty is the aggregation and interpretation of surveillance testing data at the regional level rather than the farm level. As was shown in the evaluation, this change in focus achieved highly sensitive regional detection at a low cost.

Author Contributions: Conceptualization, G.T., P.M., R.M., C.W. and J.Z.; methodology, G.T., P.M., G.S.S. and C.W.; writing, G.T., P.M., G.S.S. and J.Z.; visualization, P.N.; funding acquisition, R.M., G.S.S., G.T., C.W. and J.Z. All authors have read and agreed to the published version of the manuscript.

Funding: This research was funded by the National Pork Board, Clive, IA, U.S. (#22-028).

Institutional Review Board Statement: Not applicable.

Data Availability Statement: Current concentrated animal feeding operation (CAFO) permit data are available from the appropriate state authorities through the U.S. Freedom of Information Act.

Conflicts of Interest: The authors declare no conflicts of interest.

References

1. United States Department of Agriculture. *Changes in the U.S. Swine Industry, 1995–2012 (#678.0817)*; USDA:APHIS:VS: CEAH, National Animal Health Monitoring System: Fort Collins, CO, USA, 2017.
2. Dunn, J.W. The evolution of the U.S. swine industry. In *Changes in the Live Pig Market in Different Countries*; Szymańska, E.J., Ed.; Warsaw University of Life Sciences Press: Warsaw, Poland, 2017; pp. 19–27.
3. Food and Agriculture Organization of the United Nations. FAOSTAT Statistical Database. Available online: www.fao.org/faostat/en/#data (accessed on 12 July 2021).
4. United States Department of Agriculture. *Meat Animals Production, Disposition, and Income—2020 Summary*; USDA National Agricultural Statistics Service: Washington, DC, USA, 2021.
5. Stevenson, G.W.; Hoang, H.; Schwartz, K.J.; Burrough, E.R.; Sun, D.; Madson, D.; Cooper, V.L.; Pillatzki, A.; Gauger, P.; Schmitt, B.J.; et al. Emergence of porcine epidemic diarrhea virus in the United States: Clinical signs, lesions, and viral genomic sequences. *J. Vet. Diagn. Investig.* **2013**, *25*, 649–654. [CrossRef] [PubMed]
6. Niederwerder, M.C.; Hesse, R.A. Swine enteric coronavirus disease: A review of 4 years with porcine epidemic diarrhoea virus and porcine deltacoronavirus in the United States and Canada. *Transbound. Emerg. Dis.* **2018**, *65*, 660–675. [CrossRef] [PubMed]
7. Moura, J.A.; McManus, C.M.; Bernal, F.E.M.; de Melo, C.B. An analysis of the 1978 African swine fever outbreak in Brazil and its eradication. *Rev. Sci. Tech. Off. Int. Epiz.* **2010**, *29*, 549–563. [CrossRef] [PubMed]
8. Stegeman, A.; Elbers, A.; de Smit, H.; Moser, H.; Smak, J.; Pluimers, F. The 1997–1998 epidemic of classical swine fever in the Netherlands. *Vet. Microbiol.* **2000**, *73*, 183–196. [CrossRef] [PubMed]
9. Meuwissen, M.P.; Horst, S.H.; Huirne, R.B.; Dijkhuizen, A.A. A model to estimate the financial consequences of classical swine fever outbreaks: Principles and outcomes. *Prev. Vet. Med.* **1999**, *42*, 249–270. [CrossRef]
10. Davies, G. The foot and mouth disease (FMD) epidemic in the United Kingdom 2001. *Comp. Immunol. Microbiol. Infect. Dis.* **2002**, *25*, 331–343. [CrossRef]
11. World Organisation for Animal Health. African Swine Fever (ASF)—Situation Report 9 (7 April 2022). Available online: https://www.oie.int/en/what-we-do/animal-health-and-welfare/disease-data-collection/ (accessed on 21 April 2022).
12. Doherr, M.G.; Audigé, L. Monitoring and surveillance for rare health-related events: A review from the veterinary perspective. *Phil. Trans. R. Soc. Lond. B* **2001**, *356*, 1097–1106. [CrossRef]
13. Anderson, L.A.; Black, N.; Hagerty, T.J.; Kluge, J.P.; Sundberg, P.L. *Pseudorabies (Aujeszky's Disease) and Its Eradication. A Review of the U.S. Experience*; Technical Bulletin No. 1923; United States Department of Agriculture, Animal and Plant Health Inspection Service: Riverdale, MD, USA, 2008.
14. Hoinville, L.J.; Alban, L.; Dewe, J.A.; Gibbens, J.C.; Gustafson, L.; Häsler, B.; Saegerman, C.; Salman, M.; Stärk, K.D.C. Proposed terms and concepts for describing and evaluating animal-health surveillance systems. *Prev. Vet. Med.* **2013**, *112*, 1–12. [CrossRef]
15. Bush, E.J.; Gardner, I.A. Animal health surveillance in the United States via the National Animal Health Monitoring System (NAHMS). *Epidemiol. Sante Anim.* **1995**, *27*, 113–126.
16. Gates, M.C.; Earl, L.; Enticott, G. Factors influencing the performance of voluntary farmer disease reporting in passive surveillance systems: A scoping review. *Prev. Vet. Med.* **2021**, *196*, 105487. [CrossRef] [PubMed]
17. Mariner, J.C.; Hendrickx, S.; Pfeiffer, D.U.; Costard, S.; Knopf, L.; Okuthe, S.; Chibeu, D.; Parmley, J.; Musenero, M.; Pisang, C.; et al. Integration of participatory approaches into surveillance systems. *Rev. Sci. Tech. Off. Int. Epiz.* **2011**, *30*, 653–659. [CrossRef] [PubMed]
18. Calba, C.; Antoine-Moussiaux, N.; Charrier, F.; Hendrikx, P.; Saegerman, C.; Peyre, M.; Goutard, F.L. Applying participatory approaches in the evaluation of surveillance systems: A pilot study on African swine fever surveillance in Corsica. *Prev. Vet. Med.* **2015**, *122*, 389–398. [CrossRef]
19. Wójcik, O.P.; Brownstein, J.S.; Chunara, R.; Johansson, M.A. Public health for the people: Participatory infectious disease surveillance in the digital age. *Emerg. Themes Epidemiol.* **2014**, *11*, 7. [CrossRef]
20. Smolinski, M.S.; Crawley, A.W.; Olsen, J.M.; Jayaraman, T.; Libel, M. Participatory disease surveillance: Engaging communities directly in reporting, monitoring, and responding to health threats. *JMIR Pub. Health Surveill.* **2017**, *3*, e7540. [CrossRef] [PubMed]
21. Elbers, A.R.W.; Bouma, A.; Stegeman, J.A. Quantitative assessment of clinical signs for the detection of classical swine fever outbreaks during an epidemic. *Vet. Microbiol.* **2002**, *85*, 323–332. [CrossRef]

22. Harvey, N.; Reeves, A.; Schoenbaum, M.A.; Zagmutt-Vergara, F.J.; Dubé, C.; Hill, A.E.; Corso, B.A.; McNab, W.B.; Cartwright, C.I.; Salman, M.D. The North American Animal Disease Spread Model: A simulation model to assist decision making in evaluating animal disease incursions. *Prev. Vet. Med.* **2007**, *82*, 176–197. [CrossRef]
23. Lowe, J.; Gauger, P.; Harmon, K.; Zhang, J.; Connor, J.; Yeske, P.; Loula, T.; Levis, I.; Dufresne, L.; Main, R. Role of transportation in the spread of porcine epidemic diarrhea virus infection, United States. *Emerg. Infect. Dis.* **2014**, *20*, 872–874. [CrossRef]
24. Meyer, S. Pork Packing: Just What Is Capacity? *National Hog Farmer*, 6 October 2020; pp. 18–22.
25. R Core Team. *R: A Language and Environment for Statistical Computing*; R Foundation for Statistical Computing: Vienna, Austria, 2020; Available online: https://www.R-project.org (accessed on 30 May 2021).
26. Hagenstein, P.R.; Flocchini, R.G.; Bailar, J.C., III; Claiborn, C.; Dickerson, R.R.; Galloway, J.N.; Grossman, M.R.; Kasibhatla, P.; Kohn, R.A.; Lacy, M.P.; et al. *The Scientific Basis for Estimating Air Emissions from Animal Feeding Operations: Interim Report*; The National Academy Press: Washington, DC, USA, 2002; p. 90.
27. Bivand, R.S.; Pebesma, E.; Gómez-Rubio, V. *Applied Spatial Data Analysis with R*, 2nd ed.; Springer: New York, NY, USA, 2013; pp. 146–149.
28. Thakur, K.K.; Revie, C.W.; Hurnik, D.; Poljak, Z.; Sanchez, J. Simulation of between-farm transmission of porcine reproductive and respiratory syndrome virus in Ontario, Canada using the North American Animal Disease Spread Model. *Prev. Vet. Med.* **2015**, *118*, 413–426. [CrossRef] [PubMed]
29. Machado, G.; Galvis, J.A.; Lopes, F.P.N.; Voges, J.; Medeiros, A.A.R.; Cárdenas, N.C. Quantifying the dynamics of pig movements improves targeted disease surveillance and control plans. *Transbound. Emerg. Dis.* **2021**, *68*, 1663–1675. [CrossRef]
30. Müller, K.; Wickham, H.; James, D.A.; Falcon, S. RSQLite: SQLite Interface for R, Version 2.2.4. Available online: https://CRAN.R-project.org/package=RSQLite (accessed on 12 June 2022).
31. Thacker, S.B.; Parrish, R.G.; Trowbridge, F.L. A method for evaluating systems of epidemiological surveillance. *World Health Stat. Q.* **1988**, *41*, 11–18.
32. Nielsen, S.S.; Alvarez, J.; Bicout, D.J.; Calistri, P.; Canali, E.; Drewe, J.A.; Garin-Bastuji, B.; Gonzales Rojas, J.L.; Schmidt, C.G.; Herskin, M.; et al. Assessment of the control measures of the category A diseases of animal health law: Classical swine fever. *EFSA J.* **2021**, *19*, e06707. [CrossRef] [PubMed]
33. Nielsen, S.S.; Alvarez, J.; Bicout, D.J.; Calistri, P.; Depner, K.; Drewe, J.A.; Garin-Bastuji, B.; Gonzales Rojas, J.L.; Schmidt, C.G.; Herskin, M.; et al. Scientific opinion on the assessment of the control measures of the category A diseases of the animal health law: African swine fever. *EFSA J.* **2021**, *19*, e06402. [CrossRef]
34. Christensen, J.; Gardner, I.A. Herd-level interpretation of test results for epidemiologic studies of animal diseases. *Prev. Vet. Med.* **2000**, *45*, 83–106. [CrossRef] [PubMed]
35. Crauwels, A.P.P.; Nielen, M.; Stegeman, J.A.; Elbers, A.R.W.; Dijkhuizen, A.A.; Tielen, M.J.M. The effectiveness of routine serological surveillance: Case study of the 1997 epidemic of classical swine fever in the Netherlands. *Rev. Sci. Tech. Off. Int. Epiz.* **1999**, *18*, 627–637. [CrossRef] [PubMed]
36. Kirkland, P.D.; Le Potier, M.-F.; Finlaison, D. Pestiviruses. In *Diseases of Swine*, 11th ed.; Zimmerman, J.J., Karriker, L.A., Ramirez, A., Schwartz, K.J., Stevenson, G.W., Zhang, J., Eds.; John Wiley and Sons, Inc.: Hoboken, NJ, USA, 2019; pp. 622–640. [CrossRef]
37. Gallardo, C.; Soler, A.; Rodze, I.; Nieto, R.; Cano-Gómez, C.; Fernandez-Pinero, J.; Arias, M. Attenuated and non-haemadsorbing (non-HAD) genotype II African swine fever virus (ASFV) isolated in Europe, Latvia 2017. *Transbound. Emerg. Dis.* **2019**, *66*, 1399–1404. [CrossRef] [PubMed]
38. Bates, T.W.; Thurmond, M.C.; Hietala, S.K.; Venkateswaran, K.S.; Wilson, T.M.; Colston, B.W., Jr.; Trebes, J.E.; Milanovich, F.P. Surveillance for detection of foot-and-mouth disease. *J. Am. Vet. Med. Assoc.* **2003**, *223*, 609–614. [CrossRef]
39. Schulz, K.; Conraths, F.J.; Blome, S.; Staubach, C.; Sauter-Louis, C. African swine fever: Fast and furious or slow and steady? *Viruses* **2019**, *11*, 866. [CrossRef] [PubMed]
40. Munguía-Ramírez, B.; Armenta-Leyva, B.; Giménez-Lirola, L.; Wang, C.; Zimmerman, J. Surveillance on swine farms using antemortem specimens. In *Optimising Pig Herd Health and Production*; Maes, D., Segalés, J., Eds.; Burleigh Dodds Science Publishing: Cambridge, UK, 2023; pp. 97–138.
41. Trevisan, G.; Linhares, L.C.M.; Crim, B.; Dubey, P.; Schwartz, K.J.; Burrough, E.R.; Main, R.G.; Sundberg, P.; Thurn, M.; Lages, P.T.F.; et al. Macroepidemiological aspects of porcine reproductive and respiratory syndrome virus detection by major United States veterinary diagnostic laboratories over time, age group, and specimen. *PLoS ONE* **2019**, *10*, e0223544. [CrossRef] [PubMed]
42. Branson, B.M. Home sample collection tests for HIV infection. *J. Am. Med. Assoc.* **1998**, *280*, 1699–1701. [CrossRef]
43. Tsang, N.N.Y.; So, H.C.; Ng, K.Y.; Cowling, B.J.; Leung, G.M.; Ip, D.K.M. Diagnostic performance of different sampling approaches for SARS-CoV-2 RT-PCR testing: A systematic review and meta-analysis. *Lancet Infect. Dis.* **2021**, *21*, 1233–1245. [CrossRef] [PubMed]
44. Fatima, M.; Luo, Y.; Zhang, L.; Wang, P.-Y.; Song, H.; Fu, Y.; Li, Y.; Sun, Y.; Li, S.; Bao, Y.-J.; et al. Genotyping and molecular characterization of classical swine fever virus isolated in China during 2016–2018. *Viruses* **2021**, *13*, 664. [CrossRef] [PubMed]
45. Eichhorn, G.; Frost, J.W. Study on the suitability of sow colostrum for the serological diagnosis of porcine reproductive and respiratory syndrome (PRRS). *J. Vet. Med. B* **1997**, *44*, 65–72. [CrossRef]
46. Nijsten, R.; London, N.; van den Bogaard, A.; Stobberingh, E. Antibiotic resistance among *Escherichia coli* isolated from faecal samples of pig farmers and pigs. *J. Antimicrob. Chemother.* **1996**, *37*, 1131–1140. [CrossRef] [PubMed]

47. Newberry, K.M.; Colling, A. Quality standards and guidelines for test validation for infectious diseases in veterinary laboratories. *Rev. Sci. Tech. Off. Int. Epiz.* **2021**, *40*, 227–237. [CrossRef] [PubMed]
48. Ribeiro Miguel, A.L.; Lopes Moreira, R.P. Fernando de Oliveira. ISO/IEC 17025: History and introduction of concepts. *Química Nova* **2021**, *44*, 792–796. [CrossRef]
49. Hobbs, E.C.; Colling, A.; Gurung, R.B.; Allen, J. The potential of diagnostic point-of-care tests (POCTs) for infectious and zoonotic animal diseases in developing countries: Technical, regulatory and sociocultural considerations. *Transbound. Emerg. Dis.* **2021**, *68*, 1835–1849. [CrossRef] [PubMed]
50. U.S. Census Bureau. State Area Measurements and Internal Point Coordinates. 2010. Available online: https://www.census.gov/geographies/reference-files/2010/geo/state-area.html (accessed on 25 June 2023).
51. United Nations. *Statistical Yearbook*, 65th ed.; United Nations: New York, NY, USA, 2022; pp. 13–35.
52. Ketusing, N.; Reeves, A.; Portacci, K.; Yano, T.; Olea-Popelka, F.; Keefe, T.; Salman, M. Evaluation of strategies for the eradication of pseudorabies virus (Aujeszky's disease) in commercial swine farms in Chiang-Mai and Lampoon provinces, Thailand, using a simulation disease spread model. *Transbound. Emerg. Dis.* **2014**, *61*, 169–176. [CrossRef] [PubMed]
53. Stanojevic, S.; Valcic, M.; Stanojevic, S.; Radojicic, S.; Avramov, S.; Tambur, Z. Simulation of a classical swine fever outbreak in rural areas of the Republic of Serbia. *Vet. Med.* **2015**, *60*, 553–566. [CrossRef]
54. Hasahya, E.; Thakur, K.K.; Dione, M.M.; Wieland, B.; Oba, P.; Kungu, J.; Lee, H.S. Modeling the spread of porcine reproductive and respiratory syndrome among pig farms in Lira district of northern Uganda. *Front. Vet. Sci.* **2021**, *8*, 727895. [CrossRef] [PubMed]
55. Lee, J.; Schulz, L.L.; Tonsor, G.T. Swine producer willingness to pay for Tier 1 disease risk mitigation under multifaceted ambiguity. *Agribusiness* **2021**, *37*, 858–875. [CrossRef]

Disclaimer/Publisher's Note: The statements, opinions and data contained in all publications are solely those of the individual author(s) and contributor(s) and not of MDPI and/or the editor(s). MDPI and/or the editor(s) disclaim responsibility for any injury to people or property resulting from any ideas, methods, instructions or products referred to in the content.

animals

MDPI

Article

Performance of a Differentiation of Infected from Vaccinated Animals (DIVA) Classical Swine Fever Virus (CSFV) Serum and Oral Fluid Erns Antibody AlphaLISA Assay

Yaowalak Panyasing [1,2,*], Luis Gimenez-Lirola [3], Roongroje Thanawongnuwech [1], Phakawan Prakobsuk [4], Yanee Kawilaphan [4], Apisit Kittawornrat [4], Ting-Yu Cheng [5] and Jeffrey Zimmerman [3]

1 Department of Pathology, Faculty of Veterinary Science, Chulalongkorn University, Bangkok 10330, Thailand; roongroje.t@chula.ac.th
2 Animal Virome and Diagnostic Development Research Unit, Faculty of Veterinary Science, Chulalongkorn University, Bangkok 10330, Thailand
3 Department of Veterinary Diagnostic and Production Animal Medicine, College of Veterinary Medicine, Iowa State University, Ames, IA 50011, USA; luisggl@iastate.edu (L.G.-L.); jjzimm@isu.edu (J.Z.)
4 CPF (Thailand) Public Company Limited, Bangkok 10120, Thailand; phakawan.pra@cpf.co.th (P.P.); yanee.kaw@cpf.co.th (Y.K.); apisit.k@cpf.co.th (A.K.)
5 Department of Veterinary Preventive Medicine, College of Veterinary Medicine, The Ohio State University, Columbus, OH 43210, USA; cheng.1784@osu.edu
* Correspondence: yaowalak.p@chula.ac.th; Tel.: +66-2-218-9627

Simple Summary: Classical swine fever (CSF) is a highly contagious viral disease causing significant economic losses to swine producers in most parts of the world. Previously, pseudorabies virus (PRV) was eliminated from commercial swine herds using gene-deleted vaccines and assays capable of differentiation of infected from vaccinated animals (DIVA). Modeling on this experience, we developed an assay able to differentiate CSFV-infected pigs from pigs vaccinated with a modified live CSFV marker vaccine (Suvaxyn® CSF Marker). In brief, the CSFV Erns IgG AlphaLISA® detects specific antibodies against a CSFV envelope glycoprotein (Erns) present in wild-type viruses. In contrast, the Erns glycoprotein is absent in the CSFV marker vaccine; hence, vaccinated pigs are negative on the CSFV Erns AlphaLISA®. Importantly for ease of use in the field and in the laboratory, the assay is able to test both serum and oral fluid samples. The combined use of a marker vaccine to protect against clinical disease and a serum/oral fluids DIVA assay able to identify wild-type CSFV infections will greatly improve the capacity of the pork industry to deal quickly and decisively with CSFV.

Abstract: Classical swine fever virus (CSFV) is an OIE-listed disease that requires effective surveillance tools for its detection and control. The aim of this study was to develop and evaluate the diagnostic performance of a novel CSFV Erns IgG AlphaLISA for both serum and oral fluid specimens that would likewise be compatible with the use of CSFV E2 DIVA vaccines. Test performance was evaluated using a panel of well-characterized serum (n = 760) and individual (n = 528) or pen-based (n = 30) oral fluid samples from four groups of animals: (1) negative controls (n = 60 pigs); (2) inoculated with ALD strain wild-type CSFV (n = 30 pigs); (3) vaccinated with LOM strain live CSFV vaccine (n = 30 pigs); and (4) vaccinated with live CSFV marker vaccine on commercial farms (n = 120 pigs). At a cutoff of $S/P \geq 0.7$, the aggregate estimated diagnostic sensitivities and specificities of the assay were, respectively, 97.4% (95% CI 95.9%, 98.3%) and 100% for serum and 95.4% (95% CI 92.9%, 97.0%) and 100% for oral fluid. The Erns IgG antibody AlphaLISA combined DIVA capability with solid diagnostic performance, rapid turnaround, ease of use, and compatibility with both serum and oral fluid specimens.

Keywords: AlphaLISA; classical swine fever virus; DIVA; Erns; IgG; oral fluid; serum

1. Introduction

Classical swine fever virus (CSFV) is a member of the genus *Pestivirus* and the family *Flaviviridae* [1]. Pestiviruses infectious for pigs include bovine viral diarrhea viruses 1 and 2, border disease virus, Bungowannah virus, and atypical porcine pestivirus (APPV or Pestivirus K), but classical swine fever (CSF) is the only OIE-listed disease in the group [2]. Although its viral etiology has been recognized since 1903 [3], CSFV remains endemic in many parts of the world. The CSFV imposes a significant economic impact on pig health and international trade, and periodically emerges in CSFV-free areas, e.g., the Netherlands in 1997 [4], Great Britain in 2000 [5], South Africa in 2005 [6], Israel in 2009 [7], and Japan in 2018 [8].

Differentiation of infected from vaccinated animals (DIVA)-enabled vaccines and companion antibody assays were pioneered for the control and/or elimination of Aujeszky's disease (pseudorabies) virus and have been highly effective in this role [9–11]. Continuing on this path, research has been focused on the development of CSFV DIVA vaccines and differential tests [12]. Biologically compatible with this aim, CSFV's positive-sense, single-stranded RNA genome contains a single open reading frame encoding a polyprotein (3898 amino acids) that undergoes co- and post-translation processing to produce four structural proteins (C, Erns, E1, E2) and seven non-structural proteins (Npro, p7, NS2-3, NS4A, NS4B, NS5A, NS5B) [13–16]. Protective immunity against CSFV relies on the production of neutralizing antibodies against structural proteins E2 and Erns [14]. This fact has been used in the development of DIVA CSFV E2 vaccines and their companion Erns-based ELISAs. That is, animals vaccinated with E2 vaccines produce antibodies against CSFV E2 and not Erns, while field virus-infected animals produce antibodies against both E2 and Erns.

CSFV Erns-based ELISAs have been developed for their use with CSF E2-based DIVA vaccines [17], but studies of commercially available DIVA CSFV ELISAs underline the need for further improvement in their diagnostic sensitivity and specificity [18–21]. Alternatively, the bead-based AlphaLISA (amplified luminescent proximity homogenous assay) antibody format reportedly achieves both high analytical sensitivity and rapid turnaround because of short incubation times and the lack of wash steps [22]. The use of powerful analytical platforms such as AlphaLISA, coupled to specific target antigens and DIVA vaccines, allows for the development of highly specific and sensitive DIVA tests, which are also user-friendly and reproducible. AlphaLISAs have previously been developed for the detection of antibodies against the porcine epidemic diarrhea virus [23] and African swine fever virus [24]. The objective of the present study was to develop a CSFV Erns IgG AlphaLISA and evaluate its performance using serum and oral fluid samples from pigs of known CSFV antibody status.

2. Materials and Methods

2.1. Experimental Design

Specimens used for the analysis of the CSFV Erns IgG AlphaLISA included serum and oral fluid samples from pigs inoculated with wild-type (ALD strain) or live vaccine (LOM strain) CSF viruses in research facilities at Chulalongkorn University (Bangkok, Thailand; IACUC #1473003). In addition, to further evaluate the test's DIVA capabilities, field samples (serum and oral fluid) from commercial pigs vaccinated with a live recombinant CSFV marker live attenuated vaccine (Suvaxyn® CSF Marker, Zoetis Belgium SA, Louvain-la-Neuve, Belgium) were also included in this study. The CSFV marker vaccine is based on an infectious cDNA clone of BVDV strain CP7 in which the BVDV E2 was exchanged with the E2 from CSFV Alfort/187 [17]. The presence of CSFV antibody in samples from inoculated or vaccinated pigs was confirmed by the serum virus neutralization test (VNT). Finally, the performance of the CSFV AlphaLISA was determined by receiver operating characteristic (ROC) analyses of the serum and oral fluid testing data.

2.2. Biological Samples

A total of 1318 well-characterized serum (n = 760) and oral fluid (n = 558) samples were used in the study: Group 1. CSFV-naïve pigs (serum n = 160, individual pig oral fluid n = 160); Group 2. CSFV-naïve pigs inoculated with CSFV ALD wild-type strain (serum n = 100, individual pig oral fluid n = 184); Group 3. CSFV-naïve pigs vaccinated with live CSFV LOM strain vaccine (serum n = 100, individual pig oral fluid n = 184); Group 4. CSFV-naïve pigs vaccinated with Suvaxyn® CSFV marker vaccine (serum n = 360, pen-based oral fluids n = 30).

Groups 1, 2, and 3 samples were collected from animals housed in research facilities at Chulalongkorn University using procedures and methods described elsewhere [25]. In brief, Group 1 samples were collected prior to vaccination or inoculation of Groups 2 and 3. Groups 2 (n = 30 pigs) and 3 (n = 30 pigs) samples were collected from pigs inoculated with wild-type (ALD strain; 10^5 $TCID_{50}$/mL) CSFV or vaccinated with live vaccine virus (LOM strain; 10^3 $TCID_{50}$/mL), respectively. The serum and oral fluid samples used in the study were collected between days post-inoculation (DPIs) or days post-vaccination (DPVs) 14 and 28 [25]. Oral fluid samples were collected as described elsewhere [25]. Group 4 samples were collected from pigs in commercial swine production systems intramuscularly vaccinated with the Suvaxyn® CSF Marker vaccine (CP7_E2alf; $10^{4.8}$ to $10^{6.5}$ $TCID_{50}$/mL) at 9 weeks of age and sampled at DPVs 14, 30, and 60. At each collection, serum was collected from 120 pigs, and pen-based oral fluids were collected from 10 pens, each holding 50 to 60 pigs.

2.3. Virus Neutralization Test (VNT)

Virus neutralization testing was performed as previously described [25]. Briefly, the test was performed in 96-well microtiter plates. Minimum essential medium (MEM; 50 µL; Life Technologies, Paisley, UK) was added to all wells. Then, serum (50 µL) was added and serially 2-fold diluted. Thereafter, stock CSFV strain ALD (50 µL; 300 $TCID_{50}$/50 µL) was added to each well and the plate incubated for 1 h (37 °C, 5% CO_2). Then, SK6 cells (100 µL; 3×10^5 cells/mL) were added to each well and the plate incubated for 72 h (37 °C, 5% CO_2). Subsequently, the plate was tested with neutralizing peroxidase-liked assay (NPLA) as previously described [25] to determine the neutralizing antibody titer. Neutralizing antibody titers were measured as the reciprocal of the highest dilution of serum that completely inhibited viral infection, with neutralizing antibody titers ≥ 2 ($\log_2 1$) considered positive for CSFV antibody.

2.4. CSFV Erns Antigen

Erns antigen was produced as previously described [26]. In brief, the codified region of the truncated N-terminal CSFV Erns gene was synthetically produced (Shanghai Genery Biotech Co., Ltd., Shanghai, China) with the addition of a 5′ GP67 terminal signal peptide for expression (Bac-to-Bac™ Baculovirus Vector System, Invitrogen, Carlsbad, CA, USA), followed by a hexa-histidine tag and a "GS linker" for protein purification by affinity chromatography. The PCR-amplified Erns region was cloned into pFastBac1 (Invitrogen) using EcoRI and XhoI restriction enzymes, and confirmed by digestion and sequencing. CSFV Erns protein was purified from culture supernatant by Ni-chelating SFF affinity chromatography (GE Healthcare), according to the manufacturer's instructions. Protein elution was dialyzed and analyzed by SDS-PAGE and Western blot. Erns molecular weight measured by SDS-PAGE was 25–35 KDa. The final concentration of Erns protein was 0.75 mg/mL.

2.5. CSFV Erns AlphaLISA Bead Conjugation

The protein buffer was exchanged with PBS (0.1 M $NaKPO_4$ pH 8.0) on a Zeba™ spin column (Thermo Fisher Scientific, Rockford, IL, USA) and concentrated up to 1 mg/mL protein (BCA protein assay). AlphaLISA acceptor beads were washed with the protein buffer (pH 8) following the manufacturer's protocol. The Erns protein was conjugated to raw acceptor beads via reductive amination, as described by the manufacturer (PerkinElmer Health Sciences, Inc., Boston, MA, USA). Briefly, 0.1 mg of Erns protein (1 mg/mL, 100 µL) was mixed with 1 mg AlphaLISA® beads and reaction buffer (100 mM HEPES pH 7.4) in a 1.5 mL microtube to obtain a final reaction volume of 200 µL. Then, 1.25 µL of 10% Tween-20 (Millipore-Sigma) and 10 µL of a 400 mM solution of the reducing agent sodium cyanoborohydride (NaBH3CN; Millipore-Sigma, St. Louis, MO, USA) were added to the microtube. The tube was incubated for 18–24 h at 37 °C with mild agitation at 6–10 rpm. Thereafter, unreacted aldehyde groups on the beads were blocked by adding 10 µL of 65 mg/mL solution of carboxymethoxylamine (CMO, Millipore-Sigma) in 800 mM sodium hydroxide (NaOH; Millipore-Sigma) followed by a 60 min incubation at 37 °C under mild agitation. The uncoupled Erns protein was removed by centrifugation (16,000× *g* for 15 min at 4 °C). Then, it was washed two times with 200 µL of 100 mM Tris-HCl pH 8. The supernatant was then removed with a micropipette and the bead pellet resuspended in 200 µL of storage buffer (PBS pH 7.4 containing 0.05% Proclin-300 (Millipore-Sigma)). The bead solution was briefly sonicated with ultrasonic homogenizer (UX-050) (Mitsui-EC Corp, Tokyo, Japan) and then stored at 4 °C until assayed.

2.6. CSFV Erns IgG AlphaLISA Optimization

The CSFV Erns AlphaLISA was designed as an IgG assay (Figure 1). In brief, Erns IgG in the sample binds to Erns protein that conjugated to acceptor beads and biotinylated anti-swine IgG that conjugated to streptavidin donor beads. Binding of the Ern IgG and anti-swine IgG brings the donor and acceptor beads into proximity. Laser excitation (680 nm) of the donor beads releases singlet oxygen molecules that trigger a cascade of chemical events in acceptor beads and results in a chemiluminescent emission (615 nm) detected by the microplate reader.

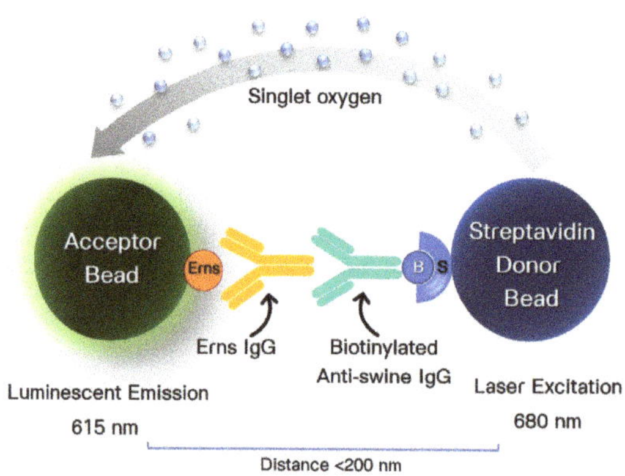

Figure 1. Classical swine fever virus (CSFV) amplified luminescent proximity homogenous assay (AlphaLISA®) platform for detection of CSFV Erns IgG antibody. In the presence of Erns IgG in a tested sample (serum, oral fluid), the Erns IgG binds Erns protein that conjugated to acceptor beads, and the biotinylated anti-swine IgG that conjugated to streptavidin donor beads. Binding of the antibody brings the donor and acceptor beads into proximity. Excitation of the donor beads at 680 nm promotes the release of singlet oxygen molecules that trigger a cascade of chemical events in the acceptor beads, resulting in a light emission at 615 nm.

To create the CSFV Erns AlphaLISA, sample dilution, incubation time, Erns protein-conjugated acceptor bead concentration, streptavidin donor bead concentration, and biotinylated goat anti-swine IgG concentration were systematically evaluated to optimize the signal and minimize background. For example, the optimal concentration of biotinylated goat anti-swine IgG was determined by evaluating concentrations of 10, 20, 40, 80, 100 nM for serum and concentrations of 20, 40, 80, 100, 160 nM for oral fluids (Figure 2).

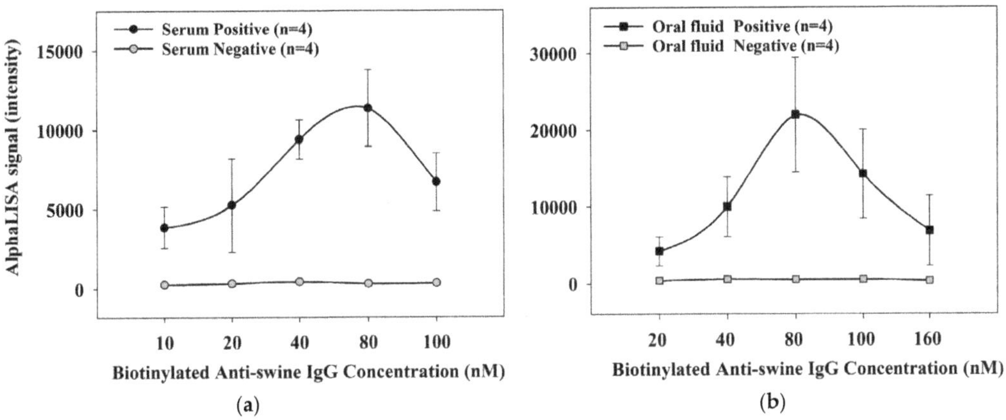

Figure 2. Optimization of biotinylated anti-swine IgG antibody concentrations for serum and oral fluid AlphaLISA®. The optimum combination of a mix containing 25 µg/mL of Erns protein-conjugated acceptor beads and 80 µg/mL of streptavidin donor beads were tested with various biotinylated anti-swine IgG concentrations: (**a**) negative and positive sera were evaluated at 10, 20, 40, 80, and 100 nM; (**b**) negative and positive oral fluids were tested at 20, 40, 80, 100, and 160 nM. The AlphaLISA® signal represents the mean values (n = 4) with error bars denoting the standard deviation.

The final one-hour procedure consisted of two steps. First, 5 µL of sample (serum or oral fluid) or control (negative and positive) was incubated for 30 min with 20 µL of a mix containing 25 µg per mL Erns protein-conjugated AlphaLISA acceptor beads and 80 nM biotinylated goat anti-swine IgG (Fc) (Bethyl Laboratories Inc., Montgomery, TX, USA) in assay buffer 1x (AlphaLISA buffer 10x; PerkinElmer, Waltham, MA, USA). In-house positive and negative controls (5 µL) were included on each plate (in duplicate) for both serum and oral fluid assays. In-house negative controls were serum samples obtained from unvaccinated and uninoculated pigs and positive controls were serum samples from pigs infected with CSFV. In the second step, 25 µL of 80 µg per mL streptavidin donor beads (PerkinElmer) was added to each well and the plates incubated for 30 min. Incubations were performed at room temperature (22–25 °C) in half-area 96-well white plates (PerkinElmer) while protected from light and held on a rotating shaker (200 rpm). Thereafter, plates were read (615 nm) using a microplate reader (Envision® 2105 multimode plate reader, PerkinElmer, Waltham, MA, USA) operated with the Envision Manager software version 1.14 (PerkinElmer). For quality purposes, negative control AlphaLISA signals (counts) were required to range from 0–400 and positive controls from 4000–7000. Sample AlphaLISA responses were converted to sample-to-positive (S/P) ratios:

$$\text{S/P ratio} = \frac{\text{Sample AlphaLISA signal} - \text{Negative control signal (mean)}}{\text{Positive control signal (mean)} - \text{Negative control signal (mean)}} \quad (1)$$

2.7. Statistical Analysis

CSFV S/P values for serum and oral fluids were analyzed by ANOVA and post hoc Tukey's pairwise comparisons. ROC analyses were done to evaluate the performance of the serum and oral fluid CSFV Erns AlphaLISAs using R software package pROC [27]. Diagnostic sensitivity and specificity were derived from the ROC analyses for specific assay cutoffs. In the ROC analyses, samples from Groups 1 and 4 were classified as Erns antibody-negative and samples from Groups 2 and 3 collected $\geq$ 14 DPI or DPV were considered positive. Associated 95% CIs for diagnostic sensitivity and specificity were estimated using a procedure for normally distributed correlated data [28]. In brief, the correlation of the data was taken account by fitting normalized data into a linear mixed model.

$$Y_{ij} = \mu + \gamma_i + \tau s_{ij} + \epsilon_{ij} \quad (2)$$

In Equation (2), Y_{ij} is the jth observation for the ith subject; μ is the overall mean for samples classified as CSFV antibody-negative; γ_i is the random effect of the ith subject; τ is the fixed effect indicating the mean difference between CSFV antibody-negative and positive groups; s_{ij} is the disease status of the jth observation for the ith subject; and ϵ_{ij} is the random error of the jth observation for the ith subject. The variances from Equation (2) were used to calculate the 95% CIs for ROC-derived diagnostic sensitivity and diagnostic specificity estimates. Logit transformation was used to prevent the estimated intervals from exceeding the range of probability, i.e., [0, 1]. Prior to performing the ROC analyses, the S/P data of serum and oral fluid AlphaLISAs were normalized by 1/5 and 1/4 power transformations, respectively.

3. Results

3.1. CSFV Antibody Responses

Serum and oral fluid CSFV VNT and Erns IgG AlphaLISA antibody responses are shown in Table 1 by group. All serum samples (n = 160) from Group 1 CSFV-naïve pigs were negative by VNT and all serum samples from Groups 2, 3, and 4 were VNT-positive (neutralizing antibody titers $\geq$ 2). Within Groups 2, 3, and 4, Tukey's pairwise comparison tests detected no difference in S/P responses between serum and oral fluids. However, comparisons between Groups 2, 3, and 4 showed higher Erns S/P responses in both serum and oral fluids from Group 2 pigs versus those in serum and oral fluids from Groups 3 or 4.

Table 1. Classical swine fever virus (CSFV) antibody responses by group and specimen.

Group and CSFV Status	Specimen (No. Samples)	Virus Neutralization Geometric Mean Titer ($\log_2$ x) (95% CI)	Erns IgG AlphaLISA Mean Sample-to-Positive (S/P) Ratio (95% CI)
1. Negative (unvaccinated and uninoculated).	Serum (n = 160) OF from individual pigs (n = 160)	negative •	−0.01 ^{a} (−0.01, 0.00) −0.00 ^{a} (−0.00, 0.00)
2. Inoculated CSFV ALD wild-type strain.	Serum (n = 100) OF from individual pigs (n = 184)	6.17 (5.82, 6.52) •	7.48 ^{b} (6.55, 8.40) 7.29 ^{b} (5.90, 8.68)
3. Vaccinated CSFV LOM live vaccine.	Serum (n = 100) OF from individual pigs (n = 184)	4.01 (3.79, 4.23) •	1.93 ^{c} (1.76, 2.11) 1.89 ^{c} (1.80, 1.99)
4. Vaccinated CSF marker live vaccine.	Serum (n = 360) OF from pens of pigs (n = 30)	5.09 (4.91, 5.27) •	0.05 ^{a} (0.04, 0.06) 0.01 ^{a} (−0.01, 0.04)

• = not tested; OF = oral fluids; $^{a–c}$ = different superscripts indicate that the mean of S/P responses differed significantly between treatment groups or specimen types ($p < 0.05$).

3.2. Evaluation of Test Diagnostic Performance

The results of the ROC curve analysis of the CSFV Erns IgG AlphaLISA based on the aggregate testing data for serum and oral fluid are shown in Table 2. At a cutoff of S/P $\geq$ 0.7, the aggregate estimated diagnostic sensitivities and specificities of the assay were, respectively, 97.4% (95% CI 95.9, 98.3) and 100% for serum and 95.4% (95% CI 92.9, 97.0) and 100% for oral fluid.

The results of the ROC analysis by treatment group are given in Table 3. Using a cutoff of S/P $\geq$ 0.7, the diagnostic specificity was 100% for both serum and oral fluid samples from the negative control (Group 1) and DIVA-vaccinated animals (Group 4). In ALD wild-type-inoculated (Group 2) pigs, the diagnostic sensitivities for serum and oral fluids were 98.0% (95% CI 93.0, 99.8) and 92.9% (95% CI 88.2, 96.2), respectively. In LOM strain-vaccinated (Group 3) pigs, the diagnostic sensitivities for serum and oral fluid tests were 96.4% (95% CI 91.9, 98.3) and 96.7% (93.0, 98.8), respectively.

Table 2. Classical swine fever virus (CSFV) Erns IgG AlphaLISA aggregate (Groups 1, 2, 3, and 4 ‡) diagnostic sensitivity (Dx Se) and diagnostic specificity (Dx Sp) by cutoff.

Cutoff (S/P)	Serum		Oral Fluids	
	Dx Se (95% CI)	Dx Sp (95% CI)	Dx Se (95% CI)	Dx Sp (95% CI)
≥0.1	100 (100, 100) *	88.3 (85.1, 90.8)	99.7 (99.2, 99.9)	100 (100, 100) *
≥0.2	100 (100, 100) *	94.8 (93.0, 96.2)	99.7 (99.4, 99.9)	100 (100, 100) *
≥0.3	97.9 (96.3, 98.8)	99.8 (98.6, 99.9)	99.7 (99.5, 99.9)	100 (100, 100) *
≥0.4	97.9 (96.5, 98.8)	99.8 (99.7, 99.9)	99.7 (99.5, 99.8)	100 (100, 100) *
≥0.5	97.4 (95.7, 98.4)	99.8 (99.7, 99.9)	99.5 (99.1, 99.7)	100 (100, 100) *
≥0.6	97.4 (95.8, 98.4)	99.8 (99.7, 99.9)	97.0 (95.2, 98.1)	100 (100, 100) *
≥0.7	97.4 (95.9, 98.3)	100 (100, 100) *	95.4 (92.9, 97.0)	100 (100, 100) *
≥0.8	97.4 (96.0, 98.3)	100 (100, 100) *	93.5 (90.2, 95.7)	100 (100, 100) *
≥0.9	97.4 (96.0, 98.3)	100 (100, 100) *	92.1 (88.5, 94.7)	100 (100, 100) *
≥1.0	97.4 (96.1, 98.3)	100 (100, 100) *	91.3 (87.5, 94.0)	100 (100, 100) *

* 95% CIs were estimated at sensitivity/specificity = 99.99% because logit (100%) is undefined. ‡ Group 1, negative controls; Group 2, inoculated with ALD strain wild-type CSFV; Group 3, vaccinated with LOM strain live CSFV vaccine; Group 4, vaccinated with CSF marker vaccine.

Table 3. Classical swine fever virus (CSFV) antibody positivity (%, 95% CI) by group, specimen type, and assay.

Group and CSFV Status	Specimen (No. Samples)	VNT	Erns IgG AlphaLISA by Cutoff		
			S/P ≥ 0.3	S/P ≥ 0.5	S/P ≥ 0.7
1. Negative (unvaccinated and uninoculated).	Serum (n = 160)	0.0 (0.0, 2.3)	0.0 (0.0, 2.3)	0.0 (0.0, 2.3)	0.0 (0.0, 2.3)
	OF from individual pigs (n = 160)	•	0.0 (0.0, 2.3)	0.0 (0.0, 2.3)	0.0 (0.0, 2.3)
2. Inoculated CSFV ALD wild-type strain.	Serum (n = 100)	100 (96.4, 100)	98.0 (93.0, 99.8)	98.0 (93.0, 99.8)	98.0 (93.0, 99.8)
	OF from individual pigs (n = 184)	•	100 (98.0, 100)	100 (98.0, 100)	92.9 (88.2, 96.2)
3. Vaccinated CSFV LOM live vaccine.	Serum (n = 100)	100 (97.4, 100)	96.4 (91.9, 98.3)	96.4 (91.9, 98.3)	96.4 (91.9, 98.3)
	OF from individual pigs (n = 184)	•	99.5 (97.0, 100)	99.5 (97.0, 100)	96.7 (93.0, 98.8)
4. Vaccinated CSF marker live vaccine.	Serum (n = 360)	100 (99.0, 100)	1.1 (0.3, 2.2)	0.3 (0.0, 1.5)	0.0 (0.0, 1.0)
	OF from pens of pigs (n = 30)	•	0.0 (0.0, 11.6)	0.0 (0.0, 11.6)	0.0 (0.0, 11.6)

• = not tested; OF = oral fluids; VNT = virus neutralization test; S/P = sample-to-positive ratio.

4. Discussion

Detectable levels of CSFV-specific antibody generally appear 10–15 days after infection or vaccination [25,29], with protective immunity conferred by the neutralizing antibody against E2 and Erns envelope glycoproteins [30]. CSFV neutralizing antibodies are measured by virus neutralization tests (VNT), which mainly detect E2-specific antibodies. The VNT is considered a confirmatory test for antibody detection and the most suitable method for discriminating CSFV-specific antibodies from antibodies against other pestiviruses that infect pigs, i.e., BVDV, BDV, APPV, and others [8]. In this study, a VNT was performed to provide information regarding the antibody status of infected or vaccinated pigs.

The results from the current study support previous reports that oral fluid CSFV antibody assays are technically possible and can provide similar diagnostic sensitivity and specificity to that achieved with serum [26]. Oral fluid samples from individual pigs or pens of pigs are easily collected and, for that reason, routinely used by pork producers and swine veterinarians for the detection of various swine diseases [31]. Oral fluid assays compatible with high-throughput testing, i.e., AlphaLISA, can enhance the effective CSFV surveillance approach at the population level.

Likewise, the ability to rapidly identify CSFV infections with an easy-to-use method would facilitate surveillance. That is, the CSFV Erns IgG AlphaLISA has a total turnaround time of one hour, with an identical protocol used for both serum and oral fluid specimens.

Serum and oral fluid Erns AlphaLISA assays likewise provide flexibility for specific CSFV situations. That is, by using different cutoffs, the rate of serum and oral fluid false negative and false positive results from CSFV Erns antibody-negative (Groups 1 and 4) and -positive (Groups 2 and 3) groups could be adjusted for the immediate surveillance objective (Table 2). Higher S/P cutoffs (e.g., $S/P \geq 0.7$) could be used to increase the test's diagnostic specificity when screening presumed negative populations. In contrast, the use of lower S/P cutoffs could be used to increase its diagnostic sensitivity, i.e., when in a "case-finding" mode after a confirmed case.

The Erns IgG AlphaLISA developed in this study provided high diagnostic sensitivity and specificity for detecting CSFV Erns antibody in serum. At a cutoff of 0.7, the assay showed a diagnostic sensitivity of 97.4% with serum from CSFV-infected pigs, and a diagnostic specificity of 100% with sera from pigs vaccinated once with CP7_E2alf. In contrast, a previous report evaluating commercial Erns antibody ELISAs from various laboratories and sample sets showed a diagnostic sensitivity of 90–98% with sera from CSFV-infected pigs and a diagnostic specificity of 89–96% with sera from CP7_E2alf-vaccinated pigs [20]. For oral fluids and using the same cutoff (0.7), the Ern IgG AlphaLISA showed a diagnostic sensitivity of 92.9% with individual oral fluids from CSFV-infected pigs and a diagnostic specificity of 100% with pen-based oral fluids from pigs vaccinated once with CP7_E2alf. There is only one prior report of a CSFV oral fluid antibody assay, i.e., an IgG Erns ELISA [27]. In that study, an IgG Erns ELISA provided a diagnostic sensitivity of 97.5% and diagnostic specificity of 100% for individual pig oral fluid samples from CSFV-infected or live virus-vaccinated pigs, and from uninfected and unvaccinated pigs, respectively [27]. However, the specificity of the oral fluid Erns ELISA was not evaluated for the DIVA properties in that study.

The AlphaLISA described in this study was specifically designed to detect CSFV-specific Erns antibodies, the basis of CSFV DIVA testing. The assay was evaluated using samples from pigs vaccinated with the marker vaccine based on a chimeric pestivirus (CP7_E2alf). The vaccine is based on the BVDV-1 strain CP7 in which the E2 protein coding sequence was replaced by the corresponding sequence of the CSFV strain Alfort/187 [32]. Thus, Erns-specific IgG antibodies were detected in serum and oral fluid samples from pigs inoculated with a wild-type CSFV isolate (Group 2) or from pigs vaccinated with a modified live vaccine (Group 3), but not in pigs vaccinated with the DIVA vaccine (Group 4). In light of the strong data presented here and given the significant animal health implications of these results, additional studies evaluating the performance of the assay are justified, e.g., repeatability, reproducibility, cross-reactivity with other pestiviruses, as well as responses in DIVA-vaccinated and -challenged pigs.

5. Conclusions

We developed and optimized an AlphaLISA system for detecting serum and oral fluid CSFV Erns IgG antibody in swine. The assay provided excellent diagnostic performance in terms of detecting Erns antibody in CSFV-infected or whole virus-vaccinated pigs and demonstrated the ability to differentiate pigs vaccinated with CSF E2-based DIVA vaccines.

Author Contributions: Y.P.; Conceptualization, Investigation, Methodology (AlphaLISA), Analysis, Writing of original draft, L.G.-L.; Conceptualization, Methodology, Review and editing, R.T.; Conceptualization, Review and editing, A.K.; Conceptualization, Methodology, P.P. and Y.K.; Methodology (virus neutralization test), T.-Y.C.; Statistical analysis, J.Z.; Conceptualization, Methodology, Review and editing. All authors have read and agreed to the published version of the manuscript.

Funding: This work was funded by Pork Checkoff funds distributed through the National Pork Board (grant number 20-148), Des Moines, Iowa, USA; the National Research Council of Thailand (NRCT): R. Thanawongnuwech Senior Scholar 2022 (grant number N42A650553).

Institutional Review Board Statement: Ethical review and approval were waived for this study. Diagnostic samples used in the study were obtained from the previous study conducted under the approval of the Chulalongkorn University institutional Animal Care and Use Committee (#1473003).

Informed Consent Statement: Not applicable.

Data Availability Statement: All data generated or analyzed during this study are included in this published article.

Acknowledgments: CSFV marker vaccines (Suvaxyn®) were kindly provided by Zoetis.

Conflicts of Interest: Phakawan Prakobsuk, Yanee Kawilaphan, and Apisit Kittawornrat are employed by CPF (Thailand) Public Company Limited. The remaining authors declare that the research was conducted in absence of any commercial or financial relationships that could be constructed as a potential conflict of interest.

References

1. Postel, A.; Smith, D.B.; Becher, P. Proposed update to the taxonomy of pestiviruses: Eight additional species with the genus *Pestivirus*, family *Flaviviridae*. *Viruses* **2021**, *13*, 1542. [CrossRef] [PubMed]
2. Smith, D.B.; Meyers, G.; Bukh, J.; Gould, E.A.; Monath, T.; Muerhoff, A.S.; Pletnev, A.; Rico-Hesse, R.; Stapleton, J.T.; Simmonds, P.; et al. Proposed revision to the taxonomy of the genus *Pestivirus*, family *Flaviviridae*. *J. Gen. Virol.* **2017**, *98*, 2106–2112. [CrossRef] [PubMed]
3. de Schweinitz, E.A.; Dorset, M.A. *A Form of Hog Cholera Not Caused the Hog Cholera Bacillus*; U.S. Bureau of Animal Industry Circular no. 41; U.S. Department of Agriculture, Government Printing Office: Washington, DC, USA, 1903.
4. Stegeman, A.; Elbers, A.; de Smit, H.; Moser, H.; Smak, J.; Pluimers, F. The 1997–1998 epidemic of classical swine fever in the Netherlands. *Vet. Microbiol.* **2000**, *73*, 183–196. [CrossRef] [PubMed]
5. Paton, D. The reappearance of classical swine fever in England in 2000. In *Trends in Emerging Viral Infections of Swine*; Morilla, A., Yoon, K.-J., Zimmerman, J., Eds.; Iowa State Press: Ames, IA, USA, 2002; pp. 153–158.
6. Sandvik, T.; Crooke, H.; Drew, T.W.; Blome, S.; Greiser-Wilke, I.; Moennig, V.; Gous, T.A.; Gers, S.; Kitching, J.A.; Bührmann, G.; et al. Classical swine fever in South Africa after 87 years' absence. *Vet. Rec.* **2005**, *157*, 267. [CrossRef] [PubMed]
7. David, D.; Edri, N.; Yakobson, B.A.; Bombarov, V.; King, R.; Davidson, I.; Pozzi, P.; Hadani, Y.; Bellaiche, M.; Schmeiser, S.; et al. Emergence of classical swine fever virus in Israel in 2009. *Vet. J.* **2011**, *190*, e146–e149. [CrossRef] [PubMed]
8. Postel, A.; Austermann-Busch, S.; Petrov, A.; Moennig, V.; Becher, P. Epidemiology, diagnosis and control of classical swine fever: Recent developments and future challenges. *Transbound. Emerg. Dis.* **2018**, *65* (Suppl. S1), 248–261. [CrossRef] [PubMed]
9. van Oirschot, J.T.; De Waal, C.A. An ELISA to distinguish between Aujeszky's disease vaccinated and infected pigs. *Vet. Rec.* **1987**, *121*, 305–306. [CrossRef] [PubMed]
10. van Oirschot, J.T.; Gielkens, A.L.; Moormann, R.J.; Berns, A.J. Marker vaccines, virus protein-specific antibody assays and the control of Aujeszky's disease. *Vet. Microbiol.* **1990**, *23*, 85–101. [CrossRef]
11. van Oirschot, J.T. Diva vaccines that reduce virus transmission. *J. Biotechnol.* **1999**, *73*, 195–205. [CrossRef]
12. Blome, S.; Moβ, C.; Reimann, I.; König, P.; Beer, M. Classical swine fever vaccines—State-of-the art. *Vet. Microbiol.* **2017**, *206*, 10–20. [CrossRef]
13. Rümenapf, T.; Unger, G.; Strauss, J.H.; Thiel, H.J. Processing of the envelope glycoproteins of pestiviruses. *J. Virol.* **1993**, *67*, 3288–3294. [CrossRef] [PubMed]
14. Meyers, G.; Thiel, H.J. Molecular characterization of pestiviruses. *Adv. Virus Res.* **1996**, *47*, 53–118. [CrossRef] [PubMed]
15. Tautz, N.; Elbers, K.; Stoll, D.; Meyers, G.; Thiel, H.J. Serine protease of pestiviruses: Determination of cleavage sites. *J. Virol.* **1997**, *71*, 5415–5422. [CrossRef] [PubMed]
16. Zhang, H.; Cao, H.W.; Wu, Z.J.; Cui, Y.D. A review of molecular characterization of classical swine fever virus (CSFV). *Isr. J. Vet. Med.* **2011**, *66*, 89–95.
17. Blome, S.; Wernike, K.; Reimann, I.; König, P.; Moβ, C.; Beer, M. A decade of research into classical swine fever marker vaccine CP7_E2alf (Suvaxyn® CSF Marker): A review of vaccine properties. *Vet. Res.* **2017**, *48*, 51. [CrossRef] [PubMed]
18. Floegel-Niesmann, G. Classical swine fever (CSF) marker vaccine. Trial III. Evaluation of discriminatory ELISAs. *Vet. Microbiol.* **2001**, *83*, 121–136. [CrossRef] [PubMed]
19. Schroeder, S.; von Rosen, T.; Blome, S.; Loeffen, W.; Haegeman, A.; Koenen, F.; Uttenthal, A. Evaluation of classical swine fever virus antibody detection assays with an emphasis on the differentiation of infected from vaccinated animals. *Rev. Sci. Tech.* **2012**, *31*, 997–1010. [CrossRef]
20. Pannhorst, K.; Frohlich, A.; Staubach, C.; Meyer, D.; Blome, S.; Becher, P. Evaluation of an Ern-based enzyme-linked immunosorbent assay to distinguish Classical swine fever virus-infected pigs from pigs vaccinated with CP7_E2alf. *J. Vet. Diagn. Investig.* **2015**, *27*, 449–460. [CrossRef]
21. Meyer, D.; Fritsche, S.; Luo, Y.; Engemann, C.; Blome, S.; Beyerbach, M.; Chang, C.-Y.; Qiu, H.-J.; Becher, P.; Poste, A. The double-antigen ELISA concept for early detection of Erns-specific classical swine fever virus antibodies and application as an accompanying test for differentiation of infected from marker vaccinated animals. *Transbound. Emerg. Dis.* **2017**, *64*, 2013–2022. [CrossRef]
22. Beaudet, L.; Rodriguez-Suarez, R.; Venne, M.-H.; Caron, M.; Bédard, J.; Brechler, V.; Parent, S.; Bielefeld-Sévigny, M. AlphaLISA® immunoassays: The no-wash alternative to ELISAs for research and drug discovery. *Nat. Methods* **2008**, *5*, 8–9. [CrossRef]

23. Kimpston-Burkgren, K.; Mora-Diaz, J.C.; Roby, P.; Bjustrom-Kraft, J.; Main, R.; Bosse, R.; Gimenez-Lirola, L.G. Characterization of the humoral immune response to porcine epidemic diarrhea virus infection under experimental and field conditions using an AlphaLISA® platform. *Pathogens* **2020**, *9*, 233. [CrossRef] [PubMed]
24. Chen, D.; Wang, D.; Wang, C.; Wei, F.; Zhao, H.; Lin, X.; Wu, S. Application of an AlphaLISA® for rapid sensitive detection of African swine fever virus in porcine serum. *Appl. Microbiol. Biotechnol.* **2021**, *105*, 4751–4759. [CrossRef] [PubMed]
25. Panyasing, Y.; Kedkovid, R.; Thanawongnuwech, R.; Kittawornrat, A.; Ji, J.; Giménez-Lirola, L.; Zimmerman, J. Effective surveillance for early classical swine fever virus detection will utilize both virus and antibody detection capabilities. *Vet. Microbiol.* **2018**, *216*, 72–78. [CrossRef] [PubMed]
26. Panyasing, Y.; Thanawongnuwech, R.; Ji, J.; Gimenez-Lirola, L.; Zimmerman, J. Detection of classical swine fever virus (CSFV) E2 and Erns antibody (IgG, IgA) in oral fluid specimens from inoculated (ALD strain) or vaccinated (LOM strain) pigs. *Vet. Microbiol.* **2018**, *224*, 70–77. [CrossRef] [PubMed]
27. Robin, X.; Turck, N.; Hainard, A.; Tiberti, N.; Lisacek, F.; Sanchez, J.-C.; Müller, M. pROC: An open-source package for R and S+ to analyze and compare ROC curves. *BMC Bioinform.* **2011**, *12*, 77. [CrossRef] [PubMed]
28. Du, Y. Choosing Cutoff Values for Correlated Continuous Diagnostic Data to Estimate Sensitivity and Specificity. Ph.D. Thesis, Iowa State University, Ames, IA, USA, 2018.
29. Moennig, V. Introduction to classical swine fever: Virus, disease, and control policy. *Vet Microbiol.* **2000**, *73*, 93–102. [CrossRef] [PubMed]
30. König, M.; Lengsfeld, T.; Pauly, T.; Stark, R.; Thiel, H.J. Classical swine fever virus: Independent induction of protective immunity by two structural glycoproteins. *J. Virol.* **1995**, *69*, 6479–6486. [CrossRef] [PubMed]
31. Henao-Diaz, A.; Gimenez-Lirola, L.; Baum, D.H.; Zimmerman, J. Guidelines for oral fluid-based surveillance of viral pathogens in swine. *Porc. Health Manag.* **2020**, *6*, 28. [CrossRef]
32. Reimann, I.; Depner, K.; Trapp, S.; Beer, M. An avirulent chimeric Pestivirus with altered cell tropism protects pigs against lethal infection with classical swine fever virus. *Virology* **2004**, *322*, 143–157. [CrossRef]

Disclaimer/Publisher's Note: The statements, opinions and data contained in all publications are solely those of the individual author(s) and contributor(s) and not of MDPI and/or the editor(s). MDPI and/or the editor(s) disclaim responsibility for any injury to people or property resulting from any ideas, methods, instructions or products referred to in the content.

animals

MDPI

Communication

Associations of Socio-Demographic Characteristics of Dairy Goat Farmers in Greece with Biosecurity-Related Practices in the Farms

Daphne T. Lianou * and George C. Fthenakis

Veterinary Faculty, University of Thessaly, 43100 Karditsa, Greece
* Correspondence: dlianou@vet.uth.gr

Simple Summary: This work studied the situation regarding biosecurity practices in goat farms in Greece and their potential associations with the socio-demographic characteristics of farmers. The findings indicated that, in goat farms, the socio-demographic characteristics of farmers are associated with the biosecurity level therein. This study identified that the presence of working staff on the farm and the length of farming experience were characteristics associated with the application of most biosecurity-related individual practices and procedures. Female farmers and the presence of working staff were associated with the overall level of biosecurity on the farms. This knowledge can be useful when developing biosecurity programs on goat farms.

Abstract: Given the significance of goat farming in Greece, our study aimed to explore biosecurity conditions on goat farms and refers to an investigation performed on 119 farms countrywide in Greece. The objective of the present work was to evaluate and report on potential associations between the socio-demographic characteristics of goat farmers and practices that play a role in biosecurity on farms. Data were obtained during a large cross-sectional study performed across Greece in 119 goat herds. The socio-demographic characteristics of farmers, as well as 16 variables related to biosecurity, were recorded; these were practices or events that could potentially enhance or compromise biosecurity on farms. A score based on all the biosecurity-related variables was devised by assigning a '1' or '−1' mark for each variable recorded on a farm that could enhance or compromise, respectively, biosecurity. Among variables potentially enhancing biosecurity on farms, the application of disinfections on the farm and the maintenance of isolation for sick animals were practiced more frequently; among variables potentially compromising biosecurity on farms, the presence of hunters in the area around the farm and grazing practices for animals were recorded more frequently. The median overall score for biosecurity-related practices for all farms in the study was 1 (interquartile range: 2.5; minimum: −4; maximum: 5). Among socio-demographic characteristics, the presence of working staff on a farm was associated with five individual biosecurity-related practices and the length of farming experience with three individual biosecurity-related practices. In the multivariable analysis, female farmers ($p = 0.007$) and the presence of working staff on the farm ($p = 0.025$) emerged as the two significant socio-demographic characteristics with an association with the overall biosecurity level on farms. This knowledge can be useful when developing biosecurity programs on goat farms. Recognition of locally applied farm-level practices enhancing biosecurity could form a basis for farmers to apply more rigorous and effective relevant plans.

Keywords: farm worker; health management; infection; quarantine; small ruminant; surveillance; transmission

1. Introduction

The World Health Organization has described biosecurity as the accumulation of strategic and integrated approaches for the management of relevant risks to human, animal, and plant health [1]. Hence, the maintenance and development of biosecurity policies, procedures, and practices aim '*to enhance the ability to protect human health, agricultural production systems, and the people and industries that depend on them*' [1]. In the context of animal health and production, biosecurity refers to a set of precautions with the purpose of protecting livestock, mainly from pathogens but also from invasive species [2–5]. Biosecurity measures and practices can be relevant at country, region, farm, or individual levels [4,6,7]. Biosecurity interventions are also applied in order to prevent the transmission of pathogens from livestock to people [8].

Goat farms are particularly prone to the transmission of pathogens, as for most of them, grazing is an integral part of their general management system. During grazing, a variety of pathogens can be transmitted to animals during direct contact with goats of other herds in the area; among these, *Mycoplasma ovipneumoniae* is transmitted most frequently through direct contact between animals [9] ('nose-to-nose' [10]), whilst *M. mucoides* and *M. capricolum* are transmitted mainly via aerosol droplets [9,11]. Pathogens can also be transmitted through ingestion during the grazing of goats on pasture, for example, the larvae of gastrointestinal nematodes (e.g., *Teladorsagia circumcincta* or *Haemonchus contortus* [12]). Finally, pathogens can also be contracted by animals during grazing itself; a notable example is *Dichelobacter nodosus*, the causal pathogen of footrot [13].

A recent (June 2024) literature evaluation performed by the authors using the topic search string [[*biosecurity*] AND [*goat** OR *caprine* OR *Capra hircus*]] revealed 113 relevant items. A detailed individual evaluation of these items revealed a total of 60 articles that covered various facets of the topic. Among these, only three have reported farmers' perspectives and attitudes on occurrence, control, and communication related to biosecurity procedures and approaches. These were carried out in Sweden [14], Kenya [15], and Australia [16] and reported the significance of farmers' experience and relevant education as determinants for the standards of biosecurity applied on goat farms. It is notable that despite the importance of the goat industry for the agricultural sector in Greece, which has the largest goat population and is the largest goat milk producer in Europe [17], the topic of biosecurity on goat farms has not been studied. This lack of studies related to biosecurity on goat farms may reflect the nature of goat farming in the country; notably, most goat farms in Greece operate under a semi-extensive or extensive management system [18], which can hinder the establishment and application of specific biosecurity measures.

Given the significance of goat farming in Greece, our study aimed to explore the biosecurity conditions on goat farms and refers to an extensive investigation performed on 119 farms countrywide in Greece. The objective of the present work was to evaluate and report on potential associations between the socio-demographic characteristics of goat farmers and practices that play a role in biosecurity on farms.

2. Materials and Methods

2.1. Visits to Goat Farms and Interviews of Farmers

The data were obtained during a large cross-sectional study performed across Greece (April 2019–June 2020) and in all 13 regional units of the country (Figure 1). The study involved 119 goat herds. The protocols for farm inclusion in the study have been presented in detail before [19].

No kind of relationship had existed between the researchers and any of the farmers prior to the visit. The researchers visited all 119 farms that participated in order to collect all relevant information. At the start, the objectives and the details of the study were presented to the farmers and discussed with them. Then, an interview was carried out by means of a structured, detailed questionnaire; the interview was always carried out by the same researcher (author D.T.L.) [19]. If farmers requested clarifications on the questions during the interview, these were provided immediately.

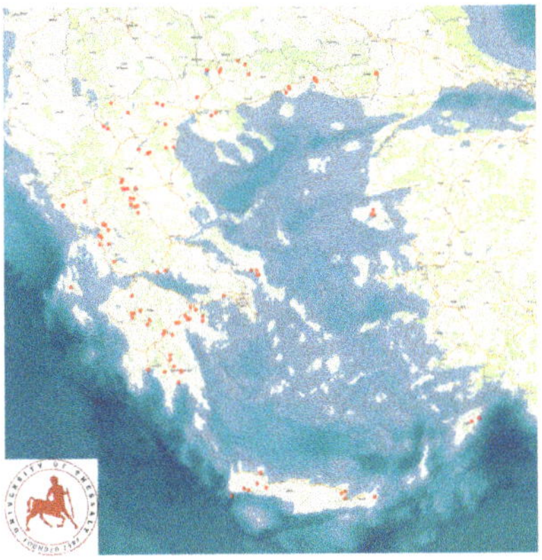

Figure 1. Map presenting locations of the 119 goat farms throughout Greece which were visited during the cross-sectional study.

2.2. Data Management and Analysis

The following socio-demographic characteristics were recorded during the interview: farmer gender, age and education level, length of farming experience, professional involvement in farming, daily period of presence on the farm, family farming tradition, and presence of working staff on the farm (Table S1). Further, a total of 16 variables related to biosecurity were additionally recorded during the interview. These were practices or events that could potentially enhance ($n = 9$) or compromise ($n = 7$) biosecurity on farms (Table S1). The general characteristics of farms (e.g., management system applied, no. of animals) were also recorded (Table S1).

An overall score based on all the above biosecurity-related variables was devised by assigning a '1' or '−1' mark for each variable recorded on a farm that could enhance or compromise, respectively, biosecurity. Thus, the overall score ranged from −7 to 9 on each farm.

Data were entered into Microsoft Excel® (Microsoft Corporation, Redmond, WA, USA) and analyzed using SPSS® v. 21 (IBM Analytics, Armonk, NY, USA). Initially, descriptive analyses were carried out.

All continuous data used in the statistical analyses were assessed by means of the Shapiro–Wilk test. The results of testing showed that the distribution of these data had a significant departure from normality: $W \leq 0.98$ ($p > 0.075$ for all assessments).

Outcomes in the form *'Occurrence of xxx'* were considered (where xxx: each of the biosecurity-related variables (Table S1)). Initially, the significance of predictors among the socio-demographic characteristics of farmers was assessed in a univariable analysis by using cross-tabulation with Pearson's chi-square test or Fisher exact test (for categorical variables) or the Mann–Whitney test (for continuous variables) with simple logistic regression. Then, a multivariable model was developed for each of the above outcomes, and parameters (i.e., the socio-demographic characteristics of farmers) found with $p < 0.2$ in the preceding univariable analyses were used in this model. The number of socio-demographic characteristics included in the models developed varied from 0 (variables: availability of footbath, availability of ditch at the main entrance, carcass disposal according to relevant regulations, application of rodenticides) to 6 (variable: grazing practice). Then, the socio-demographic characteristics were progressively removed from the model by using backwards elimination. A likelihood ratio test was performed to assess the p-value of each socio-demographic characteristic; among those found with $p > 0.2$, the one with the largest p was removed from the model. This procedure was repeated until no variable with $p > 0.2$ could be removed from the model. The socio-demographic characteristics included in the final assessment in each multivariable model are presented in Table S2. Their number varied from 0 (variables: availability of footbath, availability of ditch at the main entrance, carcass disposal according to relevant regulations, application of rodenticides) to 3 (variables: availability of fencing, application of systemic disinfections, grazing practice, presence of hunters in the area around the farm).

Subsequently, the outcome *'Overall score for biosecurity-related variables'* was considered. Initially, the significance of predictors among the socio-demographic characteristics of farmers was assessed in a univariable analysis by using Spearman's rank correlation. Then, a multivariable model was developed. Three socio-demographic characteristics were included in that model: farmer gender, length of farming experience, and presence of working staff on the farm. The same procedure described above was followed. The two variables included in the final assessment in the multivariable model were farmer gender and presence of the working staff on the farm (Table S2).

In all analyses, statistical significance was defined at $p < 0.05$.

3. Results

3.1. Descriptive Findings

Among the variables potentially enhancing biosecurity on farms, the application of disinfections on the farm and the maintenance of isolation for sick animals were the ones practiced most frequently. Among the variables potentially compromising biosecurity on farms, the presence of hunters in the area around the farm and grazing practices for animals were recorded most frequently. Details are given in Table 1.

Table 1. Descriptive findings regarding biosecurity-related variables recorded on 119 goat farms in Greece.

Variables	Goat Farms (n = 119) [1]
Processing of farm waste	0.8% (0.2–4.6%)
Availability of fencing	45.4% (36.7–54.3%)
Type of fencing available (among farms with availability of fencing)	Light wire fence: 51.9%. Strong wire fence: 48.1% [2]
Availability of footbath	8.4% (4.6–14.8%)
Availability of ditch at the main entrance	0.0% (0.0–3.1%)
Maintenance of quarantine period for new animals on the farm	51.3% (42.4–60.1%)
Maintenance of isolation for sick animals	82.4% (74.5–88.2%)
Carcass disposal according to relevant regulations	42.0% (33.5–51.0%)
Type of carcass disposal employed (among farms with carcass disposal resources according to relevant regulations)	Incineration: 18.0%. Safe burial: 78.0%. Both practices: 4.0% [2]
Application of rodenticides	42.0% (33.5–51.0%)
Application of systemic disinfections	85.7% (78.3–90.9%)
No. of systemic disinfections performed annually (among farms which applied systemic disinfections)	2 (2) occasions [3]
Grazing practice	94.1% (88.4–97.1%)
Duration of grazing annually (among farms with grazing practice)	12 (6) months [3]
Common grazing of farm animals with wildlife ruminants	9.2% (5.2–15.8%)
Presence of wild carnivore mammals [4] near the farm	82.4% (74.5–88.2%)
Presence of hunters in the area around the farm	95.0% (89.4–97.7%)
Distance from the farm (among farms with presence of hunters in the area around them)	0.5 (1.9) km [3]
Transhumance	23.5% (16.8–31.9%)
Presence of spots suitable for vector reproduction and development [5]	69.8% (61.0–77.3%)
Purchase of replacement animals	34.5% (26.5–43.4%)

[1] proportion of farm (95% confidence intervals) areas or median (interquartile range) (according to type of data); [2] proportion of farms among those that applied the practice; [3] median (interquartile range) with regard only to farms that applied the practice; [4] bears, foxes, jackals, and wolves; [5] muddy surfaces within or near the farm premises or adjacent to water troughs, manure buildups, or trenches with farm waste or manure.

The median overall score for biosecurity-related practices for all farms in the study was −1 (interquartile range: 3; minimum: −5; maximum: 4). There was a trend of higher scores on farms on islands ($p = 0.055$) (Figure 2); no other associations were found between the overall score for biosecurity-related practices and the characteristics of the farms ($p \geq 0.19$) (Table 2).

Table 2. Overall score for biosecurity-related variables recorded on 119 goat farms in Greece, in accordance with the characteristics of the farms.

Characteristics of Farms	Overall Score	p
Farms applying intensive management [2]	1 (3.0) [1]	0.23
Farms applying semi-intensive management	0 (1.0)	
Farms applying semi-extensive management	−1 (3.0)	
Farms applying extensive management	−1 (2.0)	
Farms in North Greece	−1 (2.3)	0.055
Farms in Central Greece	−1 (2.0)	
Farms in South Greece	−1 (2.5)	
Farms on the islands	1 (2.3)	
Altitude of location of farm	$r_{sp} = -0.120$	0.19
Number of animals on farm	$r_{sp} = 0.042$	0.65
Years since establishment of farm	$r_{sp} = -0.092$	0.32
Collaboration with veterinary practice	0 (1.8)	0.70
No collaboration with veterinary practice	−1 (3.0)	

[1] median (interquartile range); [2] management types applied to farms, with description given according to the European Food Safety Authority classification [20].

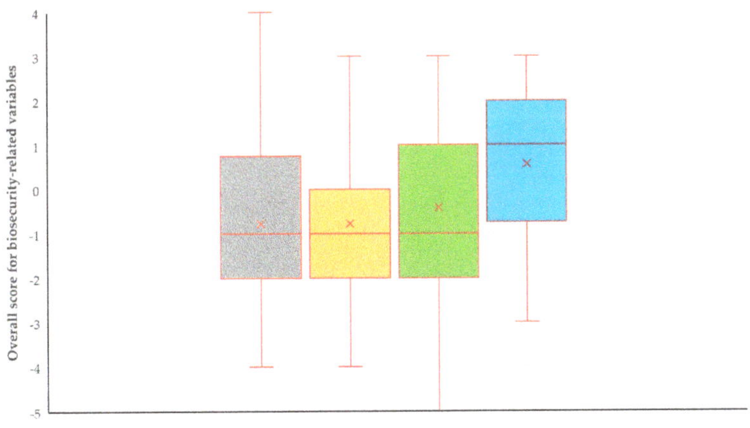

Figure 2. Box plots of the overall score for biosecurity-related practices on 119 farms in Greece, in accordance with the location of the farms (gray: farms in North Greece; yellow: farms in Central Greece; green: farms in South Greece; blue: farms on the islands).

3.2. *Associations of Socio-Demographic Characteristics with Biosecurity-Related Practices*

Details of the univariable analyses for the associations of the various socio-demographic characteristics with biosecurity-related practices are given in Tables S3–S18. The results of the subsequent multivariable analyses for each outcome are summarized in Table 3. Among the various socio-demographic characteristics, the presence of working staff on the farm was associated with five biosecurity-related practices and the length of farming experience with three ($p = 0.06$ for differences between the various characteristics) (Figure 3).

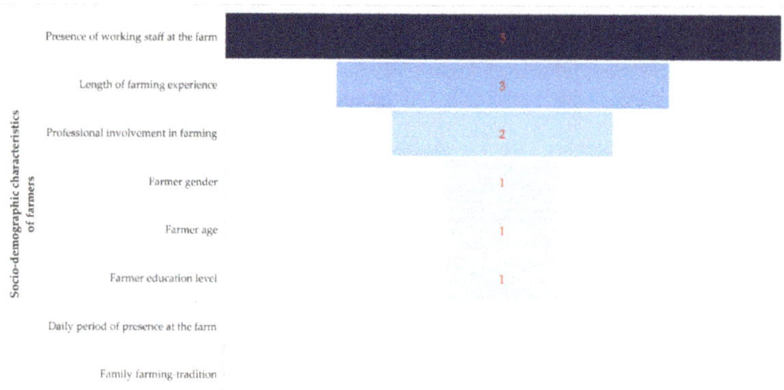

Figure 3. Funnel-type graph with the number of biosecurity-related practices with which significant associations were found with the socio-demographic characteristics of farmers.

Table 3. Summary of results of multivariable analyses for associations of socio-demographic characteristics of farmers with biosecurity-related practices on 119 goat farms in Greece.

Biosecurity-Related Practices	Socio-Demographic Characteristics of Farmers	*p*
Processing of farm waste	-	>0.08
Availability of fencing	-	0.07
Availability of footbath	-	-
Availability of ditch at the main entrance	-	-
Maintenance of quarantine period for new animals on the farm	Length of farming experience Presence of working staff on the farm	0.0003 0.001
Maintenance of isolation for sick animals	Presence of working staff on the farm	0.008
Carcass disposal according to relevant regulations	-	-
Application of rodenticides	-	-
Application of systemic disinfections	Presence of working staff on the farm	0.026
Grazing practice	Presence of working staff on the farm Length of farming experience	0.004 0.006
Common grazing of farm animals with wildlife ruminants	Farmer by profession	0.047
Presence of wild carnivore mammals near the farm	Gender	0.0007
Presence of hunters in the area around the farm	Farmer age Farmer education level Farmer by profession	0.006 0.017 0.023
Transhumance	-	>0.06
Presence of spots suitable for vector reproduction and development	-	>0.10
Purchase of replacement animals	Presence of working staff on the farm Length of farming experience	0.0003 0.003

3.3. Association of Socio-Demographic Characteristics with Overall Score for Biosecurity-Related Practices

Details of the univariable analyses for the associations of the various socio-demographic characteristics with the overall score for biosecurity-related practices are given in Table S19. In the multivariable analysis, female gender (p = 0.007) and the presence of working staff on the farm emerged as significant variables among the socio-demographic characteristics of farmers (p = 0.022) (Figure 4, Table 4).

Table 4. Results of multivariable analysis for socio-demographic characteristics with significant association with the overall score for biosecurity-related practices on 119 goat farms in Greece.

Variables	Odds Risk (±se) [1]	p
Farmer gender		0.007
Female (1 (2.0)) [2]	reference	-
Male (−1 (2.3))	4.293 ± 1.727	0.009
Presence of working staff on the farm		0.022
Yes (0 (2.0))	reference	-
No (−1 (2.0))	2.200 ± 1.424	0.028

[1] se: standard error; [2] median (interquartile range) of overall score for biosecurity-related practices.

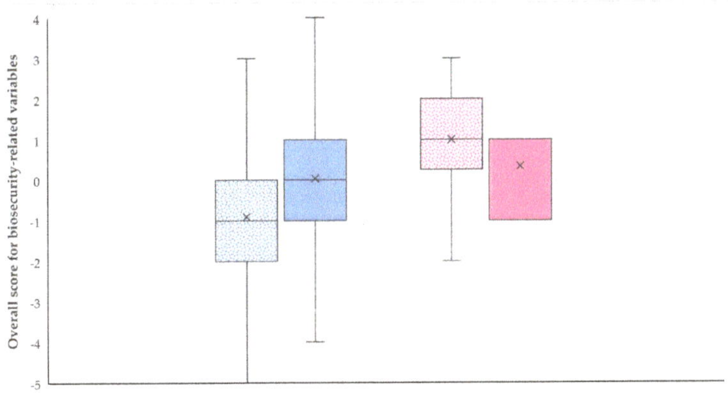

Figure 4. Box plots of the overall score for biosecurity-related practices on 119 farms in Greece in accordance with farmer gender (blue: male; pink: female) and the presence of working staff therein (no presence of working staff: motif pattern; presence of working staff: full pattern).

4. Discussion

This study focuses on the associations of the socio-demographic characteristics of farmers with biosecurity practices in goat farms, which may potentially affect the overall biosecurity level in a goat herd. This study also contributes to understanding the demographic characteristics of farmers' engagement with animal health-related biosecurity practices that should be applied on farms.

Community (i.e., social) trust can be an important component in attempts within a country to set up systems for disease prevention, as these would require a communal approach to tackling the problem [21]. However, social trust in Greece has been found to be low, with the country being ranked 17th among 18 European countries [22,23]. This may influence the willingness of local farmers to take into account and align with regulations and common rules, to collaborate with veterinary authorities, and to trust others in achieving a common goal, in this case, disease prevention, through the implementation of policies and procedures for the benefit of all. This is reflected in the fact that many farmers declared that they did not follow biosecurity rules established through legal provisions, for example, the disposal of carcasses of dead goats.

In this study, we devised and employed a binary scoring system (yes/no) for the various biosecurity-related practices taken into account. In this context, practices enhancing biosecurity were scored positively (+1) if practiced, whilst practices with negative implications for biosecurity were given a negative score (−1) if practiced. That way, within each farm, the overall level of biosecurity applied therein was scaled.

The higher biosecurity scores on farms located on the islands of the country reflect the lack of wildlife mammals on them. This contributes to minimizing threats that originate from wild mammals. Such threats include the transmission of various pathogens, for example, rabies virus from foxes and wolves or *Chlamydia* spp. from wild cervids [24,25]. Goats are particularly vulnerable to attacks by predators, as they often browse individually among bushes and scrubs, which can facilitate attacks by predators [26], and indeed, the work of Petridou et al. [27] found that the diet of wolves often includes goats.

With regard to the identification of staff as a significant determinant for the application of a higher number of individual biosecurity-related practices on the farms, this can be associated with the fact that various practices aiming to decrease disease risks require relevant manual effort and time, for example, the regular application of disinfections or the maintenance of quarantining pens in farms. Goat farms in Greece are of the dairy production type, which adds a milking routine to the daily tasks to be carried out. Hence, a lack of working staff on the farm would lead to a low prioritization of tasks related to biosecurity. This finding underlines the idea that protection against disease risk and the enhancement of biosecurity measures require increased resources at various levels.

Improved biosecurity practices by female farmers were previously reported by Backhans et al. [28] on pig farms. Indeed, on dairy farms, women farmers declared a higher frequency of intention to abide by and implement biosecurity practices [29], which may explain the current findings. In this context, it is also notable that similar findings have been reported in studies related to human health [30,31].

Moreover, the association of the length of farming experience with the level of biosecurity practices has been reported among sheep farmers in the United Kingdom [32] and among pig farmers in Europe, including Belgium, Poland, the United Kingdom, Finland [33] and Sweden [28], as well as among horticultural farmers in Australia [34]. This is interesting given the variations in animal species farmed, management systems applied, and diseases occurring in the various countries, as well as in the differences in the legislation applied between countries.

Moreover, it is notable that the higher education and training level of farmers has also been associated with better compliance with procedures and rules associated with biosecurity, according to Racicot et al. [35], possibly because of better understanding by such farmers of, for example, the transmission of microorganisms or the consequences of impaired biosecurity procedures. Within this context, farmers with a better understanding of the idea of biosecurity may feel more confident in its benefits and apply better relevant practices [36]. Hence, in addition to increased staff at farms, we also wish to indicate the importance of developing training and training material for farmers and farm workers on the topic of farm biosecurity.

The context of biosecurity is extensive [7] and can include further practices than those evaluated in the present study. Indeed, the lack of standardized biosecurity protocols for goat farms, like those devised and implemented in the pig and poultry industries, makes the formulation of concrete and precise guidelines for goat farms difficult. This reflects the lack of scientific knowledge specifically pertaining to and applicable to goat farms.

Overall, the findings of this study underline the significance of human resources in maintaining high biosecurity standards in goat herds. Potentially, veterinarians, veterinary authorities, and veterinary associations and societies could be involved in improving biosecurity standards on farms. That way, different aspects of biosecurity practices could be communicated effectively among stakeholders.

5. Conclusions

The findings indicate that, on goat farms, the socio-demographic characteristics of farmers are associated with the biosecurity level therein. This knowledge could be useful when developing biosecurity programs on goat farms. Recognition of locally applied farm-level practices enhancing biosecurity could form a basis for farmers to apply more rigorous and effective relevant plans.

Supplementary Materials: The following supporting information can be downloaded at https://www.mdpi.com/article/10.3390/ani14142136/s1: Table S1: List of variables related to biosecurity, the socio-demographic characteristics of farmers, and other characteristics recorded on 119 goat farms in Greece; Table S2: Details of multivariable models employed for the evaluation of potential associations between biosecurity-related practices and the socio-demographic characteristics of farmers on 119 goat farms in Greece; Table S3: Association between the processing of farm waste on 119 goat farms in Greece and the socio-demographic characteristics of the farmers; Table S4: Association of the availability of fencing on 119 goat farms in Greece with the socio-demographic characteristics of the farmers; Table S5: Association of the availability of a footbath on 119 goat farms in Greece with the socio-demographic characteristics of the farmers; Table S6: Association of the availability of a ditch at the main entrance on 119 goat farms in Greece with socio-demographic characteristics of the farmers; Table S7: Association of the maintenance of a quarantine period for new animals on the farm on 119 goat farms in Greece with the socio-demographic characteristics of the farmers; Table S8: Association of the maintenance of isolation for sick animals on 119 goat farms in Greece with the socio-demographic characteristics of the farmers; Table S9: Association of carcass disposal according to relevant regulations on 119 goat farms in Greece with the socio-demographic characteristics of the farmers; Table S10: Association of the application of rodenticides on 119 goat farms in Greece with the socio-demographic characteristics of the farmers; Table S11: Association of the application of systemic disinfections on 119 goat farms in Greece with the socio-demographic characteristics of the farmers; Table S12: Association of grazing practices on 119 goat farms in Greece with the socio-demographic characteristics of the farmers; Table S13: Association of common grazing of farm animals with wildlife ruminants on 119 goat farms in Greece with the socio-demographic characteristics of the farmers; Table S14: Association of the presence of wild carnivore mammals near 119 goat farms in Greece with the socio-demographic characteristics of the farmers; Table S15: Association of the presence of hunters in the area around the farm on 119 goat farms in Greece with the socio-demographic characteristics of the farmers; Table S16: Association of transhumance on 119 goat farms in Greece with the socio-demographic characteristics of the farmers; Table S17: Association of the presence of spots suitable for vector reproduction and development on 119 goat farms in Greece with the socio-demographic characteristics of the farmers; Table S18: Association of the purchase of replacement animals on 119 goat farms in Greece with the socio-demographic characteristics of the farmers; Table S19: Association of the overall score for biosecurity-related practices on 119 goat farms in Greece with the socio-demographic characteristics of the farmers.

Author Contributions: Conceptualization, D.T.L. and G.C.F.; methodology, D.T.L. and G.C.F.; formal analysis, D.T.L.; investigation, D.T.L.; resources, G.C.F.; data curation, D.T.L.; writing—original draft preparation, D.T.L.; writing—review and editing, D.T.L. and G.C.F.; visualization, D.T.L. All authors have read and agreed to the published version of the manuscript.

Funding: This research received no external funding.

Institutional Review Board Statement: The protocols of this study were approved by the academic board of the Veterinary Faculty of the University of Thessaly, meeting 34/03.04.19 (items 5a & 5b).

Data Availability Statement: Detailed results associated with this study are presented in the Supplementary Materials.

Conflicts of Interest: The authors declare no conflicts of interest.

References

1. World Health Organization. *Biosecurity: An Integrated Approach to Manage Risk to Human, Animal and Plant Life and Health*; Information Note No. 1/2010; World Health Organization: Geneva, Switzerland, 2010; p. 4.
2. van der Merwe, P. Prevention and preparedness: Biosecurity, early warning and contingency planning. *Rev. Scientif. Techniq. Off. Internat. Epiz.* **2020**, *39*, 551–559. [CrossRef] [PubMed]
3. Msimang, V.; Rostal, M.K.; Cordel, C.; Machalaba, C.; Tempia, S.; Bagge, W.; Burt, F.J.; Karesh, W.B.; Paweska, J.T.; Thompson, P.N. Factors affecting the use of biosecurity measures for the protection of ruminant livestock and farm workers against infectious diseases in central South Africa. *Transbound. Emerg. Dis.* **2022**, *69*, e1899–e1912. [CrossRef] [PubMed]
4. Militzer, N.; Mclaws, M.; Rozstalnyy, A.; Li, Y.S.; Dhingra, M.; Auplish, A.; Mintiens, K.; Sabirovic, M.; von Dobschuetz, S.; Heilmann, M. Characterising biosecurity initiatives globally to support the development of a progressive management pathway for terrestrial animals: A scoping review. *Animals* **2023**, *13*, 2672. [CrossRef] [PubMed]
5. Ngom, R.V.; Laconi, A.; Mouiche, M.M.M.; Ayissi, G.J.; Akoussa, A.M.M.; Ziebe, S.D.; Tilli, G.; Zangue, H.A.; Piccirillo, A. Methods and tools used for biosecurity assessment in lvestock farms in Africa: A scoping review. *Transbound. Emerg. Dis.* **2024**, *2024*, 5524022.
6. Laanen, M.; Persoons, D.; Ribbens, S.; De Jong, E.; Callens, B.; Strubbe, M.; Maes, D.; Dewulf, J. Relationship between biosecurity and production/antimicrobial treatment characteristics in pig herds. *Vet. J.* **2013**, *198*, 508–512. [CrossRef] [PubMed]
7. Huber, N.; Andraud, M.; Sassu, E.L.; Prigge, C.; Zoche-Golob, V.; Käsbohrer, A.; D'Angelantonio, D.; Viltrop, A.; Żmudzki, J.; Jones, H.; et al. What is a biosecurity measure? A definition proposal for animal production and linked processing operations. *One Health* **2022**, *15*, 100433. [CrossRef] [PubMed]
8. Youssef, D.M.; Wieland, B.; Knight, G.M.; Lines, J.; Naylor, N.R. The effectiveness of biosecurity interventions in reducing the transmission of bacteria from livestock to humans at the farm level: A systematic literature review. *Zoonoses Publ. Health* **2021**, *68*, 549–562. [CrossRef] [PubMed]
9. Martin, W.B.; Aitken, I.D. Mycoplasma respiratory infections. In *Diseases of Sheep*, 3rd ed.; Martin, W.B., Aitken, I.D., Eds.; Blackwell Science: Oxford, UK, 2000; pp. 198–201.
10. Plant and Animal Health Branch of the Ministry of Agriculture of Canada. Domestic and Wild Sheep and Goats, and the Risk of Mycoplasma ovipneumoniae. 2024. Available online: https://www2.gov.bc.ca/gov/content/industry/agriculture-seafood/animals-and-crops/animal-production/sheep-and-goats/mycoplasma-ovipneumoniae (accessed on 25 May 2024).
11. MacKay, E. Mycoplasma pneumonias in goats. In *MSD Veterinary Manual*; Hess, L., Ed.; Merck & Co.: Rahway, NJ, USA, 2022.
12. Taylor, M.A.; Coop, R.L.; Wall, R.L. *Veterinary Parasitology*, 3rd ed.; Blackwell Publishing: London, UK, 2007.
13. Gelasakis, A.I.; Kalogianni, A.I.; Bossis, I. Aetiology, risk factors, diagnosis and control of foot-related lameness in dairy sheep. *Animals* **2019**, *9*, 509. [CrossRef]
14. Frössling, J.; Nöremark, M. Differing perceptions-Swedish farmers' views of infectious disease control. *Vet. Med. Sci.* **2016**, *2*, 54–68. [CrossRef]
15. Nyokabi, S.; Birner, R.; Bett, B.; Isuyi, L.; Grace, D.; Güttler, D.; Lindahl, J. Informal value chain actors' knowledge and perceptions about zoonotic diseases and biosecurity in Kenya and the importance for food safety and public health. *Trop. Anim. Health Prod.* **2018**, *50*, 509–518. [CrossRef]
16. Hernández-Jover, M.; Higgins, V.; Bryant, M.; Rast, L.; McShane, C. Biosecurity and the management of emergency animal disease among commercial beef producers in New South Wales and Queensland (Australia). *Prev. Vet. Med.* **2016**, *134*, 92–102. [CrossRef] [PubMed]
17. Pulina, G.; Milan, M.J.; Lavin, M.P.; Theodoridis, A.; Morin, E.; Capote, J.; Thomas, D.L.; Francesconi, A.H.D.; Caja, G. Current production trends, farm structures, and economics of the dairy sheep and goat sector. *J. Dairy Sci.* **2018**, *101*, 6715–6729. [CrossRef] [PubMed]
18. Lianou, D.T.; Michael, C.K.; Fthenakis, G.C. Data on mapping 444 dairy small ruminant farms during a countrywide investigation performed in Greece. *Animals* **2023**, *13*, 2044. [CrossRef] [PubMed]
19. Lianou, D.T.; Chatziprodromidou, I.P.; Vasileiou, N.G.C.; Michael, C.K.; Mavrogianni, V.S.; Politis, A.P.; Kordalis, N.G.; Billinis, C.; Giannakopoulos, A.; Papadopoulos, E.; et al. A detailed questionnaire for the evaluation of health management in dairy sheep and goats. *Animals* **2020**, *10*, 1489. [CrossRef]
20. European Food Safety Authority. Scientific opinion on the welfare risks related to the farming of sheep for wool, meat and milk production. *EFSA J.* **2014**, *12*, 3933–4060.
21. Wollebæk, D.; Wallman Lundåsen, S.; Trägårdh, L. Three forms of interpersonal trust: Evidence from Swedish municipalities. *Scand. Polit. Stud.* **2012**, *35*, 319–346. [CrossRef]
22. Anon. Survey Finds Low Levels of Social Trust in Greece. Kathimerini 2018. Available online: https://www.ekathimerini.com/news/232870/survey-finds-low-levels-of-social-trust-in-greece/ (accessed on 24 May 2024).
23. Ervasti, H.; Kouvo, A.; Venetoklis, T. Social and institutional trust in times of crisis: Greece, 2002–2011. *Soc. Indic. Res.* **2019**, *141*, 1207–1231. [CrossRef]
24. Cuteri, V.; Diverio, S.; Carnieletto, P.; Turilli, C.; Valente, C. Serological survey for antibodies against selected infectious agents among fallow deer (*Dama dama*) in central Italy. *J. Vet. Med. B* **1999**, *46*, 545–549. [CrossRef]

25. Salinas, J.; Caro, M.R.; Vicente, J.; Cuello, F.; Reyes-Garcia, A.R.; Buendia, A.J.; Rodolakis, A.; Gortazar, C. High prevalence of antibodies against Chlamydiaceae and *Chlamydophila abortus* in wild ungulates using two "in house" blocking-ELISA tests. *Vet. Microbiol.* **2009**, *135*, 46–53. [CrossRef]
26. Cozza, K.; Fico, R.; Battistini, M.L.; Rogers, E. The damage–conservation interface illustrated by predation on domestic livestock in central Italy. *Biol. Conserv.* **1996**, *78*, 329–336. [CrossRef]
27. Petridou, M.; Youlatos, D.; Lazarou, Y.; Selinides, K.; Pylidis, C.; Giannakopoulos, A.; Kati, V.; Iliopoulos, Y. Wolf diet and livestock selection in central Greece. *Mammalia* **2019**, *83*, 530–538. [CrossRef]
28. Backhans, A.; Sjölund, M.; Lindberg, A.; Emanuelson, U. Biosecurity level and health management practices in 60 Swedish farrow-to-finish herds. *Acta Vet. Scand.* **2015**, *57*, 14. [CrossRef] [PubMed]
29. Nöremark, M.; Sternberg Lewerin, S.; Ernholm, L.; Frössling, J. Swedish farmers' opinions about biosecurity and their intention to make professionals use clean protective clothing when entering the stable. *Front. Vet. Sci.* **2016**, *3*, 46. [CrossRef]
30. Sax, H.; Uckay, I.; Richet, H.; Allegranzi, B.; Pittet, D. Determinants of good adherence to hand hygiene among healthcare workers who have extensive exposure to hand hygiene campaigns. *Infect. Control Hosp. Epidemiol.* **2007**, *28*, 1267–1274. [CrossRef] [PubMed]
31. Wright, J.G.; Jung, S.; Holman, R.C.; Marano, N.N.; McQuiston, J.H. Infection control practices and zoonotic disease risks among veterinarians in the United States. *J. Am. Vet. Med. Assoc.* **2008**, *232*, 1863–1872. [CrossRef] [PubMed]
32. Jack, C.; Hotchkiss, E.; Sargison, N.D.; Toma, L.; Milne, C.; Bartley, D.J. Determining the influence of socio-psychological factors on the adoption of individual 'best practice' parasite control behaviours from Scottish sheep farmers. *Prev. Vet. Med.* **2022**, *200*, 105594. [CrossRef]
33. Chantziaras, I.; Dewulf, J.; Van Limbergen, T.; Stadejek, T.; Niemi, J.; Kyriazakis, I.; Maes, D. Biosecurity levels of pig fattening farms from four EU countries and links with the farm characteristics. *Liv. Sci.* **2020**, *237*, 104037. [CrossRef]
34. Duong, T.T.; Brewer, T.D.; Luck, J.; Zander, K.K. Farmers' assessment of plant biosecurity risk management strategies and influencing factors: A study of smallholder farmers in Australia. *Outlook Agric.* **2021**, *48*, 48–57. [CrossRef]
35. Racicot, M.; Venne, D.; Durivage, A.; Vaillancourt, J.P. Evaluation of the relationship between personality traits, experience, education and biosecurity compliance on poultry farms in Québec, Canada. *Prev. Vet. Med.* **2012**, *103*, 201–207. [CrossRef]
36. Laanen, M.; Maes, D.; Hendriksen, C.; Gelaude, P.; De Vliegher, S.; Rosseel, Y.; Dewulf, J. Pig, cattle and poultry farmers with a known interest in research have comparable perspectives on disease prevention and on-farm biosecurity. *Prev. Vet. Med.* **2014**, *115*, 1–9. [CrossRef]

Disclaimer/Publisher's Note: The statements, opinions and data contained in all publications are solely those of the individual author(s) and contributor(s) and not of MDPI and/or the editor(s). MDPI and/or the editor(s) disclaim responsibility for any injury to people or property resulting from any ideas, methods, instructions or products referred to in the content.

animals

MDPI

Article

Farmers' Biosecurity Awareness in Small-Scale Alpine Dairy Farms and the Crucial Role of Veterinarians

Marica Toson [1], Manuela Dalla Pozza [1] and Piera Ceschi [2,*]

1 Istituto Zooprofilattico Sperimentale delle Venezie, 35020 Padova, Italy
2 Department of Prevention—Hygiene and Public Health Service, South Tyrol Health Service, 39042 Bressanone/Brixen, Italy
* Correspondence: piera.ceschi@sabes.it

Simple Summary: Biosecurity measures are necessary to the development of a family alpine farming system. In this study, good practices and areas of improvement have been investigated. In particular, practices related to quarantine, cleaning, waste milk feeding to calves, immunization, and diagnostics could be improved to implement optimal biosecurity measures. Farmers confirm that they are willing to adopt effective preventive measures under the recommendation of the veterinarian, and this attitude may be crucial to designing multisectoral participatory activities and strategies. Further investigations and the development of customized applications and digital toolkits are now required to better understand farmers' drivers toward biosecurity and to provide support.

Abstract: Background: Biosecurity measures are essential to improve animal health and welfare, tackle antimicrobial resistance (AMR), minimize the burden of infectious diseases, and enhance the safety, security, and quality of sustainable agricultural products. However, the implementation of these measures in small mountain farms can be influenced by several factors, which probably differ from the main variables that affect intensive farming systems. Methods: This study investigated the awareness among farmers regarding the implementation of biosecurity measures at the local level in small dairy farms located in the Autonomous Province of Bolzano/Bozen (Italy). In order to determine to what extent these measures are in line with the recommendations, a questionnaire was conceptualized and sent by post to a representative group of local farmers. The initiative was voluntary and anonymous. Results: A total of 50 farmers responded to the questionnaire, resulting in a response rate of 27.5%. This study confirms that most of the farmers are familiar with biosecurity, and 38% of them know its comprehensive definition. The results indicate that 77% of respondents are willing to implement preventive measures to reduce the use of antimicrobials, and 76% of them acknowledge that they would follow the recommendations provided by veterinarians. In agreement with other studies, the role of the veterinarian as a person of trust among farmers is confirmed. Conclusions: Main strengths and areas of improvement have been identified. Additional data and effective tools are needed to better investigate drivers toward biosecurity and to implement impactful and practical measures for mountain farming.

Keywords: biosecurity; mountain farming; disease prevention; AMR; vaccination

1. Introduction

Biosecurity is defined by the European Regulation 2016/429 ("Animal Health Law") as "the sum of management and physical measures designed to reduce the risk of the introduction, development, and spread of diseases to, from, and within an animal population or an establishment, zone, compartment, means of transport, or any other facilities, premises, or location" [1].

This definition emphasizes the importance of prevention tools to keep infection pressure as low as possible in order to protect animal health and welfare. Importantly, implementing stronger biosecurity on dairy farms also has an impact on food safety, food security, and public and even environmental health [2].

The implementation of biosecurity measures has been strengthened in the last decade at the national level in Italy, and the development of strategies for prevention of animal infectious diseases represents one of the seven pillars of the National Action Plan on Antimicrobial Resistance (Piano di Contrasto all'Antimicrobico Resistenza—PNCAR), adopted for the first time in 2017 [3]. Antimicrobial resistance (AMR) is a major public health concern [4], and the association between biosecurity and antimicrobial use in farmed animals is the subject of intense research [5]. To translate the principles of the National Plan into action at the local level, it is necessary to implement appropriate measures and good practices to prevent diseases that otherwise have to be treated with antimicrobials. For this reason, local stakeholders' awareness about biosecurity and potential areas for improvement needs to be addressed.

During the last three years, the Directorate General of Animal Health and Veterinary Medicines of the Italian Ministry of Health has implemented an innovative information system, named ClassyFarm, which allows farms to be categorized according to risk. It is available to official veterinarians, farm veterinarians, and farmers, and it allows the processing of data related to the following assessment areas, i.e., biosafety, animal welfare, health and production parameters, animal feeding, antimicrobial consumption, and lesions found at slaughterhouses [6]. However, in 2023, an accurate investigation of Classyfarm's implementation in the Autonomous Province of Bolzano/Bozen was published, and the authors raised concerns about the appropriateness of the questionnaires developed, as the peculiarities of small mountain dairy farms are not taken into consideration [7].

In accordance with paragraph 43 of the European "Animal Health Law", measures need to be applied with a certain degree of flexibility. In particular, specific productions, species or categories of animals, and local circumstances must be considered. Therefore, a study about biosecurity was designed, focusing on specific variables that may have an impact on the implementation of these measures in alpine dairy farms.

In order to not only collect specific additional data about this farm system but also to highlight strengths in disease prevention and to provide practical evidence-based recommendations to local farmers, if required, a questionnaire consisting of 74 questions related to farm characteristics, general breeding management, selling and purchasing policy, management of calves, management of adult cattle, and relevant opinions was conceptualized. A representative subgroup of farms was selected, and questionnaires were sent by post. Enrollment into the initiative was voluntary and anonymous.

2. Materials and Methods

2.1. Study Area and Selection of the Farms

The area of the Autonomous Province of Bolzano/Bozen covers 7400 square kilometers; it has 535,774 inhabitants according to the 2021 census [8], and it is entirely located in the Alps.

This province is divided into four main local health districts, which are subdivided into 20 subdistricts. The district of Bressanone/Brixen was chosen to perform this study. This area includes three subdistricts, i.e., Alta Valle Isarco/Wipptal (with 5 municipalities), Bressanone/Brixen (with 6 municipalities), and Chiusa/Klausen (with 6 municipalities). In this province there are three official languages, i.e., German, Italian, and Ladin.

According to the last agricultural census [8], the total number of cattle registered in the province was 132,784 and the number of registered farms was 8315. The median herd size was 15 animals per dairy farm.

A sample of 800 dairy farms was selected from a total of 3285 dairy farms located in the province of Bolzano/Bozen that were registered in the national animal register. Then, from this sample, all the farms located in the district of Bressanone/Brixen were included in the investigation, and, finally, a sample of 182 farms was randomly selected in order to achieve an assumed 25% response rate (with at least 44 expected respondents), based on the existing literature [9]. The selection of farms involved simple random sampling performed through the SAS procedure (PROC SURVEYSELECT PROCEDURE of SAS 9.4 software (Copyright @ 2002–2012 by SAS Institute Inc., Cary, NC, USA), stratified by municipality, and categorization of the number of animals (on the basis of median animal capacity: median = 15 animals).

2.2. Questionnaire Development, Distribution, and Collection

A draft questionnaire was developed in 2021, and four farmers, two veterinary practitioners, and an expert in mountain agriculture and milk production tested it and provided feedback.

To facilitate its completion and the subsequent analyses, the questionnaire was divided into 6 main sections as follows: general information (n = 25 questions), general breeding management (n = 6 questions), selling and purchasing of animals (n = 9 questions), management of calves (n = 8 questions), management of adult cattle (n = 5 questions), and opinions (n = 21 questions). The questions referred to the sanitary situation in the last 12 months. Both fill-in-the-blank and multiple-choice questions were presented to the farmers. A 5-point Likert scale was used to design the last part of the questionnaire to assess farmers' opinions towards biosecurity.

Questionnaires were sent by post at the beginning of February 2022, and farmers had two months to return them to their local veterinary service. As the aim of the initiative was neither to assess the compliance of the farmers' behaviors and practices with the legislation in force nor to provide individual counseling, participation was voluntary and anonymous. An invitation letter was added to the questionnaire to inform the farmers about the initiative. The questionnaire was written in German.

2.3. Data Analysis (Statistical Methods, Multivariable Models, and Associations)

Data collected through the paper survey were stored and validated in an Excel file. Validation has been performed through a consistency check of the answers reported for different variables and questions but logically related. Descriptive statistics were applied to the data. The association between categorical variables was evaluated through Pearson's chi-squared test or Fisher's exact test. Data were analyzed using STATA 17.0, and results with a p-value < 0.05 were considered significant.

An ordered regression model, as conducted by Robichaud et al., 2019 [10], was performed to assess the variables that were most closely associated with the rank of agreement with some items. However, the marginal analysis did not show any significant associated variables to put into the multivariable ordinal model, even if we considered a 5-point Likert scale or a 3-point Likert scale, created by grouping the extreme values. The Likert data have been presented as categorical data through histogram.

3. Results

3.1. General Data

A total of 50 farmers filled in and returned the questionnaire to their local Office of Veterinary Services, representing a response rate of 27.5% (Table 1). Questionnaires were returned from 14 of the 17 municipalities included in the study. One farmer omitted to specify the municipality where their farm was located. Figure 1 shows the area of study in the Autonomous Province of Bolzano/Bozen and the geographical distribution of the respondents in the district of interest.

Three questionnaires were considered not eligible for analysis. Two of them were uncompleted, and one was filled in by a farmer who reported having only two bovines (one cow and one calf), and the milk was kept for domestic use only. Among the respondents, two of them were women younger than 60 years old, 27 of them were men younger than 60 years old, and 15 of them were men over 60 years old. Two respondents did not answer these questions. Data are displayed in Table 2.

Table 1. Total respondents (N = 50).

Geographical Distribution of Participants			
Sub-Districts	**No of Selected Farms**	**No of Respondent Farms**	**Respondents/Selected**
Alta Valle Isarco/Wipptal	62	12	19.3%
Bressanone/Brixen	53	13	24.5%
Chiusa/Klausen	67	24	35.8%
Unknown		1	-
Total	**182**	**50**	**27.5%**

Table 2. General information (questionnaires included in the analysis, N = 47).

Variable	No	% [1]
Groups (Age and Gender)		
Women < 60 years old	2	4.3%
Men < 60 years old	27	57.4%
Men > 60 years old	15	31.9%
Unknown	3	6.4%
On-farm Employment		
Full-time	29	61.7%
Part-time	17	36.2%
Unknown	1	2.1%
Farm type		
Holiday accommodation	13	27.7%
Private	34	72.3%
Location		
Urban area	7	14.9%
Countryside	39	83.0%
Unknown	1	2.1%
Altitude		
Above 1000 masl	32	68.1%
Under 1000 masl	15	31.9%
Milk Production		
Conventional milk	28	59.6%
Organic, hay milk, organic hay milk	12	25.5%
Only dry cows (at the moment)	7	14.9%
Production and Delivery		
Dairy plant	33	70.2%
Dairy plant, small quantities of milk, and milk products to end consumers	6	12.8%
Only dry cows (at the moment)	7	14.9%
Unknown	1	2.1%

[1] Sum of the categories within a variable and within a column equals 100%.

Figure 1. Area of study and geographical distribution of the respondents.

Most of the farmers (61.7%) reported working full-time on-farm, and almost one-third of them (28%) offered on-farm holiday accommodation. Overall, 83% of the farms were located in the countryside, and 66% of the farms were located above 1000 m above sea level (masl). No significant association (chi-squared test: p-value = 0.335) was found between the husbandry system (tie- vs. free-stall) and the location of the farm (in terms of sea level and urban or rural areas).

The median size of the herd was 15 cows per farm; 62% of the respondents had more than 15 cattle (26.8 animals on average), 36% had fewer, and 2% omitted to provide this information.

Almost two-thirds of the farms (60%) produced conventional milk, while 26% adhered to specific production systems (organic, hay milk, and organic hay milk). Among the farms, 70% sold milk exclusively to a dairy plant, while 13% also sold small quantities of milk and milk products to end consumers (2% unknown). Overall, 15% of the farmers did not sell milk during the period of study (dry period). Data are shown in Table 2.

3.2. Health Literacy

One question about the definition of biosecurity was included in the questionnaire. The correct answer included both measures in place to avoid the spread of pathogens within the farm (internal measures) and to prevent pathogens from entering the farms (external measures). Overall, 38% of the farmers identified the comprehensive definition of biosecurity (including both internal and external measures); 15% of the respondents associated the meaning of this concept with either the internal or the external biosecurity definition; 9% indicated that it relates only to organic farms; and 38% did not know the definition.

3.3. Breeding Management

At least 68% of the farmers reported that they kept cattle in tie-stalls. Most of the farmers (78%) separated animals by age and/or production group. Both artificial insemination (83%) and natural breeding (17%) were practiced.

3.4. Diseases

A positive association between herd size and the presence of reported diseases was observed (Kruskal–Wallis test: p value = 0.0025 for mastitis; p-value = 0.0246 for diarrhea); in fact, the median number of animals was higher in those farms that experienced cases of mastitis or diarrhea (median number of animals: 22 vs. 9 for mastitis; 20 vs. 8 for diarrhea).

Whereas no significant association between the husbandry system and mastitis was found (p-value = 0.428), there was a significant association between the free-stall system and the presence of diarrhea (p-value = 0.002).

3.5. Working Organization and Cooperations

Overall, 6% of the farmers reported the presence of employees on their farm. Also, cooperation with auxiliary forces during the peak summer mowing season was reported. These people are usually recruited on a voluntary basis, not paid, and not allowed entry to the stables where animals are kept. More than 50% of the farmers stated that only family members were allowed to enter the stables where animals are kept, with some exceptions (veterinarians, authorities, etc.) (Figure 2). Stable entry is usually allowed for veterinarians (96%), followed by delegates of the local farmers' association, people in charge of the identification of animals (77%), and the evaluation of animals (47%), respectively (Figure 2). Most of the respondents reported regularly attending animal trade or exhibition fairs (70%) and stables belonging to other farms (53%). The practice of tool and equipment sharing was adopted by 21% of the respondents. Overall, 42% of farmers attended both markets/auctions and other farms, while 13% of them followed more restrictive behavior (Figure 3).

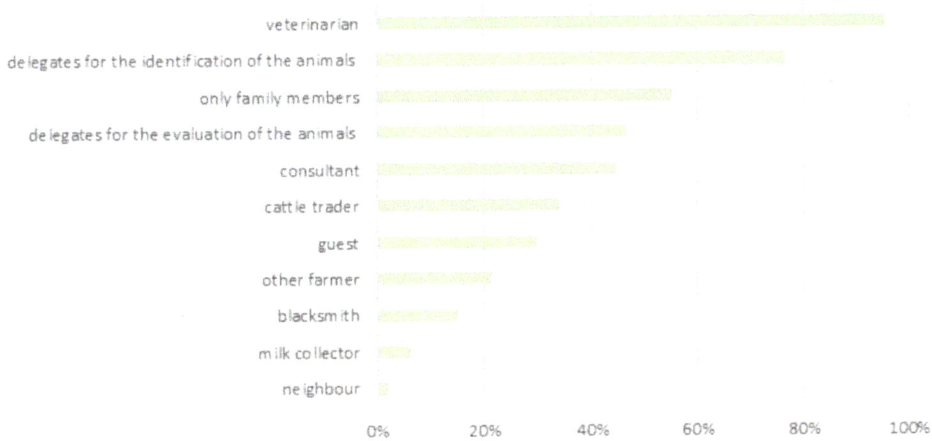

Figure 2. People allowed to enter the stables.

Figure 3. Attendance of facilities where animals are kept.

3.6. Maintenance and Cleaning Procedures

Milking equipment and installation maintenance occurred regularly in 51% of the farms, while 30% of the farmers indicated that service operations took place every 3 to 5 years or as required. Overall, 19% did not answer this question, and, among them, 15% declared keeping non-lactating cows. Meanwhile, 42% of the respondents indicated that they clean drinking troughs when they are dirty.

Around 60% of the farmers complied with regular preventive hygiene practices (cleaning of the vehicles used for animal transport, of water troughs, and calf shelters). No statistically significant differences between full- and part-time farmers were detected.

3.7. Other Potential Risk Factors

Dead animals are removed from the farm by a specialized company in 62% of cases, while 38% of the farmers reported using their own vehicles to transport carcasses to the local disposal plants.

Manure management is usually mechanical (66%). Only 2% of the farmers use manure for biogas production.

Natural water spring access is allowed to cattle in 62% of farms.

Overall, 64% of the farmers confirmed that only cattle were kept on their farms. However, 27% of them cannot exclude contact with other animals (both domesticated and wild). At least 91% of respondents acknowledged that animals are kept in summer pastures, where the presence of other species (pets, wild and farmed animals) cannot be ruled out.

More than 60% of the farmers reported selling cattle only if necessary, and, when they sell them, they usually contact professional salesmen (89%). The majority of farmers usually purchased cattle at the local trade fair, from other local farmers or from salesmen (Figure 4). Moreover, 4% of farmers (N = 2) reported applying a quarantine period after purchasing animals. Farmers report that they do not place animals in quarantine, either because it is not possible (21%), or because it is not considered to be necessary (55%). Usually, cattle are purchased inside the province (55%), from abroad (Austria and Germany), or the neighboring Autonomous Province of Trento.

Figure 4. Purchase of cattle (private, local trade, etc.).

3.8. Calf Management

Calf shelters are usually cleaned only if necessary (38%). Most of the farmers avoid feeding with milk of cows treated with antibiotics during the withdrawal period (60%), while some of them feed with mastitic milk (45%). During the last 12 months, 43% reported administering antibiotic treatments, and 12% reported administering vaccinations. Regarding diseases, in 23% of cases, diagnostic testing was performed.

3.9. Cattle Management

During the last 12 months, 68% of the farmers reported administering antibiotic treatments, and 4% reported administering vaccinations. Regarding disease, in 17% of cases, diagnostic testing was performed, and 8% of the farmers used antimicrobial susceptibility testing. Overall, 51% of the farmers reported fewer than five cases per year of mastitis, and 21% reported no cases at all. Overall, 26% of farmers did not answer this question.

3.10. Opinions

Overall, 53% of the farmers agreed that their adult cattle may be at risk of infectious diseases, while 34% of them were confident that they were not. Diarrhea in calves (89%) and mastitis in cows (43%) are considered to be the most relevant diseases. Meanwhile, 49% of the respondents were confident that contact with other animals represents the most relevant risk factor, followed by unknown variables (28%), contact with people, feed, the purchasing of cattle, viral agents, and poor hygienic conditions (Figure 5).

Figure 5. Factors that are considered to be relevant in the transmission and spread of infectious diseases at farm level.

As regards biosecurity, the most effective actions taken to prevent infectious disease were the cleaning of clothes and boots, preventive measures, and the isolation of sick animals. Farmers who do not take any action explained that these measures are too expensive, not applicable, or even not necessary (as animals are healthy). In 28% of these cases, the farmers asserted that they did not know what more they should do.

The main drivers of implementing preventive measures were to minimize the use of antimicrobials (74%), and to save money (68%).

Some farmers stated that they will adopt preventive measures only if they are effective on other farms (60%), if necessary (59%), or when forced (17%). Overall, 77% of them were willing to adopt additional measures under the recommendations of a veterinarian. Moreover, 62% of them confirmed that a veterinarian had already provided information about biosecurity, and at least 51% showed interest in this topic. Farmers' answers are displayed in Figure 6. For clarity, data are displayed on a three-point Likerts scale where the group "agree" reports the answers of farmers who strongly agree and agree with the questions, the group "disagree" reports the answers of people who are in strong disagreement and disagreement, and the group "uncertain" represents the respondents who are not sure about the answer.

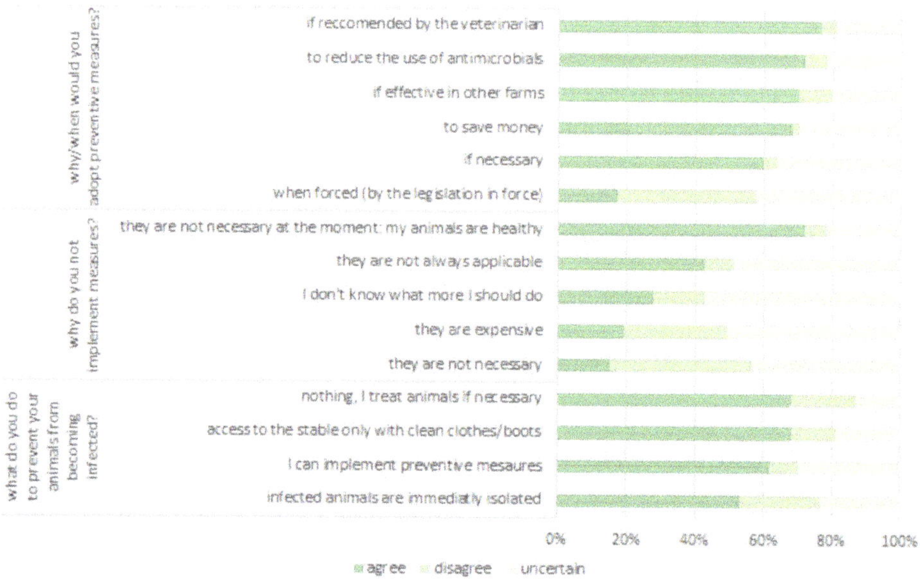

Figure 6. Opinions about biosecurity.

4. Discussion

Due to its position, Italy presents different climatic conditions, varying from Alpine to the Po Valley, Mediterranean, and Peninsular climate. Dairy farming in Italy varies within the country also according to different local settings, traditions, and geographic conditions. In Italy, four main regions located in the northern part of the country deliver 66% of the national milk [11]. These regions are not entirely located in the Alps; most of their farms are in the flatland, and they keep a large number of animals in free stalls all the year. On the contrary, the Autonomous Province of Bolzano/Bozen is entirely located in the Alps, and it is characterized by small dairy farms, where animals are often kept in tie-stalls. However, summer grazing in the Alpine pastures is a typical practice that allows cattle to move outside the farms in better climate conditions and helps reduce the workload for farmers. This practice is also linked to the local traditions and the development of tourism. On the other hand, these characteristics may influence the burden of infectious diseases, their spread, and the measures taken to contrast them. The discussion highlights peculiarities that have been investigated.

4.1. The Context and the Response Rate

The response rate is relevant to assessing the quality of a survey [12]. In this study, 27.5% of the questionnaires were filled in and returned. As expected, this rate is lower in comparison to similar studies that included an additional reminder for selected farmers and/or a visit to the farm, which reported response rates of over 40% [13,14], but higher than expected (25%) when compared to other studies and online surveys [9]. In this study, factors that may have had a positive impact on the response rate include the selection of a well-defined target population (farmers that live in the same district and breed cattle) and the relevance of the topic for this category of people. Also, the choice to avoid contacting the participants in advance and reminding them about the survey probably affected this result. However, it is not possible to ascertain if this had a positive or a negative effect. The authors decided to prioritize the anonymity of the respondents in order to gain insights into the opinions and practices without any possible influence of intermediary persons before or during the completion of the questionnaires. As a higher number of potential participants and the use of incentives do not necessarily guarantee an increase in response rate (11), the study was not extended to other districts and no incentives were offered.

The debate is still open: whether postal surveys can result in higher or lower response rates compared to e-mail surveys [15]. We preferred to send the survey per post, also assuming that every farmer would receive it, in order to reduce the potential impact of respondents' attitude towards technology on the non-response bias.

A low number of farmers were women, and most of the respondents were men over 40 working full-time on farms located outside the villages, above 1000 masl, producing conventional milk. These data provide insights into the demographics of farmer society; however, they are not sufficient to understand the underlying factors that characterize the reality of local dairy farms. Importantly, women play an essential role, making invaluable contributions to livestock, family, and, if present, farm holiday management. Moreover, often more than one generation live together on a farm; youths have a job in the main town, and, in addition, they provide regular support to their parents, who work full-time on their farm. This approach is crucial to allow the existence of a traditional alpine farming system.

4.2. Health Literacy and Awareness on Preventive Measures

In accordance with other studies [16], most of the farmers already knew the complete definition of biosecurity or related it to the definition of external or internal biosecurity. Moreover, it is pertinent to mention that, as the German word for organic farm is "biologisch", the suffix "bio" could have been misleading.

Among the participants, the majority declared being interested in additional information about biosecurity measures, and this attitude is a valuable strength. As a matter of fact, the improvement of health literacy is extremely relevant in the veterinary and agricultural sectors, as this could lead to improved compliance and outcomes [17]. This study was also designed to provide farmers with relevant information to share with them their strengths and areas for improvement, to raise awareness on this topic and to improve their knowledge. To meet the demands of farmers and to address their willingness to learn more about preventive measures, on the basis of the results of this study, an awareness initiative can be promoted.

Behaviors and motivators among farmers have been investigated [18,19], and it has been shown that if preventive measures are not perceived as relevant by farmers, they might end up causing a higher risk of the spread of diseases [18,20,21]. The current study found a low willingness to vaccinate calves (12%), while an even lower percentage of farmers (only 4% of the respondents) had immunized their adult cattle during the last 12 months. As determinants such as age, education, and family size have been associated with the willingness to vaccinate cattle [22], it would be worthwhile to assess which factors and drivers may have an impact on this choice. Moreover, in the future, it will be possible to detailed obtain data from the National Electronic Veterinary Recipe Platform. This tool has recently been implemented by the Ministry of Health [23] to monitor the use of vaccinations and, if required, to design and recommend effective immunization strategies to reduce the incidence of relevant vaccine-preventable diseases in cattle, especially those that can be transmitted to humans.

On the other hand, one of the factors that motivated the participant to this study to improve preventive measures was the possibility to reduce antimicrobial consumption. As vaccination is one of the most cost-effective biosecurity measures to prevent infectious diseases in cattle and, consequently, to reduce the need for antimicrobial treatments [4,5,24], it would be reasonable to inform thoroughly the stakeholders about the benefits of immunization.

Nevertheless, it must be considered that some vaccinations are not available. Moreover, the adoption of some immunization programs is not allowed or required, as this province has achieved the status of being officially free from several major animal diseases, including bovine tuberculosis, enzootic bovine leukosis (EBL), and bovine brucellosis. Moreover, additional compulsory eradication programs have been adopted for the control of relevant transmittable diseases such as bovine viral diarrhea virus (BVDV) [25] and infectious bovine rhinotracheitis [26]. The high level of the cattle health status in this province was achieved thanks to veterinary strategic decisions that promoted and strengthened effective surveillance, control, and eradication programs in close collaboration with farmers and their local associations. Appropriate policies were implemented to protect both animal and human health and to guarantee the supply of local safe and sustainable products of animal origin for human consumption. Anyway, surveillance programs for early detection of infectious diseases have to be permanently ensured [27]. These aspects are relevant in this province because the alpine farming, with its prevalent traditional, family-based management system, could be vulnerable to the spread of re-emerging diseases, and the successful marketing of products with high biological value proteins could be threatened. On the other hand, this farming system could be intrinsically resilient to threats. In fact, in this study, we observed a significant association between herd size and the presence of reported cases of mastitis and diarrhea, which represent the most relevant diseases that may occur from the participants' point of view. This can be explained by the fact that, very often, farmers' entire families put a particular effort into the daily care of a relatively low number of cattle, in conjunction with the awareness among local farmers of the relevance of both diseases. Under these circumstances, if farmers can be trained in disease detection, they will probably be willing to cooperate with the local veterinary authorities to enhance the local warning system for the prompt detection of signs of other relevant infectious diseases.

4.3. The Role of the Veterinarians, Strengths, and Areas of Improvement

In agreement with other studies [2,16,21,28–30], veterinarians have been acknowledged to be trustworthy informants, and the decision to implement preventive measures can be strongly influenced by them. In this study, 62% of respondents had already been informed by a veterinarian about these measures, and, importantly, 77% of the respondents reported that they were willing to improve their biosecurity measures under the recommendations of a veterinarian.

These results can suggest additional research questions for further studies to better investigate, for example, whether veterinarians are consulted on a regular basis or just for unusual cases, as reported by other authors [31], and how veterinarians can support the change in the long run [32].

This study allows us to acknowledge good practices and habits that need to be further encouraged and how the veterinarians can play a pivotal role. For instance, stable entry is usually allowed only to family members who have the primary interest of keeping their animals healthy and to staff who is trained and already has comprehensive knowledge about preventive measures (among them, veterinarians, and delegates of the local farmer association). Under these circumstances, the risk of diseases' transmission can be considered low, provided that hygiene good practices are respected. In fact, the risks linked with the daily attendance of many farms where both healthy and sick animals are kept and the limited time available for changing the clothes and disinfecting the boots cannot be underestimated.

Other good practices that are already in place and must be further promoted are wearing clean clothes and boots when entering a pen, the early isolation of sick animals, and the allocation of animals by age or production group.

On the other hand, areas for improvement also emerged. Only approximately half of respondents confirmed that they perform regular inspections and replacement of milking equipment, while less than half of them cleaned drinking troughs and calf shelters regularly. Even though a high number of farmers purchase cattle, only 4% quarantine their cattle, either because it is not possible or because this practice is not considered to be necessary. At this point, it is crucial to emphasize that mountain farming is demanding and that there are restrictions and limits. In particular, it can be challenging or even impossible to renovate buildings to provide space for quarantine pens or to convert tie-stalls into animal-friendly free-stalls equipped with quarantine boxes. Nevertheless, in this study, it was not possible to find an association between the lack of quarantine practices and the incidence of diseases due to the low number of farmers that do apply quarantine. Interestingly, in accordance with other studies [33], we did not find any association between the tie-stall system and a higher incidence of diseases. On the contrary, a positive association between the free-stall system and the presence of diarrhea was confirmed.

During the 12 months before this study, 43% of farmers reported the use of antimicrobials. However, diagnostic practices are not common; in fact, only 8% of the farmers indicated the use of antimicrobial susceptibility tests. Moreover, 45% of them reported feeding calves with mastitic milk and 40% with milk of cows treated with antimicrobials during the withdrawal period, even though this practice should be avoided as it is detrimental to calves' health [34]. Of course, shifting from a curative approach to a preventive strategy requires time and training. However, if farmers and veterinarians agree on shared objectives and perceive counseling as an opportunity, it will be possible to improve both diagnostic and feeding practices [2,13].

This study shows that the majority of the farmers purchase cattle at trade fairs from other farmers and salesmen. Even though animals' movement must comply with the applicable law and, usually, animals are moved from areas and countries with the same health status as this province, purchasing cattle can be a risk in terms of the introduction of diseases into a farm, especially those that are not targets of official control programs. Several factors, such as the frequency and number of animal purchases but also transport practices, play a crucial role. For these reasons, strict hygiene measures at purchase and during transport must be ensured. In particular, salespersons and farmers who regularly attend cattle markets must be aware of potential risks and be trained in preventive measures to minimize the risk of disease spread.

Other potential factors that need to be monitored are the use of artificial insemination (AI), the sharing of equipment, carcass removal, and manure management. We found that reproduction strategies are mainly the responsibility of veterinarians or trained staff, so this practice should not pose a significant risk. Equipment sharing is not very common; however, the situation needs to be monitored, as it may change in the future due to the need to split the costs of buying and maintaining machines and instruments. Carcasses can either be removed and disposed of by specialized companies or transported by the farmer to a disposal plant. Considering the small herd sizes and the location of the farms (most of them located outside of villages), disinfection baths and gates at farm entrances are not common. As vehicles usually have to enter the premises of several farms, either when they collect milk or when they remove carcasses, truck cleaning and disinfection at the available vehicle washing facilities before leaving them is essential in order to minimize risks. Another peculiarity that must be considered for small alpine farms is the truck traffic, which is not always possible as streets can be too narrow. For this reason, many farmers bring milk tanks directly to local collection points, and this can be considered a useful practice to reduce the daily entry of vehicles into the premises.

Summer pastures characterize alpine farming, and as expected, we found that a high percentage of farmers (91%) adopt this practice. This means that cattle may have contact with other animals (pets, wild and farmed animals) and access water springs. Even if local farmers identify the variable "contact with animals" as the most relevant risk factor for disease transmission, unfortunately, they lack control over this factor, as animals in pastures are usually kept outdoors and are not supervised all the time.

The transmission of waterborne diseases represents a specific topic that goes behind the scope of this study. However, we asked farmers two questions about cattle water intake. As potable water is available in local stables, the first question was whether drinking troughs are cleaned, and in 42% of the farms this happens only if they are dirty. The second question addressed the potential access to natural water springs, i.e., as 91% of the farmers bring their cattle to summer pastures, this possibility cannot be ruled out. In the first case, cattle are exposed to risk, and this can be minimized by adopting good cleaning practices. On the other hand, the second scenario may have implications from a public health perspective, as water can be contaminated with pathogens from cattle feces [35]. However, this risk should be negligible, as access to natural water springs is limited by fences.

4.4. Perspectives

In this study, the majority of the farmers showed interest in biosecurity (51%), and they confirmed that they would be willing to adopt preventive measures if they are effective on other farms (60%) and if necessary (59%), in order to reduce the use of antibiotics (74%), and to save money (68%). However, it should be further investigated as to which preventive measures are not applicable, why some respondents prefer to wait and treat animals if necessary, and how farmers can be better supported, as some of them stated that they do not know what more they should do. A high number of farmers said that they do not apply biosecurity measures because their animals are healthy. The value of some good practices that are already in place (cleaning of boots, separation of animals, stall entry allowed only to a restricted number of persons) should be acknowledged, and the intervention of a veterinarian would be crucial to explain why some measures are crucial and should be maintained.

Mountain farming can be considered as a model for local sustainable development, even if challenges have to be taken into account [36–38]. To avoid the risk of decline and abandonment, local farmers have been able to diversify their activities, and they can generate revenue from tourism, which is of vital importance for small-scale farming [39–41]. Even though a low number of women and farmers under 40 responded to the questionnaire, this does not necessarily mean that women and young people are not involved in working activities on farms. Unless proven otherwise, we suggest that women can feel empowered and help to develop touristic activities on farms, thus contributing to farmers' families' income. Furthermore, mountain agriculture may be attractive for young people who can adapt innovative concepts to ensure the modernization and sustainability of this sector [42,43].

This study highlights the importance of drivers (such as a reduction in the use of antimicrobials and cost savings) and of the main actors (in particular veterinarians), who may play a crucial role in the implementation of preventive measures.

Prevention measures are associated both with costs (for example, vaccinations and testing of animals) and benefits (healthier animals are more productive). In this study we included two questions, one related to the willingness to save money in the long term as a driver to improve biosecurity, and one to better understand whether these measures are considered too expensive. Most of the farmers confirm that improving biosecurity brings more benefits in terms of saving money, and less than 20% of the respondents consider these measures expensive. These findings confirm an optimistic attitude toward biosecurity. As interest in this topic is growing and promising projects about biosecurity's cost-effectiveness are being promoted at the international level, in the future it will be possible to provide evidence-based practical guidance also from this perspective.

As the enrollment was voluntary and anonymous, no farm interviews or reminder calls were planned. Because of this, it was not possible to address specific answers to better investigate some factors that could influence the outcomes or ask for clarifications for some missing or unclear answers. As studies on farmers' perspectives, attitudes, beliefs, and motivations regarding biosecurity implementation are effective in order to understand why some measures are not implemented and to optimize the undertaking of good practices [44,45], further specific surveys focusing on behavioral changes are recommended. In addition, as the role of self-evaluation tools like *Classyfarm* [6] and Biocheck® [19] has been consolidated in improving the awareness of biosecurity and strengthening the implementation of specific preventive measures, we would recommend the development of additional applications and digital toolkits for self-evaluations customized to small-scale family alpine farms.

5. Conclusions

This study highlights several strengths of small-scale alpine dairy farms and those practices related to quarantine, cleaning, waste milk feeding to calves, immunization and diagnostics that could be improved to implement optimal biosecurity measures. As farmers confirmed that they were willing to adopt effective preventive measures under the recommendations of a veterinarian, this attitude will be crucial to designing multisectoral participative initiatives. Further investigations and the development of customized applications and digital toolkits are now required to better understand local farmers' drivers toward biosecurity and to provide support.

Author Contributions: Conceptualization, M.T. and P.C.; Methodology, M.T. and P.C.; Software, M.T.; Validation, M.T.; Formal analysis, M.T.; Investigation, P.C.; Resources, P.C.; Data curation, M.T.; Writing—original draft, P.C.; Writing—review & editing, M.T. and P.C.; Visualization, M.T.; Supervision, M.T. and M.D.P.; Project administration, P.C.; Funding acquisition, P.C. All authors have read and agreed to the published version of the manuscript.

Funding: This project was funded by the Department of Prevention of the South Tyrol Health Service, project number 18/2021. No compensation or incentives have been paid for participating in the survey.

Institutional Review Board Statement: The study was carried out in compliance with Directive 86/609/EEC. The information collected cannot identify the participants or place them at risk of damaging them.

Informed Consent Statement: This paper, based on anonymous and voluntary collection of data, did not require a signed consent form. We provided the farmers with an information document that explained that the participation in the survey was voluntary, that the purpose of the study was to obtain an insight into the biosecurity awareness, and that the data collection was anonymous.

Data Availability Statement: Relevant questions are reported in the manuscript. The full text questionnaire (in German) is available upon request to interested researchers.

Acknowledgments: We would like to thank all the farmers who voluntarily took time to complete the questionnaire, Maria Grazia Zuccaro for authorizing the project, Anne Marie Kaser for her valuable advice, the veterinarians, and the farmers who revised the questionnaire. We are very grateful to Antonio Fanolla and Sabine Weiss for providing the map of the area of study.

Conflicts of Interest: The authors declare no conflicts of interest.

References

1. The European Parliament and the Council of the European Union. Regulation (EU) 2016/429 of the European Parliament and of the Council of 9 March 2016 on Transmissible Animal Diseases and Amending and Repealing Certain Acts in the Area of Animal Health ('Animal Health Law'). *Off. J. Eur. Union* **2016**, *59*, 1–208.
2. Renault, V.; Humblet, M.F.; Pham, P.N.; Saegerman, C. Biosecurity at cattle farms: Strengths, weaknesses, opportunities and threats. *Pathogens* **2021**, *10*, 1315. [CrossRef]
3. Ministero della Salute. PNCAR. 2023. Available online: https://www.salute.gov.it/portale/antibioticoresistenza (accessed on 31 October 2023).
4. EFSA Panel on Animal Health and Welfare (AHAW); Nielsen, S.S.; Bicout, D.J.; Calistri, P.; Canali, E.; Drewe, J.A.; Garin-Bastuji, B.; Rojas, J.L.G.; Schmidt, C.G.; Herskin, M.; et al. Assessment of animal diseases caused by bacteria resistant to antimicrobials: Cattle. *EFSA J.* **2021**, *19*, e06955. [CrossRef] [PubMed]
5. Dhaka, P.; Chantziaras, I.; Vijay, D.; Bedi, J.S.; Makovska, I.; Biebaut, E.; Dewulf, J. Can Improved Farm Biosecurity Reduce the Need for Antimicrobials in Food Animals? A Scoping Review. *Antibiotics* **2023**, *12*, 893. [CrossRef] [PubMed]

6. Instituto Zooprofilattico Sperimentale della Lombardia e dell'Emilia-Romagna. Classyfarm. Available online: https://www.classyfarm.it/ (accessed on 4 January 2022).
7. Holighaus, L.; Zanon, T.; Kemper, N.; Gauly, M. First evaluation of the practicability of the CLASSYFARM welfare assessment protocol in Italian small-scale mountain dairy farms—A case study. *Ital. J. Anim. Sci.* **2023**, *22*, 995–1007. [CrossRef]
8. Istituto Provinciale di Statistica, ASTAT. Alto Adige in Cifre 2021. 2021. Available online: https://astat.provincia.bz.it (accessed on 1 May 2024).
9. Klein-Jöbstl, D.; Arnholdt, T.; Sturmlechner, F.; Iwersen, M.; Drillich, M. Results of an online questionnaire to survey calf management practices on dairy cattle breeding farms in Austria and to estimate differences in disease incidences depending on farm structure and management practices. *Acta Vet. Scand.* **2015**, *57*, 44. [CrossRef]
10. Robichaud, M.V.; Rushen, J.; de Passillé, A.M.; Vasseur, E.; Orsel, K.; Pellerin, D. Associations between on-farm animal welfare indicators and productivity and profitability on Canadian dairies: I. On freestall farms. *J. Dairy Sci.* **2019**, *102*, 4341–4351. [CrossRef] [PubMed]
11. ISTAT. Latte e Prodotti Lattiero Caseari. Available online: https:////dati.istat.it/Index.aspx?DataSetCode=DCSP_LATTE (accessed on 18 June 2024).
12. Wu, M.J.; Zhao, K.; Fils-Aime, F. Response rates of online surveys in published research: A meta-analysis. *Comput. Hum. Behav. Rep.* **2022**, *7*, 100206. [CrossRef]
13. Sahlström, L.; Virtanen, T.; Kyyrö, J.; Lyytikäinen, T. Biosecurity on Finnish cattle, pig and sheep farms—Results from a questionnaire. *Prev. Vet. Med.* **2014**, *117*, 59–67. [CrossRef]
14. Bernaerdt, E.; Dewulf, J.; Verhulst, R.; Bonckaert, C.; Maes, D. Purchasing policy, quarantine and acclimation practices of breeding gilts in Belgian pig farms. *Porc. Health Manag.* **2021**, *7*, 25. [CrossRef]
15. Zahl-Thanem, A.; Burton, R.; Vik, J. Should we use email for farm survey? A comparative study of email and postal survey response rate and non-response bias. *J. Rural. Stud.* **2021**, *87*, 352–360. [CrossRef]
16. Brennan, M.L.; Wright, N.; Wapenaar, W.; Jarratt, S.; Hobson-West, P.; Richens, I.F.; Kaler, J.; Buchanan, H.; Huxley, J.N.; O'Connor, H.M. Exploring attitudes and beliefs towards implementing cattle disease prevention and control measures: A qualitative study with dairy farmers in Great Britain. *Animals* **2016**, *6*, 61. [CrossRef] [PubMed]
17. Ba, K.C.; Kaewkungwal, J.; Pacheun, O.; Nguyen Thi To, U.; Lawpoolsri, S. Health Literacy Toward Zoonotic Diseases Among Livestock Farmers in Vietnam. *Environ. Health Insights* **2020**, *14*, 1178630220932540. [CrossRef]
18. Lestari, V.S.; Rahardja, D.P.; Mappigau, P.; Rohani, S.T.; Sirajuddin, S.N. Beef cattle farmers behavior toward biosecurity. *J. Indones Trop. Anim. Agric.* **2019**, *44*, 204–212. [CrossRef]
19. Damiaans, B.; Sarrazin, S.; Heremans, E.; Dewulf, J. Perceptie, drijfveren en hindernissen van bioveiligheid in de rundveeproductie. *Vlaams Diergeneeskd Tijdschr.* **2018**, *87*, 150–163. [CrossRef]
20. Kuster, K.; Cousin, M.E.; Jemmi, T.; Schüpbach-Regula, G.; Magouras, I. Expert opinion on the perceived effectiveness and importance of on-farm biosecurity measures for cattle and swine farms in Switzerland. *PLoS ONE* **2015**, *10*, e0144533. [CrossRef] [PubMed]
21. Moya, S.; Tirado, F.; Espluga, J.; Ciaravino, G.; Armengol, R.; Diéguez, J.; Yus, E.; Benavides, B.; Casal, J.; Allepuz, A. Dairy farmers' decision-making to implement biosecurity measures: A study of psychosocial factors. *Transbound. Emerg. Dis.* **2020**, *67*, 698–710. [CrossRef]
22. Lestari, V.S.; Rahardja, D.P.; Sirajuddin, S.N. The willingness of beef cattle farmers to adopt vaccination from the animal health service unit in Bone regency, South Sulawesi province. In *IOP Conference Series: Earth and Environmental Science, Proceedings of the 5th International Conference on Agriculture, Environment, and Food Security, Medan, Indonesia, 18 November 2021*; Institute of Physics: London, UK, 2022. [CrossRef]
23. Ministero della Salute. REV M. Available online: https://www.ricettaveterinariaelettronica.it/ (accessed on 4 January 2022).
24. Dewulf, J.; Joosten, P.; Chantziaras, I.; Bernaerdt, E.; Vanderhaeghen, W.; Postma, M.; Maes, D. Antibiotic Use in European Pig Production: Less Is More. *Antibiotics* **2022**, *11*, 1493. [CrossRef]
25. Luzzago, C.; Decaro, N. Epidemiology of Bovine Pestiviruses Circulating in Italy. *Front. Vet. Sci.* **2021**, *8*, 669942. [CrossRef]
26. Bettini, A.; Stella, M.; Precazzini, F.; Degasperi, M.; Colorio, S.; Tavella, A. Infectious Bovine Rhinotracheitis Post-Eradication Program in the Autonomous Province of Bolzano, Italy: A Retrospective Study on Potential Bovine Herpesvirus Type 2 Cross-Reactivity. *Animals* **2023**, *13*, 3502. [CrossRef]
27. Schüpbach, G.; Silva, L.C.; Buzzell-Hatav, A. Surveillance plan proposal for early detection of zoonotic pathogens in ruminants. *EFSA Support. Publ.* **2023**, *20*, 7887E. [CrossRef]
28. Durán, S.M.; Casal, J.; Ciaravino, G.; Diéguez, F.J. Influence of official veterinary services on the implementation of biosecurity measures on dairy cattle farms in Spain National surveillance on vaccine preventable diseases (ISCIII-Madrid, Spain) View project "Portraits of Bios in the Contemporary Societies View project". In Proceedings of the 4th International Conference on Animal Health Surveillance, Copenhagen, Denmark, 3–5 May 2022. [CrossRef]
29. Moya, S.; Navea, J.; Casal, J.; Ciaravino, G.; Yus, E.; Diéguez, F.J.; Benavides, B.; Tirado, F.; Allepuz, A. Government veterinarians' perceptions of routine biosecurity focused on dairy cattle farms in north-western and north-eastern Spain. *Front. Vet. Sci.* **2023**, *10*, 1043966. [CrossRef]
30. Pritchard, K.; Wapenaar, W.; Brennan, M.L. Cattle veterinarians' awareness and understanding of biosecurity. *Vet. Rec.* **2015**, *176*, 546. [CrossRef]

31. Cobo-Angel, C.; LeBlanc, S.J.; Roche, S.M.; Ritter, C. A Focus Group Study of Canadian Dairy Farmers' Attitudes and Social Referents on Antimicrobial Use and Antimicrobial Resistance. *Front. Vet. Sci.* **2021**, *8*, 645221. [CrossRef] [PubMed]
32. Bard, A.M.; Main, D.; Roe, E.; Haase, A.; Whay, H.R.; Reyher, K.K. To change or not to change? Veterinarian and farmer perceptions of relational factors influencing the enactment of veterinary advice on dairy farms in the United Kingdom. *Dairy Sci.* **2019**, *102*, 10379–10394. [CrossRef]
33. Witkowska, D.; Ponieważ, A. The Effect of Housing System on Disease Prevalence and Productive Lifespan of Dairy Herds—A Case Study. *Animals* **2022**, *12*, 1610. [CrossRef] [PubMed]
34. Penati, M.; Sala, G.; Biscarini, F.; Boccardo, A.; Bronzo, V.; Castiglioni, B.; Cremonesi, P.; Moroni, P.; Pravettoni, D.; Addis, M.F. Feeding Pre-weaned Calves with Waste Milk Containing Antibiotic Residues Is Related to a Higher Incidence of Diarrhea and Alterations in the Fecal Microbiota. *Front. Vet. Sci.* **2021**, *8*, 650150. [CrossRef] [PubMed]
35. Antunes, P.O.; Óhuallacháin, D.; Dunne, N.; Kelly-Quinn, M.; O'sullivan, M.; Bragina, L.; Jennings, E. Cattle access to small streams increases concentrations of *Escherichia coli* in bed sediments. *Hydrobiologia* **2023**, *850*, 3273–3291. [CrossRef]
36. European Commission, Directorate-General for Agricultural and Rural Development. *New Insight into Mountain Farming in the European Union*; Commission Staff Working Document: Brussels, Belgium, 2009.
37. Schneider, A. Land's Constraints and Possibilities–High-Altitude Farmers in the Eastern Alps. *Cult. Agric. Food Environ.* **2022**, *44*, 18–26. [CrossRef]
38. Wanner, A.; Pröbstl-Haider, U.; Feilhammer, M. The future of Alpine pastures – Agricultural or tourism development? Experiences from the German Alps. *J. Outdoor Recreat. Tour.* **2021**, *35*, 100405. [CrossRef]
39. Giaccio, V.; Mastronardi, L.; Marino, D.; Giannelli, A.; Scardera, A. Do rural policies impact on tourism development in Italy? A case study of agritourism. *Sustainability* **2018**, *10*, 2938. [CrossRef]
40. Lun, L.M.; Pechlaner, H.; Volgger, M. Rural Tourism Development in Mountain Regions: Identifying Success Factors, Challenges and Potentials. *J. Qual. Assur. Hosp. Tour.* **2016**, *17*, 389–411. [CrossRef]
41. Streifeneder, T.; Dax, T. *Agritourism in Europe: Enabling Factors and Current Developments of Sustainable On-Farm Tourism in Rural Areas. Global Opportunities and Challenges for Rural and Mountain Tourism*; Kala, D., Bagri, S.C., Eds.; IGI Global: Hershey, PA, USA, 2020; pp. 40–58. [CrossRef]
42. Marescotti, M.E.; Demartini, E.; Filippini, R.; Gaviglio, A. Smart farming in mountain areas: Investigating livestock farmers' technophobia and technophilia and their perception of innovation. *J. Rural. Stud.* **2021**, *86*, 463–472. [CrossRef]
43. White, B. Agriculture and the Generation Problem: Rural Youth, Employment and the Future of Farming. *IDS Bull.* **2012**, *43*, 9–19. [CrossRef]
44. Ellis-Iversen, J.; Cook, A.J.; Watson, E.; Nielen, M.; Larkin, L.; Wooldridge, M.; Hogeveen, H. Perceptions, circumstances and motivators that influence implementation of zoonotic control programs on cattle farms. *Prev. Vet. Med.* **2010**, *93*, 276–285. [CrossRef]
45. Laanen, M.; Maes, D.; Hendriksen, C.; Gelaude, P.; De Vliegher, S.; Rosseel, Y.; Dewulf, J. Pig, cattle and poultry farmers with a known interest in research have comparable perspectives on disease prevention and on-farm biosecurity. *Prev. Vet. Med.* **2014**, *115*, 1–9. [CrossRef]

Disclaimer/Publisher's Note: The statements, opinions and data contained in all publications are solely those of the individual author(s) and contributor(s) and not of MDPI and/or the editor(s). MDPI and/or the editor(s) disclaim responsibility for any injury to people or property resulting from any ideas, methods, instructions or products referred to in the content.

animals

MDPI

Article

Risk Awareness and Attitude of German Farmers towards Biosecurity Measures

Anna Herrmann [1], Katja Schulz [1], Natalie Wischnewski [1,2,*], Jule Brüssau [1], Eva Zeiler [2] and Carola Sauter-Louis [1]

1 Institute of Epidemiology, Friedrich-Loeffler-Institut, Südufer 10, 17493 Greifswald-Insel Riems, Germany; anna.herrmann@fli.de (A.H.); katja.schulz@fli.de (K.S.); j.bruessau@web.de (J.B.); carola.sauter-louis@fli.de (C.S.-L.)

2 Faculty of Sustainable Agricultural and Energy Systems, University of Applied Sciences Weihenstephan-Triesdorf, Am Staudengarten 1, 85354 Freising, Germany; eva.zeiler@hswt.de

* Correspondence: natalie.wischnewski@fli.de

Simple Summary: Biosecurity measures are crucial for protecting farms against the introduction of animal diseases, given the dynamic nature of outbreaks. Livestock farmers play a key role in the success of these strategies. To understand German farmers' perspectives on biosecurity, an anonymous online survey was conducted. Alongside general farm information, data were collected on existing biosecurity protocols and potential disease entry pathways, along with corresponding protective measures. The evaluation of the survey results indicates farmers' awareness of the importance of biosecurity, evidenced by the presence of biosecurity concepts on their farms. While farmers recognize potential disease entry points, there is a lack of knowledge on improving biosecurity in these areas. Overall, farmers' high commitment to biosecurity suggests routine implementation of measures. However, their lack of knowledge about the effectiveness of biosecurity measures needs to be further investigated in future studies.

Abstract: The implementation of management activities depends on both the attitude of the people performing the work and their understanding of why the work should be performed. In the context of animal husbandry, the implementation of such practices is crucial for the functionality of biosecurity. Therefore, it is important to know how farmers perceive biosecurity as a whole. An anonymous online survey was conducted among German farmers. In addition to general data about their farm, information about their existing concept of biosecurity, as well as about the assessment of possible introduction routes for animal diseases into the farm with regard to their likelihood, was gathered. Furthermore, information on measures to protect their farm against disease introduction were retrieved. Analysis showed that in general, farmers were aware of the importance of biosecurity and consequently had concepts of biosecurity on their farms. However, awareness about dangerous introduction routes for animal diseases into a farm was associated with a lack of knowledge of how to improve the measures in these areas. The role of the veterinarian in the context of biosecurity was highlighted and further problematic areas were indicated. Overall, the high level of commitment from farmers indicated a good implementation of daily practices.

Keywords: biosecurity; farm management; animal husbandry; farmer's attitude; online-survey; Germany

1. Introduction

Pathogens infecting animals kept for food production pose a threat on multiple layers. Not only is the health and therefore the wellbeing of the livestock in threat [1,2], but also the health of the humans consuming the produced food [3–5]. Furthermore, losses of animals along the production chain reduce the efficient use of resources, leading to an economic loss as well as a higher impact on the environment per produced unit [6]. The concept of biosecurity generally deals with the elimination or at least the minimization of biological contamination [7]. With regard to animal health and production systems, the term biosecurity is defined by the World Organization for Animal Health (WOAH) as "a set of management and physical measures designed to reduce the risk of introduction, establishment and spread of animal diseases, infections or infestations to, from and within an animal population" [8]. Other definitions focus more on the link of animal diseases to public health and thereby highlight the importance of the One Health concept [9]. Biosecurity on a farm can be divided into external biosecurity, ensuring that no pathogens enter the farm, and internal biosecurity, hindering the spread of pathogens within the farm [10]. In addition, a distinction can be made between operational biosecurity, which describes management practices, and constructional biosecurity, which includes building measures [11]. While constructional biosecurity can be controlled by national authorities, the actual daily implementation of management practices on the farm cannot. Therefore, the success of biosecurity measures is strongly dependent on their implementation on site. Thus, the people in charge of the animal facility represent the key figures in a successful biosecurity concept [12]. Certain socio-psychological factors affect the choice of a farmer of whether to implement biosecurity measures on their farm or not [13,14]. The awareness of the problem as well as the perceived effectiveness and feasibility of a measurement are thereby the main factors of the probability of its adaption [15,16].

In Germany, animal production systems are currently threatened by the circulation of highly pathogenic avian influenza (HPAI) [17] as well as African swine fever (ASF) [18]. Therefore, functioning biosecurity on a farm level is crucial in order to prevent outbreaks. Evaluating the current situation regarding biosecurity measures in Germany, an online survey was conducted among farmers. The aim was to assess the risk awareness in regards to pathogen introduction to the farm, the self-perception of the farm's biosecurity status and the general attitude of farmers towards biosecurity. The collected data allow for an estimation about the functionality of the operational biosecurity on farms in Germany.

Even though biosecurity measures are implemented to hinder the introduction and spread of all kinds of diseases into and within the farm, some infectious animal diseases like HPAI and ASF are listed by the European Union (EU) law (Regulation (EU) 2016/429 (Animal Health Law)) and are associated with a specific series of control measures. The term 'Tierseuchen' used in the original German questionnaire was translated as 'animal diseases' for the writing of the present paper, but implies that the animal disease is listed by the EU. The focus of the presented paper lies therefore on these listed animal diseases.

2. Materials and Methods

2.1. Recruitment of Participants

An anonymous online survey was conducted among farmers that keep farm animals for the production of food. More precisely, people that owned a farm or were hired to oversee an animal facility were encouraged to participate. All German farmers and agricultural workers who were involved in biosecurity issues and reachable through the selected distribution system were encouraged to participate in the survey. There was no selection based on regions or production types. All completed datasets were considered in the survey analysis. Due to significant differences in husbandry and management practices, fish and bee holdings were not included in this questionnaire. The survey was accessible starting from the 1st of March and until the 31st of May 2023, leading to a conscience sampled data pool [19]. The recruitment of participants was achieved through indirect snowball sampling [20,21]. In snowball sampling, participants or survey respondents were requested to invite additional individuals who belonged to the target group to participate in the survey. Altogether, three social media influencers that focused on agricultural topics with a focus on political themes, three governmental institutions, three industrial companies, fourteen media outlets in the field of agricultural news and thirty farmers' associations were addressed in the beginning of March 2023 simultaneously with the start of the survey. Nine addressees responded, of which eight released a call for participation on various social media platforms or sent an informational letter to their members. Further announcements were posted on various platforms of the contacted institutions, even though they did not respond to the authors' initial request. The organizations contacted were only involved in the distribution of the survey. Nevertheless, it was possible for individual employees of the organizations or agricultural influencers who, for example, managed a farm and were confronted with the issue of biosecurity in their day-to-day operations to take part in the survey.

2.2. Questionnaire

The survey was designed using the ScoSci Survey software (version 3.4.10; https://www.soscisurvey.de; [22] accessed on 13 July 2023) and was made available via this website. It consisted of ten questions about the structure of the farm, the state of its biosecurity and the awareness of potential threats as well as potential measures of improvement. Pretesting was performed by seven veterinarians working in the field of epidemiology and three agronomists working in the field of animal husbandry. The questionnaire can be found in Supplementary S1.

The questions were presented on four pages headlined and categorized by topics. First, general information was collected via four questions. Information about the respondent's number of years of experience in the husbandry of farm animals, the type of production system, the type of animal species kept on the farm and the location of the farm on a federal state level was collected. The question about the species kept on the farm had respective subcategories concerning the type of production or the species. For each subcategory, except for the subcategories of the 'Ungulates', the number of animals kept in that category was asked, using three categories of ranges.

The second category was labelled state of biosecurity. It contained a question with five statements about biosecurity on their farm that participants were asked to evaluate according to a five-point Likert scale [23]. Statements are abbreviated in the following text as described in Table 1.

Table 1. Statements about biosecurity assessed according to a five-point Likert scale.

Statement	Abbreviation
'I consider biosecurity measures on my farm important.'	'Importance'
'The biosecurity concept on my farm is well thought out and offers the best possible protection against the introduction of animal diseases.'	'Concept'
'The adherence of the biosecurity concept by all employees is controlled frequently.'	'Adherence'
'I am familiar with the measures officially applied in case of an animal disease outbreak.'	'Official measures'
'In case of an animal disease outbreak, my farm's existence is under threat.'	'Threat'

Subsequently, the participants were asked from which source they obtained the information for the development and maintenance of the biosecurity concept on their farm. Furthermore, they were asked about the revision of their biosecurity plan and the tools used in that context.

In the third category, a picture with twelve pictograms combined with respective subtitles of twelve introduction routes for pathogens into the farm was displayed. Two statements were given, accompanied by a distinctive symbol: 'Here I see a high risk of disease introduction into my farm!' and 'Here my farm is very well positioned in terms of biosecurity!'. Participants were asked to allocate each statement to, at most, three of the displayed introduction routes by clicking into the respective pictogram (Supplementary S1).

In the last category, one question was asked about possible offers that would interest the participants for improved biosecurity on their farm. The opt-out option was that no offers were needed. Finally, an open text field question was displayed for further thoughts on the topic of on-farm biosecurity.

2.3. Data Inclusion

In order to raise the participation rate, answering was made optional for all questions. All interviews that were submitted by pressing the button "send data" at the last page of the survey were considered as datasets. For the final analysis, datasets that did not include any answers and datasets that indicated that the farm was located in Austria or in Switzerland were excluded, as well as farms that indicated that they kept fish or bees only.

2.4. Data Analysis

Data were analyzed descriptively as well as exploratively. For descriptions of farm sizes, the number of animals per animal species kept was analyzed and categorized. The median was calculated with the mean absolute deviation (MAD). Farms that kept at least one species subgroup of the highest given number category were categorized as large. Every other farm that had information about at least one animal species kept was categorized as small. In parts, animal species other than cattle, swine and chicken were summarized under the term 'other animal species' for a more comprehensive description. Similarly, the answers 'Does rather apply' and 'Does apply' were summarized as 'agreed' for parts of the interpretation.

The correlation between the location of the participating farms and the total number of farms keeping animals in Germany was analyzed by a simple linear regression model. In addition, a correlation analysis was performed with a subset of data for each animal species kept. The data concerning the total farms registered in Germany were obtained from the Federal Statistical Office of Germany. For the overall number of farms as well as the number of farms keeping chicken, goats and poultry, data from March 2020 were used [24]. For the number of registered farms keeping cattle, pigs and sheep, more recent data from November 2022 were available [25]. The correlation of answers given to the statements about attitudes towards biosecurity were analyzed using Spearman's rank correlation coefficient (ρ) [26] and interpreted according to Cohen, whereby correlation coefficients larger than 0.5 were considered as strong correlations [27]. The p-values for Spearman correlation were calculated using asymptotic t approximation with Edgeworth series approximation. Analysis included all datasets that dealt with all five statements. In all correlation analyses, a p-value of less than 0.05 was considered significant. Thematic text analysis was performed on the answers from the open question, coding the information given in the text answer according to Braun and Clarke [28]. Therefore, after familiarization with the data, codes were generated according to features found in the data. These codes were then combined into overarching themes. Statistical analysis was performed in R [29] using the package psych [30]. For the creation of figures, the packages ggplot2 [31], pheatmap [32], corrplot [33] and UpSetR [34,35] were used.

3. Results

3.1. Properties of Participating Farms

Overall, 239 complete datasets were acquired. Farms were described to be located in 13 out of the 16 German federal states, with most farms located in Baden-Württemberg (n = 51, 21.3%), North Rhine-Westphalia (n = 45, 18.8%) and Bavaria (n = 41, 17.2%). None of the farms were located in the three city-states (Berlin, Hamburg, Bremen) (Figure 1).

The proportion of farms participating in the study complied (p = 0.00, R2 = 0.58) with the proportion of farms in Germany (Figure 2). Furthermore, the proportion of participating farms keeping cattle (p = 0.02, R2 = 0.35), pigs (p = 0.00, R2 = 0.79) and poultry (p = 0.05, R2 = 0.24) complied with the proportion of registered German farms keeping the respective animal species. In contrast, farms that stated that they kept sheep (p = 0.14, R2 = 0.11), other poultry (p = 0.78, R2 = −0.08), ungulates (p = 0.23, R2 = 0.05) and goats (p = 0.12, R2 = 0.13) did not comply with the proportion of these farms in Germany.

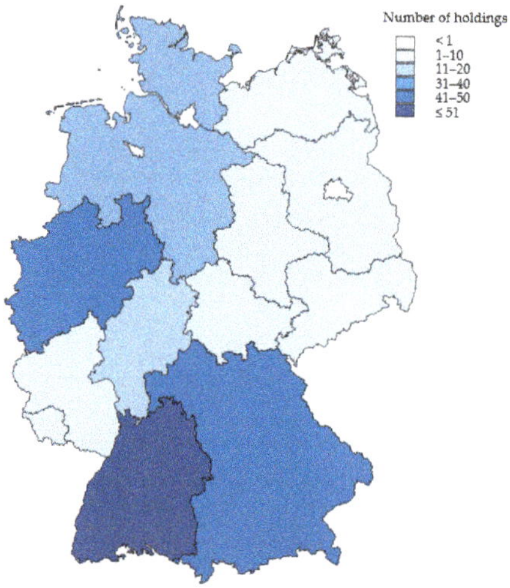

Figure 1. Number of farms per federal state. Four Austrian and two Swiss holdings also took part. Five holdings did not provide any information on their situation.

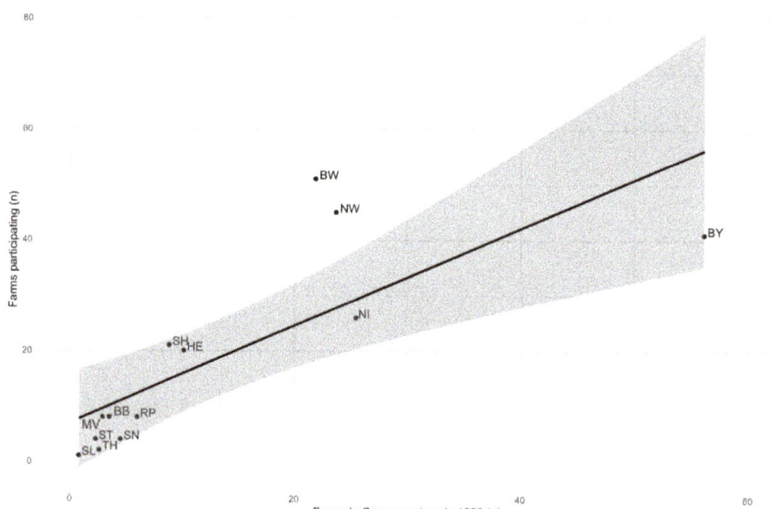

Figure 2. Univariable linear regression model of participating farms from each German federal state in dependence of farms registered in Germany. Line fits represent linear regression and 95% confidence interval. BB = Brandenburg, BW = Baden-Württemberg, BY = Bavaria, HE = Hesse, MV = Mecklenburg-Western Pomerania, NI = Lower Saxony, NW = North Rhine-Westphalia, RP = Rhineland-Palatinate, SH = Schleswig-Holstein, SL = Saarland, SN = Saxony, ST = Saxony-Anhalt, TH = Thuringia.

Participants stated in most cases that they had 'more than 20 years' of experience in animal husbandry (n = 116, 48.5%). This was followed by '11 to 20 years' (n = 45, 18.8%) and '5 to 10 years' (n = 35, 14.6%). Lastly, 17 participants stated that they had 'less than 5 years' of experience (n = 17, 7.1%). Twenty-six participants did not answer this question (n = 26, 10.9%). Use of 'conventional husbandry with outdoor access' was the most common type of production system (n = 125, 52.3%), followed by 'conventional husbandry' (n = 92, 38.5%), 'conventional husbandry with outdoor climate stimulus'

(n = 48, 20.1%) and lastly 'biological/ecological animal husbandry' (n = 43, 17.6%). The majority of farms kept their animals in only one type of animal production system (n = 182, 76.2%). The most common combination of production systems was 'conventional husbandry' and 'conventional husbandry with outdoor access' (n = 16, 6.7%), followed by the addition of a 'conventional husbandry with outdoor climate stimulus' system to the former combination on 15 farms (6.3%). Two participants did not answer the question concerning their production system type (n = 2, 0.8%). Chicken (n = 141, 59.0%), cattle (n = 106, 44.4%) and swine (n = 70, 29.3%) were the most prevalent species kept on the farms. More than half of the farms kept more than one animal species (n = 133, 55.6%), with a median of 2 (MAD = 1) animal species per farm. The most prevalent combinations were chicken and cattle (n = 52, 21.8%), chicken and swine (n = 29, 12.1%) and swine and cattle (n = 24, 10.0%) (Figure 3). Three farmers did not state which animal species were being kept (n = 3, 1.3%).

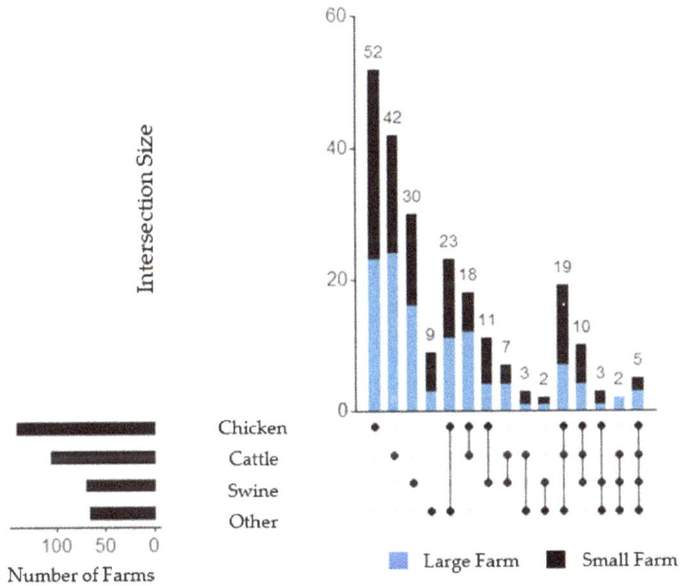

Figure 3. Animal species kept on the farms of participants. The number of farms keeping the respective animal species is given on the vertical columns on the right side. The vertical columns give the number of farms that kept animals of a specific species or resp. multiple species, as indicated underneath the bars by the black dot(s) in combination with the black lines in between them, indicating a combination of these species. The number of farms categorized as large in each category is highlighted in blue. Classification for cattle and other animals: S = 1–99 animals, L = more than 100 animals; for swine and chicken: S = 1–999 animals, L = more than 1000 animals.

More than two thirds of all cattle holdings were categorized as beef cattle (n = 72, 67.9%), with 16 (15.1%) farms simultaneously keeping milk cattle and 49 (46.2%) keeping at least one other animal species on the farm. The majority of chicken holdings kept layer chicken (n = 121, 85.8%). Twenty farms also kept broiler chicken (n = 20, 14.2%) and 79 (56.0%) some kind of other animal species. Ninety percent of all farms stating to keep pigs categorized their animals as fattening pigs (n = 63, 90.0%). Half of them only kept pigs for fattening (n = 35, 50.0%), while the rest kept pigs of earlier stages of the production chain. Over half of the farms keeping pigs kept also another animal species (n = 37, 52.9%).

3.2. State of Biosecurity

All five statements about the importance of biosecurity on the farm were categorized as 'rather applies' or 'applies' by the majority of farmers. The strongest agreement was found for the statement 'Importance' (n = 199, 83.3%), followed by 'Official measures' (n = 176, 73.6%). This was followed by the statements 'Concept' (n = 160, 66.9%), 'Adherence' (n = 147, 61.5%) and 'Threat' (n = 126, 52.7%). Three participants did not rate any of the statements (n = 3, 1.3%) and one farmer each did not evaluate one, three and four answers, respectively (n = 1, 0.4%) (Figure 4).

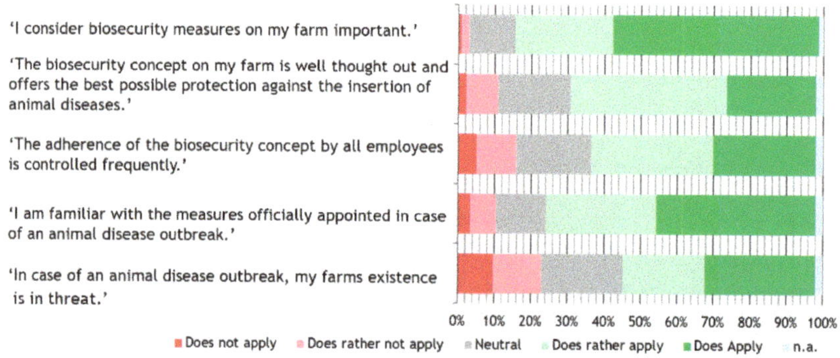

Figure 4. Likert scale for statements concerning the importance of biosecurity on the farm as stated by the participants.

Two-hundred-and-thirty-three datasets (n = 233, 97.5%) were included in the Spearman's ρ calculation. Spearman's rank correlation showed significant correlation between each statement combination. However, a strong positive association was only found for the statement combinations 'Adherence' and 'Concept', with an ρ of 0.68, and 'Importance' and 'Concept', with an ρ of 0.51 (Figure 5).

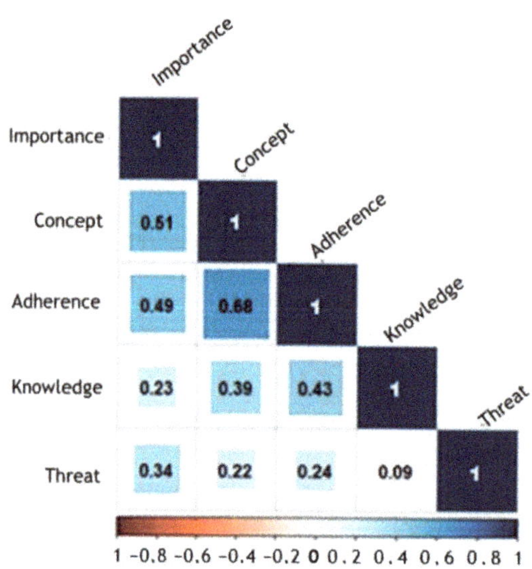

Figure 5. Spearman's rank correlation coefficient (ρ) between answers given to statements about biosecurity in a Likert-scale format.

Overall, the majority (n = 203, 84.5%) of farmers stated that there was a concept of biosecurity on their farm. The farmers received the information for their conception and development of biosecurity, in the largest percentage, from their veterinarian (n = 151, 63.2%). In addition, information was gathered from other farmers (n = 78, 32.6%), the animal health service (n = 73, 30.5%) and the chamber of agriculture (n = 62, 25.9%). Farmers consulted a median of two (MAD = 1) sources for the conception of the biosecurity concept. One farmer did not respond to the question (n = 1, 0.4%). Nearly two thirds of farmers (n = 152, 63.6%) stated that they evaluated their concept of biosecurity with the help of their veterinarian (n = 71, 29.7%), an online-tool (n = 47, 19.7%) or an institution (n = 34, 14.2%). Seven farmers did not answer this question (n = 7, 2.9%).

3.3. *Assessment of Hazardous Situation*

The highest risk for an introduction of an animal disease into the farm was seen in the introduction routes 'wild birds' (n = 129, 54.0%), 'humans' (n = 109, 45.6%) and 'pests' (n = 106, 44.4%). Furthermore, the introduction routes 'transporter' (n = 76, 31.8%) and 'purchase of new animals' (n = 74, 31.0%) were selected by approximately one third of farmers. With three introduction routes being the maximum allowed in the questionnaire, farmers selected a medium of 3 (MAD = 0) answers, resulting in 625 introduction routes being selected all together. Sixteen farmers did not select any of the given introduction routes (n = 16, 6.7%) (Figure 6A).

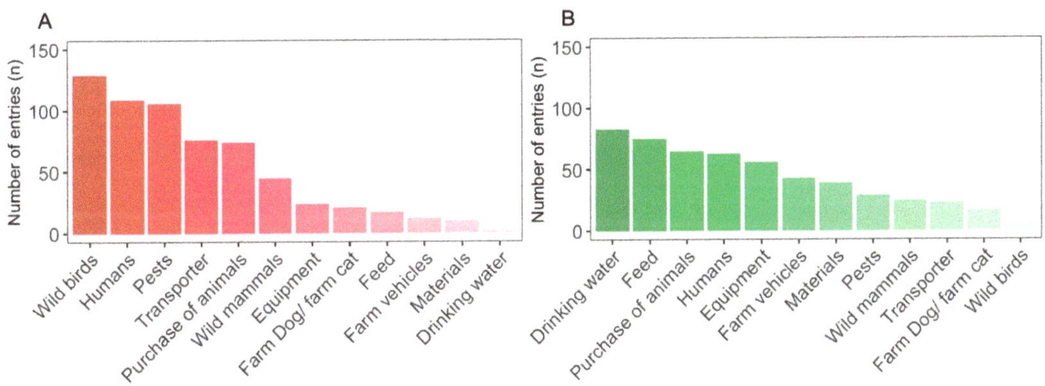

Figure 6. Pareto charts of the possible introduction routes for a disease into the farms (**A**) as well as the introduction routes perceived to have a high protection status (**B**). Number of entries are represented as bars.

The introduction routes that were described as best protected by the existing biosecurity measures were 'drinking water' (n = 83, 34.7%), 'feed' (n = 75, 31.4%), 'purchase of new animals' (n = 65, 27.2%) and 'humans' (n = 63, 26.4%). Overall, the introduction routes were selected 523 times, with the five most frequently selected introduction routes making up two thirds of all answers (n = 342, 65.4%). Farmers selected a median of three (MAD = 0) answers, and 57 farmers did not choose any introduction routes (n = 57, 23.9%) (Figure 6B).

In most cases, the introduction routes that were labeled as highest risk were not marked as best protected by the same farmer. This was only the case for the introduction route 'humans' for twelve farmers (n = 12, 5.0%), 'pests' as well as 'purchase of new animals' (n = 8, 3.3%), 'transporters' (n = 5, 2.1%), 'equipment' (n = 2, 0.8%) and 'wild mammals' as well as 'farm vehicles' (n = 1, 0.4%) (Figure 7).

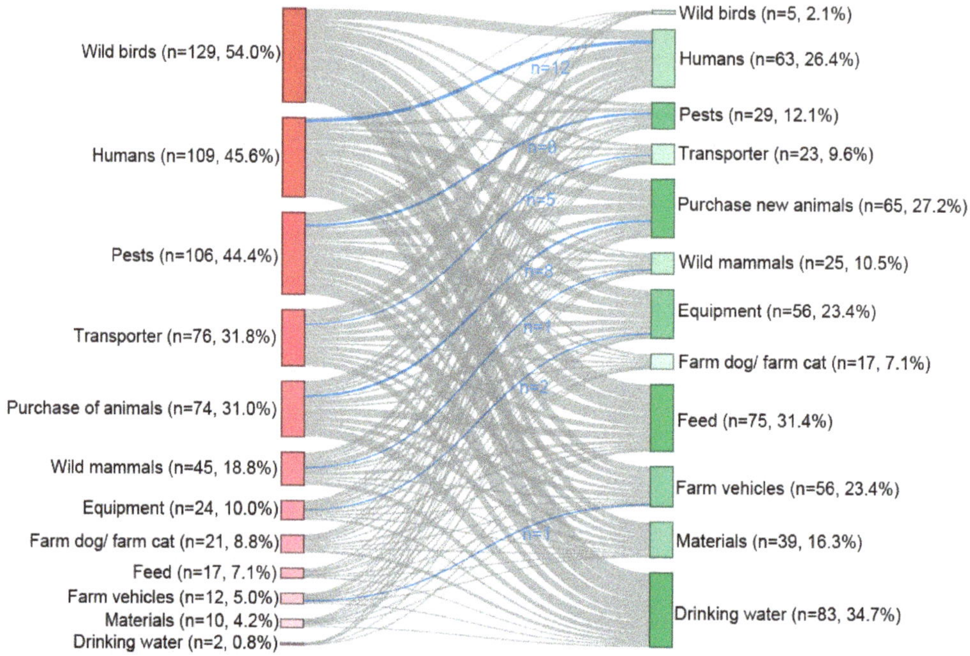

Figure 7. Risk factors (red), as well as introduction routes that were perceived as well protected (green) by the participants. Gray lines indicate their mutual selection by one participant, with the blue lines highlighting the selection of one introduction route in both categories by one participant. The color gradients of red and green change with increasing percentages (dark red: highest risk, dark green: best protected).

3.4. *Options for Improvement*

Farmers chose a median of two (MAD = 1) answers of the given nine options on possible offers for improved biosecurity on their farm. More than half of the farmers wanted regular information on the current state of animal disease outbreaks (n = 139, 58.2%) and short fact sheets (n = 131, 54.8%), as well as a mobile app (n = 81, 33.9%). Twenty-one farmers did not see a need for such offers (n = 21, 8.8%) and three did not answer the question (n = 3, 1.3%).

Twenty-seven participants filed out the open text field (n = 27, 11.3%). The information was classified into 13 themes (Supplementary S2). The majority of statements (n = 20, 74.1%) contained information that was assigned to a single theme. In addition, the information of six statements (n = 5, 18.5%) and two statements (n = 2, 7.4%) was assigned to two and three themes, respectively. Most statements discussed a general uncertainty about biosecurity (n = 6, 2.5%), focusing on the lack of absolute safety and the need for reliable instructions. The personal responsibility of the farmer was mentioned (n = 4, 1.7%) by highlighting the fact that everyone should start with themselves. Furthermore, the feasibility of biosecurity measures (n = 3, 1.3%) was discussed by mentioning structural and financial impediments. Some statements highlighted free-range holdings (n = 3, 1.3%) as well as hobby husbandry (n = 1, 0.4%) as the main risk source for animal disease outbreaks affecting larger, commercial farms in their area. In contrast, two statements mentioned the good information about biosecurity provided from the mobile stable association (n = 2, 0.8%). The role of veterinary officer was described (n = 4, 1.6%) as both helpful and difficult. In addition, two participants (n = 2, 0.8%) mentioned the psychological stress that farmers are exposed to as a result of an animal disease outbreak in the form of self-blame.

Two farmers mentioned that there was unauthorized entry happening on their farm ($n = 2$, 0.8%). One farmer each highlighted the importance of biosecurity respective discrepancies in the family ($n = 1$, 0.4%). Four statements added further information about their farm ($n = 4$, 1.7%) and three criticized the design of the questionnaire ($n = 3$, 1.3%). Codes used in the generation of themes can be found in Supplementary S2.

4. Discussion

The implementation of daily management practices is essential for functioning biosecurity on a farm level. It relies on the attitude and the risk awareness of the people on site. Therefore, it is important that farmers actively engage with the topic of biosecurity. This fact is also reflected in the current European legislation. In the Animal Health Law, a direct focus is put on the responsibility of the farmer in terms of biosecurity. It states that the operators of farms are responsible for 'minimizing the risk of the spread of diseases'. The large majority of farmers in the present study declared that they had a concept of biosecurity on their farm and thereby implied that they were actively taking charge on this topic. Furthermore, nearly two-thirds of farmers reevaluated their biosecurity concept with the help of an external institution, indicating a strong willingness to implement the best available biosecurity measures on their farm. In both the creation and the evaluation of their biosecurity plan, the consultation of a farm's veterinarian was by far the most prevalent form of external input. The role of the veterinarian in terms of biosecurity has been shown before [36–38]. In a series of interviews with German pig holders on biosecurity in the context of ASF, a similar strong focus on trust in regards to their respective farm veterinarian was found, thus highlighting the practicability of their advice [39]. Veterinarians seem to be aware of their influence on farmers, as the majority state that they advise farmers on on-farm biosecurity [40–44]. However, studies suggest veterinarians do encounter problems in communicating relevant information to farmers [45,46]. Therefore, veterinarians should be educated during their studies, as well as later on through information campaigns, about biosecurity on the farm level and effective strategies on how to communicate with farmers [42,47]. In addition, veterinarians working in the field should be actively involved in the evaluation of mandatory biosecurity measures.

The attitude of farmers towards biosecurity was further explored by five statements about biosecurity. As the statement about the importance of biosecurity found a nearly one-hundred percent agreement, farmers seemed to recognize the importance of the topic. The strong correlation between this statement and the statement about the existence of a well thought-through concept of biosecurity seems to be reasonable, since the importance of a topic directly impacts the motivation to act on it. In addition, a strong correlation was found between the latter statement and the control of adherence to the biosecurity plan. The carrying out of daily practices according to an existing biosecurity concept was found to be, in general, poor as well as error-prone in execution [48,49]. Therefore, the active control of adherence to the concept by all people working on a farm seems to be essential for effective biosecurity on the farm. Since the majority of farmers agreed to this statement, this shows that participants in the present study were not only aware of the importance of biosecurity and a defined concept, but also about the need for active control of the implementation on their farm. In contrast, the weakest correlation was found between farmers that stated that they had a good knowledge of official appointed measures in the case of an animal disease outbreak and farmers that thought their farm's existence would be under threat in such a scenario. Although no negative correlation was found, probably due to the small percentage of participants disagreeing to the statement 'Knowledge', this hints at a lack of knowledge in this area, which should be explored in further surveys. An outbreak of an animal disease listed in the European Animal health law requires the application of certain measures to contain the outbreak, such as culling of the whole stock [50]. Even though the loss is compensated for by the responsible authorities, this reparation only includes the animal's market value and no other costs arising during an outbreak on the farm [51]. Recently, in parts of the federal state Lower Saxony, this compensation was partially reduced when breaches of the biosecurity regulations were identified on the affected farm by the authorities [52]. Therefore, it seems reasonable that over half of the farmers participating did fear for their farm's existence in case of an outbreak happening on their farm.

Introduction routes of animal diseases to the farms perceived by farmers as dangerous showed a uniform picture. This was shown by the fact that the four most frequently named introduction routes made up the majority of answers given and that nearly all farmers answered this question. The ranking of farmers in the present study showed similarities with the results of expert panels that were asked to weigh subcategories of biosecurity against each other in terms of importance [53,54]. Since the human aspect, animal trade and pest control were, among others, the factors with the greatest weight assigned, it seems that farmers perceived the risk posed by the different introduction routes similarly to experts in the field. This further supports the already discussed engagement of farmers in the topic of biosecurity. In contrast, the question about introduction routes that were well protected by existing measures was not answered by a quarter of farmers, and there was much more variation in this question than in the previous one. Both the diversity of selected introduction routes as well as the large group of farmers not selecting any introduction route hints at a general insecurity about the effectiveness of biosecurity. Rating the introduction routes as best protected implies a certain knowledge about the effectiveness of existing measures. Even when looking into the scientific literature, it is hard to quantify the effectiveness of biosecurity measures, which is a situation that is acknowledged by veterinary experts [55,56]. Due to the lack of quantitative data, existing scores designed to rate the biosecurity of a farm as a whole rely on expert opinions for their weights of the importance of different measures [53,54,57,58]. In that context, the discrepancy between the selection of dangerous introduction routes and the selection of good protected introduction routes was obvious, both at a general level and on a single farmer's choice level. Additionally, when looking at a single farmer's choice, introduction routes considered dangerous to the farm were only considered rarely as well-protected introduction routes. This further emphasizes the problem that even though farmers identified introduction routes of concern in their biosecurity concept, they apparently lacked knowledge about possibilities for improvement. The survey did not investigate the underlying causes of farmers' uncertainties regarding the effectiveness of biosecurity measures. Due to the ongoing dynamics of the animal disease situation in Germany, farmers were increasingly being sensitized to the significance and importance of biosecurity via information events, veterinary consultations and political influences. One reason for the uncertainty with regard to effectiveness could be that, although training has emphasized the importance of biosecurity, the selection, prioritization and implementation of targeted biosecurity measures has been neglected. Furthermore, biosecurity measures are very farm-specific and dependent on the actual implementation, which makes it difficult to compare the measures applied across farms. This could make it difficult for farmers to assess whether measures are relevant for their farm or not, due to their type of production. In addition, the aim of biosecurity is not to introduce an animal disease into the farm; thus, as long as the farm has a high health status, the success of the biosecurity measures and their effectiveness are only indirectly visible and therefore difficult for farmers to quantify. Furthermore, it may be difficult for farmers to identify whether the health status of their herd is due to the biosecurity measures implemented or whether the farm is currently not exposed to any potential pathogens.

Lastly, open text field question entries hinted at further problems that concern farmers in the context of biosecurity. Foremost, the mentioned uncertainty of biosecurity in combination with the feasibility of biosecurity measures on farms further highlight the need for both further research on the topic as well as clear communication from the authorities to farmers. This is in line with the conclusions found in the aforementioned series of interviews with German pig holders [39] and a similar study with UK broiler farm workers [59]. Furthermore, the open free-text answers suggested tensions between conventional farmers and small-holder farms concerning biosecurity on free-range holdings. Since the already described compensation in case of a listed animal disease outbreak is only paid to the affected farm and not to farms in the region affected by further measures, such as restriction zones with a ban on animal transportation, farms are directly impacted by the situations of their neighboring farms. A similar conflict seems to be happening in the UK [60] and should be further explored in the German context. The resulting peer pressure also relates to the psychological impact on farmers that was mentioned in the free-text section. In that context, it should not be forgotten that farmers are mostly people building their and their family's existence around the existence and profitability of their farm [61]. Keeping the high suicide rate of farmers in mind [62], authorities should aim for clear communication of reasonable measures and see farmers as partners in the common goal of achieving the best biosecurity possible.

A limitation of this study might be that the participating group of farmers showed some deviations from the overall group of farmers in Germany, as can be estimated from the available data from official state sources. However, the location of participating farms complied with farms in Germany on the level of the three main species kept. Furthermore, as the average age of farmers in Germany was estimated to be 53 in 2013 [63] and recent data indicate that 85% of farm managers in Germany are above the age of 40 [64], the overall high experience of animal husbandry in the participating group seems reasonable. In contrast to these two general information criteria, animal species kept did not represent the situation on German farms. The most prevalent animal species kept on German farms for food production are cattle (64%) followed by chicken (30%) and pigs (19%) [65]. In the present study, those three species were also by far the most frequently species kept. However, chicken and pigs were found in a higher percentage than expected. This might be due to the two most threatening animal diseases circulating in Germany right now, being ASF [66] and HPAI [67], which target these particular species and lead to high economic losses. Farmers keeping those species might be, in general, more engaged in the topic of biosecurity, raising their likelihood of participating in a survey about the topic [68]. Similar, the percentage of farms with organic husbandry exceeded the estimated 10% in Germany [69]. The percentage of conventional husbandry with outdoor contact is not systematically collected for pigs [70] and cattle in Germany. For layer chicken, which is considered to be the sector of animal production with the highest percentage of husbandries with outdoor contact, 21% of all farms hold layer chicken in a free-range system [71], which is considerably less than the more than half of our participants stated to have a conventional husbandry with outdoor access on their farm. Both organic and conventional husbandries with outdoor access have a higher risk of contracting a disease into their animal stock since they are more exposed to the outdoor world and biosecurity measures are harder to implement [72–74]. Therefore, as mentioned above in the context of the over-representation of pigs and chicken, an increased engagement with the topic due to the higher risk in these holdings with outdoor contact might have led to a higher participation rate [68].

Overall, the participating subgroup of German farmers keeping animals for food production constituted a regionally representative cross-section with a focus on farmers keeping pigs and/or chickens in husbandries with contact to the outdoor world. Since this group of farmers is actually the group that is most challenged by the current animal disease situation in Germany, the collected information is very valuable in order to develop targeted strategies to help these farmers improve the biosecurity of their husbandry.

5. Conclusions

The high percentage of farmers actively engaging in the topic of biosecurity concerning animal husbandry indicates a general willingness to implement management practices on a daily basis. However, the underlying discrepancy of the perception of dangerous as well as well-protected introduction routes for pathogens hints at a potential lack of knowledge of the effectiveness of biosecurity measures. This displays a need for further research on this effectiveness and clear recommendations for farmers, which were also asked for by the participants.

Supplementary Materials: The following supporting information can be downloaded at: https://www.mdpi.com/article/10.3390/ani14071102/s1, Supplementary S1: Survey questionnaire, Supplementary S2: Themes and corresponding codes from the open text field question about further thoughts on biosecurity.

Author Contributions: Conceptualization, A.H., C.S.-L. and E.Z.; methodology, A.H., C.S.-L., K.S., J.B. and N.W.; formal analysis, A.H. and K.S.; writing, A.H. and N.W.—original draft preparation, A.H. and N.W.—review and editing, A.H., K.S., J.B., N.W., E.Z. and C.S.-L.; funding acquisition, C.S.-L. and E.Z. All authors have read and agreed to the published version of the manuscript.

Funding: This research was funded by the Federal Ministry of Food and Agriculture as part of the German Climate Protection Programme 2022, grant number 123-65405/0004.

Institutional Review Board Statement: Not applicable.

Informed Consent Statement: Informed consent was obtained from all subjects involved in the study.

Data Availability Statement: The data presented in this study are available upon reasonable request from the authors.

Acknowledgments: We thank all farmers participating in the study as well as the scientific experts pre-testing the survey in the preliminary stages. Furthermore, we thank all persons and institutions helping in promoting the online survey.

Conflicts of Interest: The authors declare no conflicts of interest. The funders had no role in the design of the study; in the collection, analyses, or interpretation of data; in the writing of the manuscript; or in the decision to publish the results.

References

1. Ducrot, C.; Bed'Hom, B.; Béringue, V.; Coulon, J.-B.; Fourichon, C.; Guérin, J.-L.; Krebs, S.; Rainard, P.; Schwartz-Cornil, I.; Torny, D.; et al. Issues and special features of animal health research. *Vet. Res.* **2011**, *42*, 96. [CrossRef]
2. Oxford Analytica. Animal Health and Sustainability: A Global Data Analysis: A Report Produced for HealthforAnimals. Available online: https://www.healthforanimals.org/wp-content/uploads/2023/04/Animal-health-and-Sustainability-A-Global-Data-Analysis-FINAL.pdf (accessed on 23 June 2023).
3. Nørrung, B.; Buncic, S. Microbial safety of meat in the European Union. *Meat Sci.* **2008**, *78*, 14–24. [CrossRef]
4. Chousalkar, K.K.; Khan, S.; McWhorter, A.R. Microbial quality, safety and storage of eggs. *Curr. Opin. Food Sci.* **2021**, *38*, 91–95. [CrossRef]
5. Fusco, V.; Chieffi, D.; Fanelli, F.; Logrieco, A.F.; Cho, G.-S.; Kabisch, J.; Böhnlein, C.; Franz, C.M. Microbial quality and safety of milk and milk products in the 21st century. *Compr. Rev. Food Sci. Food Saf.* **2020**, *19*, 2013–2049. [CrossRef]
6. OIE. Feeding the World Better by Controlling Animal Diseases. Available online: https://www.woah.org/en/feeding-the-world-better-by-controlling-animal-diseases/ (accessed on 23 July 2023).
7. Bayot, M.L.; Limaiem, F. *Biosafety Guidelines*; StatPearls Publishing: Treasure Island, FL, USA, 2020.
8. WOAH. Terrestrial Animal Health Code. 2021. Available online: https://www.woah.org/en/what-we-do/standards/codes-and-manuals/terrestrial-code-online-access/?id=169&L=1&htmfile=sommaire.htm (accessed on 27 June 2023).
9. Renault, V.; Humblet, M.-F.; Saegerman, C. Biosecurity concept: Origins, evolution and perspectives. *Animals* **2022**, *12*, 63. [CrossRef]
10. FAO/OIE/WB. Good Practices for Biosecurity in the Pig Sector—Issues and Options in Developing and Transition Countries. Available online: http://www.fao.org/docrep/012/i1435e/i1435e00.pdf (accessed on 30 July 2023).
11. Bucherer, M.; Holzhausen, J.; Conraths, F.J.; Probst, C. Infrastructure of animal farms: Key constructional elements in terms of biosecurity based on experience from Germany. *Berl. Münch. Tierärztl. Wochenschr.* **2021**, *134*, 1–12. [CrossRef]
12. Mankad, A. Psychological influences on biosecurity control and farmer decision-making. A review. *Agron. Sustain. Dev.* **2016**, *36*, 40. [CrossRef]
13. Jack, C.; Hotchkiss, E.; Sargison, N.D.; Toma, L.; Milne, C.; Bartley, D. Determining the influence of socio-psychological factors on the adoption of individual 'best practice' parasite control behaviours from Scottish sheep farmers. *Prev. Vet. Med.* **2022**, *200*, 105594. [CrossRef]
14. Delpont, M.; Racicot, M.; Durivage, A.; Fornili, L.; Guerin, J.-L.; Vaillancourt, J.-P.; Paul, M.C. Determinants of biosecurity practices in French duck farms after a H5N8 Highly Pathogenic Avian Influenza epidemic: The effect of farmer knowledge, attitudes and personality traits. *Transbound. Emerg. Dis.* **2021**, *68*, 51–61. [CrossRef]

15. Ritter, C.; Jansen, J.; Roche, S.; Kelton, D.F.; Adams, C.L.; Orsel, K.; Erskine, R.J.; Benedictus, G.; Lam, T.J.; Barkema, H.W. Invited review: Determinants of farmers' adoption of management-based strategies for infectious disease prevention and control. *J. Dairy. Sci.* **2017**, *100*, 3329–3347. [CrossRef]
16. Merrill, S.C.; Moegenburg, S.; Koliba, C.J.; Zia, A.; Trinity, L.; Clark, E.; Bucini, G.; Wiltshire, S.; Sellnow, T.; Sellnow, D.; et al. Willingness to comply with biosecurity in livestock facilities: Evidence from experimental simulations. *Front. Vet. Sci.* **2019**, *6*, 156. [CrossRef]
17. King, J.; Staubach, C.; Lüder, C.; Koethe, S.; Günther, A.; Stacker, L.; Rubbenstroth, D.; Dietze, K.; Grund, C.; Conraths, F.J.; et al. Connect to protect: Dynamics and genetic connections of highly pathogenic avian influenza Outbreaks in Poultry from 2016 to 2021 in Germany. *Viruses* **2022**, *14*, 1849. [CrossRef]
18. Sauter-Louis, C.; Schulz, K.; Richter, M.; Staubach, C.; Mettenleiter, T.C.; Conraths, F.J. African swine fever: Why the situation in Germany is not comparable to that in the Czech Republic or Belgium. *Transbound. Emerg. Dis.* **2022**, *69*, 2201–2208. [CrossRef]
19. Leiner, D.J. Our research's breadth lives on convenience samples. A case study of the online respondent pool "SoSci Panel". *Int. J. Internet Sci.* **2016**, *5*, 367–396. [CrossRef]
20. Baltar, F.; Brunet, I. Social research 2.0: Virtual snowball sampling method using Facebook. *Internet Res.* **2012**, *22*, 57–74. [CrossRef]
21. Goodman, L.A. Snowball Sampling. *Ann. Inst. Stat. Math.* **1961**, *32*, 148–170. [CrossRef]
22. Leiner, D.J. ScSci Survey. [Computer Software]. Available online: https://www.soscisurvey.de (accessed on 13 July 2023).
23. Joshi, A.; Kale, S.; Chandel, S.; Pal, D.K. Likert scale: Explored and explained. *Br. J. Appl. Sci. Technol.* **2015**, *7*, 396. [CrossRef]
24. Statistisches Bundesamt. *Viehhaltung der Betriebe: Land- und Forstwirtschaft, Fischerei*; 2.1.3 No. 2030213209005; Statistisches Bundesamt: Wiesbaden, Germany, 2021.
25. Statistisches Bundesamt. *Viehbestand: Land und Forstwirtschaft, Fischerei*; 4.1 No. 2030410225324; Statistisches Bundesamt: Wiesbaden, Germany, 2023.
26. Spearman, C. The proof and measurement of association between two things. *Am. J. Psychol.* **1904**, *15*, 72–101. [CrossRef]
27. Cohen, J. *Statistical Power Analysis for the Behavioral Sciences*, 2nd ed.; Lawrence Erlbaum Associates: New York, NY USA, 1988.
28. Braun, V.; Clarke, V. Using thematic analysis in psychology. *Qual. Res. Psychol.* **2006**, *3*, 77–101. [CrossRef]
29. R Development Core Team. *R: A Language and Environment for Statistical Computin*; R Foundation for Statistical Computing: Vienna, Austria, 2010.
30. Revelle, W. psych: Procedures for Psychological, Psychometric, and Personality Research: R Package Version 2.3.6. Available online: https://CRAN.R-project.org/package=psych (accessed on 14 August 2023).
31. Wickham, H. *ggplot2: Elegant Graphics for Data Analysis*; Springer: New York, NY, USA, 2016; ISBN 978-3-319-24277-4.
32. Kolde, R. pheatmap: Pretty Heatmaps. Available online: https://CRAN.R-project.org/package=pheatmap (accessed on 14 August 2023).
33. Wei, T.; Simko, V. R Package 'corrplot': Visualization of a Correlation Matrix. (Version 0.92). Available online: https://github.com/taiyun/corrplot (accessed on 14 August 2023).
34. Lex, A.; Gehlenborg, N.; Strobelt, H.; Vuillemot, R.; Pfister, H. UpSet: Visualization of intersecting sets. *IEEE Trans. Vis. Comput. Graph.* **2014**, *20*, 1983–1992. [CrossRef]
35. Conway, J.R.; Lex, A.; Gehlenborg, N. UpSetR: An R package for the visualization of intersecting sets and their properties. *Bioinformatics* **2017**, *33*, 2938–2940. [CrossRef]
36. Garforth, C.J.; Bailey, A.P.; Tranter, R.B. Farmers' attitudes to disease risk management in England: A comparative analysis of sheep and pig farmers. *Prev. Vet. Med.* **2013**, *110*, 456–466. [CrossRef]
37. Laanen, M.; Maes, D.; Hendriksen, C.; Gelaude, P.; de Vliegher, S.; Rosseel, Y.; Dewulf, J. Pig, cattle and poultry farmers with a known interest in research have comparable perspectives on disease prevention and on-farm biosecurity. *Prev. Vet. Med.* **2014**, *115*, 1–9. [CrossRef]
38. Malakauskas, A.; Schulz, K.; Kukanauskaitė, I.; Masiulis, M.; Conraths, F.J.; Sauter-Louis, C. African Swine Fever Outbreaks in Lithuanian Domestic Pigs in 2019. *Animals* **2022**, *12*, 115. [CrossRef]
39. Klein, L.; Hessling-Zeinen, S.; Adler, F.; Gerdes, U.; Blome, S.; Beilage, E.g.; Campe, A. Exploring pig farmers' decision-making concerning biosecurity measures against african swine fever. *Prev. Vet. Med.* **2023**, *217*, 105949. [CrossRef]
40. Pritchard, K.; Wapenaar, W.; Brennan, M.L. Cattle veterinarians' awareness and understanding of biosecurity. *Vet. Rec.* **2015**, *176*, 546. [CrossRef]
41. Agrawal, I.; Bromfield, C.; Varga, C. Assessing and improving on-farm biosecurity knowledge and practices among swine producers and veterinarians through online surveys and an educational website in Illinois, United States. *Front. Vet. Sci.* **2023**, *10*, 1167056. [CrossRef]
42. Valeeva, N.I.; van Asseldonk, M.; Backus, G. Perceived risk and strategy efficacy as motivators of risk management strategy adoption to prevent animal diseases in pig farming. *Prev. Vet. Med.* **2011**, *102*, 284–295. [CrossRef]
43. Moya, S.; Tirado, F.; Espluga, J.; Ciaravino, G.; Armengol, R.; Diéguez, J.; Yus, E.; Benavides, B.; Casal, J.; Allepuz, A. Dairy farmers' decision-making to implement biosecurity measures: A study of psychosocial factors. *Transbound. Emerg. Dis.* **2020**, *67*, 698–710. [CrossRef]
44. Sayers, R.G.; Good, M.; Sayers, G.P. A survey of biosecurity-related practices, opinions and communications across dairy farm veterinarians and advisors. *Vet. J.* **2014**, *200*, 261–269. [CrossRef]

45. Denis-Robichaud, J.; Kelton, D.; Bauman, C.; Keefe, G.; Dubuc, J. Gap between producers and veterinarians regarding biosecurity on Quebec dairy farms. *Can. Vet. J.* **2020**, *61*, 757–762.
46. Shortall, O.; Ruston, A.; Green, M.; Brennan, M.; Wapenaar, W.; Kaler, J. Broken biosecurity? Veterinarians' framing of biosecurity on dairy farms in England. *Prev. Vet. Med.* **2016**, *132*, 20–31. [CrossRef]
47. Merrill, S.C.; Trinity, L.; Clark, E.M.; Shrum, T.R.; Koliba, C.J.; Zia, A.; Bucini, G.; Sellnow, T.L.; Sellnow, D.D.; Smith, J.M. Message delivery strategy influences willingness to comply with biosecurity. *Front. Vet. Sci.* **2021**, *8*, 667265. [CrossRef]
48. Racicot, M.; Venne, D.; Durivage, A.; Vaillancourt, J.-P. Description of 44 biosecurity errors while entering and exiting poultry barns based on video surveillance in Quebec, Canada. *Prev. Vet. Med.* **2011**, *100*, 193–199. [CrossRef]
49. Racicot, M.; Venne, D.; Durivage, A.; Vaillancourt, J.-P. Evaluation of strategies to enhance biosecurity compliance on poultry farms in Québec: Effect of audits and cameras. *Prev. Vet. Med.* **2012**, *103*, 208–218. [CrossRef]
50. Degeling, C.; Lederman, Z.; Rock, M. Culling and the Common Good: Re-evaluating Harms and Benefits under the One Health Paradigm. *Public Health Ethics* **2016**, *9*, 244–254. [CrossRef]
51. Bätza, H.J. The German system of compensating animal keepers in cases of outbreaks of animal diseases. In *The Economics of Livestock Disease Insurance: Concepts, Issues and International Case Studies*; CABI Pub: Oxfordshire, UK, 2006; pp. 141–144, ISBN 9780851990774.
52. Niedersächsische Tierseuchenkasse. Kürzung von Leistungen bei Verstößen. Available online: https://www.ndstsk.de/uebersicht/tierkoerperbeseitigung/biosicherheit/1050_information-zu-entsch-auml-digungsleistungen-und-beihilfen-im-tierseuchenfall-k-uuml-rzung-der-leistungen-bei-verst-ouml.html (accessed on 14 September 2023).
53. Gelaude, P.; Schlepers, M.; Verlinden, M.; Laanen, M.; Dewulf, J. Biocheck.UGent: A quantitative tool to measure biosecurity at broiler farms and the relationship with technical performances and antimicrobial use. *Poult. Sci.* **2014**, *93*, 2740–2751. [CrossRef]
54. Laanen, M.; Persoons, D.; Ribbens, S.; de Jong, E.; Callens, B.; Strubbe, M.; Maes, D.; Dewulf, J. Relationship between biosecurity and production/antimicrobial treatment characteristics in pig herds. *Vet. J.* **2013**, *198*, 508–512. [CrossRef]
55. Kuster, K.; Cousin, M.-E.; Jemmi, T.; Schüpbach-Regula, G.; Magouras, I. Expert opinion on the perceived effectiveness and importance of on-farm biosecurity measures for cattle and swine farms in Switzerland. *PLoS ONE* **2015**, *10*, e0144533. [CrossRef]
56. EFSA Panel on Animal Health and Welfare; More, S.; Bicout, D.; Bøtner, A.; Butterworth, A.; Calistri, P.; Depner, K.; Edwards, S.; Garin-Bastuji, B.; Good, M.; et al. Urgent request on avian influenza. *EFSA J.* **2017**, *15*, e04687. [CrossRef]
57. Grabkowsky, B.; Conraths, F.; Globig, A.; Wilke, A.; Denzin, N. A self-assessment tool to improve poultry farm biosecurity regarding avian influenza. *Berl. Münch. Tierärztl. Wochenschr.* **2020**. [CrossRef]
58. Sasaki, Y.; Furutani, A.; Furuichi, T.; Hayakawa, Y.; Ishizeki, S.; Kano, R.; Koike, F.; Miyashita, M.; Mizukami, Y.; Watanabe, Y.; et al. Development of a biosecurity assessment tool and the assessment of biosecurity levels by this tool on Japanese commercial swine farms. *Prev. Vet. Med.* **2020**, *175*, 104848. [CrossRef]
59. Royden, A.; Christley, R.; Prendiville, A.; Williams, N.J. The role of biosecurity in the control of campylobacter: A qualitative study of the attitudes and Perceptions of UK Broiler Farm Workers. *Front. Vet. Sci.* **2021**, *8*, 751699. [CrossRef]
60. Shortall, O.; Sutherland, L.-A.; Ruston, A.; Kaler, J. True cowmen and commercial farmers: Exploring vets' and dairy farmers' contrasting views of 'good farming' in relation to biosecurity. *Sociol. Rural.* **2018**, *58*, 583–603. [CrossRef]
61. Graskemper, V.; Feil, J.-H. Values of farmers—Evidence from Germany. *J. Rural. Stud.* **2022**, *89*, 13–24. [CrossRef]
62. Steck, N.; Junker, C.; Bopp, M.; Egger, M.; Zwahlen, M. Time trend of suicide in Swiss male farmers and comparison with other men: A cohort study. *Swiss Med. Wkly.* **2020**, *150*, w20251. [CrossRef]
63. Deutscher Bauernverband. *Zukunft wächst auf dem Land: Geschäftsbericht des Deutschen Bauernverbandes 2017/2018*; Deutscher Bauernverband e.V.: Berlin, Germany, 2018.
64. eurostat. Farm Indicators by Age and Sex of the Manager, Economic Size of the Farm, Utilised Agricultural Area and NUTS2 Region [EF_M_FARMANG__custom_7103129]. Available online: https://ec.europa.eu/eurostat/databrowser/bookmark/81051733-a4f8-4b74-ab1e-3cfc4ada9e1b?lang=en&page=time:2020 (accessed on 7 August 2023).
65. Statistisches Bundesamt. Landwirtschaftliche Betriebe mit Viehhaltung und Zahl der Tiere-Stichtag-Regionale Ebenen. Available online: https://www.destatis.de/DE/Themen/Branchen-Unternehmen/Landwirtschaft-Forstwirtschaft-Fischerei/Tiere-Tierische-Erzeugung/Publikationen/Downloads-Tiere-und-tierische-Erzeugung/viehhaltung-2030213209004.html?nn=207832 (accessed on 7 August 2013).
66. Sauter-Louis, C.; Conraths, F.J.; Probst, C.; Blohm, U.; Schulz, K.; Sehl, J.; Fischer, M.; Forth, J.H.; Zani, L.; Depner, K.; et al. African swine fever in wild boar in Europe—A review. *Viruses* **2021**, *13*, 1717. [CrossRef]
67. Conraths, F.J.; Sauter-Louis, C.; Globig, A.; Dietze, K.; Pannwitz, G.; Albrecht, K.; Höreth-Böntgen, D.; Beer, M.; Staubach, C.; Homeier-Bachmann, T. Highly Pathogenic Avian Influenza H5N8 in Germany: Outbreak Investigations. *Transbound. Emerg. Dis.* **2016**, *63*, 10–13. [CrossRef]
68. Keusch, F. Why do people participate in Web surveys? Applying survey participation theory to Internet survey data collection. *Manag. Rev. Q.* **2015**, *65*, 183–216. [CrossRef]
69. Statistisches Bundesamt. Viehbestand in Betrieben mit Konventionellem und ökologischem Landbau. Available online: https://www.destatis.de/DE/Themen/Branchen-Unternehmen/Landwirtschaft-Forstwirtschaft-Fischerei/Tiere-Tierische-Erzeugung/Tabellen/oekologischer-landbau-viehbestand.html (accessed on 24 August 2013)
70. Nielsen, S.S.; Alvarez, J.; Bicout, D.J.; Calistri, P.; Canali, E.; Drewe, J.A.; Garin-Bastuji, B.; Gonzales Rojas, J.L.; Herskin, M.; Miranda Chueca, M.Á.; et al. African swine fever and outdoor farming of pigs. *EFSA J.* **2021**, *19*, e06639. [CrossRef]

71. BMEL. Legehennenhaltung nach Haltungsformen. Available online: https://www.bmel-statistik.de/fileadmin/daten/SJT-3102000-0000.xlsx (accessed on 7 August 2023).
72. Delsart, M.; Pol, F.; Dufour, B.; Rose, N.; Fablet, C. Pig Farming in Alternative Systems: Strengths and Challenges in Terms of Animal Welfare, Biosecurity, Animal Health and Pork Safety. *Agriculture* **2020**, *10*, 261. [CrossRef]
73. Artois, M.; Bicout, D.; Doctrinal, D.; Fouchier, R.; Gavier-Widen, D.; Globig, A.; Hagemeijer, W.; Mundkur, T.; Munster, V.; Olsen, B. Outbreaks of highly pathogenic avian influenza in Europe: The risks associated with wild birds. *Rev. Sci. Tech. Off. Int. Epiz* **2009**, *28*, 69–92. [CrossRef]
74. Gonzales, J.L.; Stegeman, J.A.; Koch, G.; de Wit, S.J.; Elbers, A.R.W. Rate of introduction of a low pathogenic avian influenza virus infection in different poultry production sectors in the Netherlands. *Influenza Other Respir. Viruses* **2013**, *7*, 6–10. [CrossRef]

Disclaimer/Publisher's Note: The statements, opinions and data contained in all publications are solely those of the individual author(s) and contributor(s) and not of MDPI and/or the editor(s). MDPI and/or the editor(s) disclaim responsibility for any injury to people or property resulting from any ideas, methods, instructions or products referred to in the content.

animals

MDPI

Communication

Risk Assessments and Risk Mitigation to Prevent the Introduction of African Swine Fever into the Danish Pig Population

Jan Dahl

Danish Agriculture & Food Council, DK-1609 Copenhagen, Denmark; jd@lf.dk; Tel.: +45-24635877

Simple Summary: The Danish pig industry, in collaboration with the authorities, has implemented several risk-reducing measures to prevent the introduction of African Swine Fever into Danish pig production. Ongoing risk assessments are conducted to adjust or establish new measures. The first outbreaks in a new area in Northern Europe often occur in wild boar, likely due to discarded food waste at rest areas and in forested areas. A fence has been established on the Danish–German border to limit the migration of wild boar from Germany to Denmark, and the limited population of wild boar in Denmark has been culled. All trucks that transport pigs or cattle are cleaned and disinfected at the border if they have been abroad, according to the DANISH Transportstandard. Through continuous contact with the feed industry, changes in the importation of feedstuffs and other auxiliaries are analyzed, and risk assessments for each feed item are conducted. Based on this, according to the Danish Product Standard, it is not permitted to store or use feed items such as hay, straw, silage, or bedding imported from areas with outbreaks of African Swine Fever on Danish properties with pigs. These types of feed items can contain parts of dead animals in which the virus can survive for a long time. Oil cakes, fats, and oils imported from countries with African Swine Fever may only be used after heat treatment.

Abstract: African Swine Fever poses a significant threat to pig production. An outbreak in Denmark would have severe economic consequences, potentially resulting in a loss between 335 million and 670 million euros. To mitigate the major risk factors, the Danish authorities and the Danish pig industry have implemented several risk-reducing measures. The small wild boar population in Danish nature has been culled, and a fence between Denmark and Germany has been constructed to stop or reduce the risk of migrating wild boar from Germany. All trucks arriving from abroad, intended for the transport of Danish pigs, are inspected, washed, and disinfected at facilities near the border before being allowed to transport pigs between herds or from herds out of Denmark. Cross-border trade of feedstuffs and potentially risky materials is continuously monitored. Based on risk assessments, feed types or bedding materials from countries with African Swine Fever that could contain parts of wild boar are banned from Danish pig herds. Certain types of fats and oils from countries with African Swine Fever can only be used after heat treatment. The import of whole kernels of corn, wheat, or barley is not considered a risk.

Keywords: African swine fever; truck wash; risk mitigation; wild boar

1. Introduction

In June 2007, African Swine Fever (ASF) was confirmed in the Caucasus region of Georgia after many years of being confined to Africa and the Italian island of Sardinia. It gradually spread north and west from the Caucasus, primarily through wild boar, and due to the transport of infected pigs, or feeding pigs or wild boar contaminated pork products. Since 2014, ASF has been circulating in wild boar populations in the Baltic countries and Poland, and it has been introduced into several Eastern European countries, reaching as far west as Belgium, Italy, the Czech Republic, and Germany, and as far north as Sweden.

Table 1 summarizes the number of outbreaks in Europe from 2022 to August 2024, based on official statistics from the EU [1].

Table 1. Registered outbreaks in domestic pigs and wild boar in Europe.

Country	2024 1 January–7 August		2023		2022	
	Domestic Pigs	Wild Boar	Domestic Pigs	Wild Boar	Domestic Pigs	Wild Boar
Estonia		12	2	53		57
Latvia	4	516	8	730	6	913
Lithuania	6	376	3	439	16	302
Poland	35	1165	30	2744	14	2152
Italy	10	1106	17	1047	4	277
Ukraine	34	11	37	9	7	2
Czech Republic		26		56		1
Romania	122	108	740	292	329	465
Hungary		245		407		550
Bulgaria	1	95	3	322	2	387
Slovakia	1	97		546	5	550
Serbia	116	82	991	213	107	146
Moldova	8	6	19	6	14	3
Montenegro		1				
Bosnia-Hercegovina	27	36	1508	22		
Croatia	3	38	1124	11		
North Macedonia	2	36	15	41	30	9
Kosovo			9	4		
Belgium						
Germany	9	279	1	899	3	1628
Sweden		8		60		
Greece	5	17	6	2		
Albania		2				
Total	383	4262	4513	7903	537	7442

Danish pig production is export-based, with an annual production of between 28 and 32 million pigs over the last 10 years [2]. Between 30% and 50% of the pigs produced are exported as weaners or growers to many European countries, and 50% to 70% are finished and slaughtered in Denmark. With an annual export of more than 1 million tons of pork, Denmark would be severely affected by an outbreak that hinders the exportation of pigs and pork.

The Department of Food and Resource Economics, Copenhagen University [3], estimated that an outbreak of African Swine Fever would cause an economic loss for Danish pig production of EUR 335 million if the outbreak is in a pig herd, and EUR 670 million if wild boar become infected. An outbreak in wild boar would affect exportation to third countries for more than a year [4].

This makes it vital for the Danish pig industry to keep African Swine Fever out of Denmark. To achieve this, formal and informal risk assessments are conducted continuously to mitigate the risks of introduction, considering the changing epidemiological picture and changes in the trade patterns of potential fomites, like feed and other materials, crossing borders between countries.

The Danish Veterinary and Food Administration can mitigate some of these risks, but some risks can only be handled by the industry. This does not imply a conflict between the industry and the authorities; rather, there are different responsibilities and opportunities for risk mitigation for both parties.

The Danish pig industry has developed the Danish Product Standard and the Danish Transport Standard [5], which cover more than 95% of production. An important part of these standards is the implementation and regulation of several biosecurity measures. One-third of all herds are visited each year by a third-party auditor to ensure compliance.

2. Risk Factors

The World Organisation of Animal Health (WOAH) has identified several pathways for the introduction of African Swine Fever into pig herds or wild boar populations, including direct contact between infected and non-infected pigs, feeding garbage containing infected pork, as well as via vehicles, clothes, and implements [6]. Contaminated feed is another possible pathway, as shown by Dee [7]. The Danish pig industry and the Danish veterinary authorities have taken several steps to mitigate these risks, including mandatory washing and disinfection of returning livestock trucks, eliminating free-living wild boar from Danish territory, establishing a fence at the Danish–German border, and establishing industry rules for the handling of feed items that might constitute a risk.

In Northern Europe, wild boar has been a major factor in the spread of African Swine Fever. The first cases in many Northern European countries occurred in wild boar (Belgium, Germany, Sweden, and the Czech Republic) [8]. In several European countries, the infection has persisted in wild boar for years [9], demonstrating that successful eradication of African Swine Fever in wild boar populations is difficult once the infection has become widespread. However, eradication was achieved in some countries when the outbreak had a limited geographical extent (the Czech Republic and Belgium) [10]. Soft ticks (Ornithodoros) play an important role in the epidemiology of African Swine Fever in warmer climates, but are not part of the epidemiology in Northern Europe [11]. This subject is not discussed further herein.

3. Wild Boar in Denmark

Wild boar became extinct in Denmark at the beginning of the 19th century due to a combination of hunting, cold winters, and lack of suitable habitats [11]. However, since 2008, wild boars have been observed in low numbers in Denmark [11].

In 2018, the Danish Ministry of Food, Agriculture and Fisheries and the Danish parliament decided, in close collaboration with the Danish pig industry, that the small population of wild boar should be removed, and a fence should be built at the border between Denmark and Germany to prevent the migration of wild boar from Germany to Denmark. The Danish Nature Agency estimated the total population of wild boar in Denmark as 150–200 [12].

4. The Wild Boar Fence between Denmark and Germany

In 2019, the construction of the 68 km-long fence was completed, with 20 openings for traffic and numerous openings that can be passed by smaller animals and humans (ladders, small holes, and cattle grids) (Figure 1). In 2021, the last wild boar in the free range was shot; in total, 157 were either trapped or shot between 2018 and 2021. Since then, there have been fewer than a handful of sightings of wild boar; some were shot, and others likely migrated south again.

Figure 1. The wild boar fence between Denmark and Germany. The fence is 1.5 m high and has been dug 0.5 m into the ground (photo courtesy of Dr. Gordon Spronk).

The cost of the fence was approximately EUR 6 million, according to the Danish Nature Agency [12].

5. Mitigating the Risk of Introduction through Importing or Exporting Pigs

Denmark has a limited importation of live pigs, averaging less than 200 pigs annually, mostly boars or gilts for breeding. These are subject to normal importation procedures according to EU regulations [13]. Additionally, the Danish pig industry has implemented a voluntary quarantine and testing program, with high compliance so far.

In 2023, Denmark exported more than 15,000,000 pigs, mainly to Poland and Germany, including areas where African Swine Fever was present. More than 25,000 trucks were used for these exports, posing a risk when returning to Denmark to pick up pigs again. EU regulations [14] demand that all means of transport used for animal transport must be cleaned and disinfected after unloading. However, controlling the efficiency of cleaning and disinfection is challenging.

After the foot-and-mouth outbreak in the UK in 2001, Danish authorities required returning trucks to undergo additional inspection, cleaning, and disinfection at authorized washing facilities after crossing the Danish border [15,16]. Ten years later, the EU commission intervened, claiming the extra control, cleaning, and disinfection violated EU regulations on free trade.

Based on experiences in the washing facilities, Danish pig and cattle organizations decided that this control, cleaning, and disinfection were necessary to ensure an adequate level of hygiene. Thus, the pig industry formed the DANISH Transportstandard [5] as an adjunct to the DANISH Produktstandard to replace the regulation from Danish authorities.

All herds delivering pigs to major Danish slaughterhouses are part of the Danish Product Standard, which is compliant with the German QS scheme (Quality Scheme for Food). This means pigs exported to Germany are also covered by the Product Standard and the Transport Standard. More than 95% of Danish production is covered by the Danish Product Standard.

Trucks returning to Denmark must go through one of three authorized facilities at the border. Each truck is inspected on arrival, and if no potentially contaminated material is found inside, the truck is washed on the outside and disinfected inside and out. GPS data from the truck is uploaded and analyzed. Based on the GPS information, a certificate is issued electronically and on paper.

The Danish Transport Standard divides Europe into three zones: infected, buffer, and free zones. If GPS data show a truck has been in an infected zone, the certificate indicates a 7-day quarantine after leaving the washing facility before it can load pigs from a Danish farm. If a truck has been in a buffer zone, it has a 48 h quarantine. If it has only been in free zones, it can pick up pigs for export directly from a herd.

Trucks with 48 h or 7-day quarantines can always pick up pigs from a collection point, where pigs are unloaded from a truck only used for domestic transport and then loaded onto a truck transporting pigs across the border. If an export truck is used for domestic transport, it always has a 48 h quarantine, even if it has only been in a free zone. If it has been in an infected zone, the quarantine is 7 days.

All movements of live pigs in Denmark must be registered in the official Danish pig movement database. The Danish Agriculture & Food Council cross-references this database with certificates issued at washing facilities at the border to ensure that quarantine requirements have been fulfilled.

The cost of extra control, washing, and disinfection for Danish pig production is estimated to be approximately EUR 3 million.

6. Mitigating the Feed Risk for the Introduction of African Swine Fever

Swill feeding with infected, non-heat-treated pork products is likely the main pathway for the introduction of African Swine Fever into new areas. Although not clearly proven, it is probable that this was the main introductory pathway into the Dominican Republic, several countries in Eastern and Central Europe, and Southeast Asia.

EU regulation 1069/2009 [17] prohibits the use of food and kitchen waste or swill for feeding pigs and other farm animals. In large parts of Northern and Western Europe, there have been no introductions into farmed pigs by this route. However, unintentional introduction through some feed matrices could occur. African Swine Fever can survive for long periods in pig tissues [6]. Hay, silage, wrapped grass, and similar feed types can be contaminated by tissues of dead pigs or wild boar during harvest. Using these feed types could infect pig herds if introduced into the herd. To mitigate this risk, the use, or even presence, of these feed types is not allowed on pig farms, according to the Danish Product Standard, if the feed is imported from areas with African Swine Fever.

Dee [7] showed that the African Swine Fever virus can remain infectious under conditions simulating realistic transport conditions for trans-Atlantic and trans-Pacific passage in, for example, soybean meal and soy oil cake. The Danish pig industry closely monitors changing trade patterns of feed ingredients in collaboration with feed mills. Whenever changes in trade patterns are suspected or revealed, representatives from the feed mills cooperation (DAKOFO) and representatives from the pig production industry meet and discuss the circumstances.

Denmark is normally self-sufficient in grain production, but in 2018, a drought reduced Danish grain production, necessitating the importation of 1 to 1.5 million tons of wheat or corn, potentially from countries with African Swine Fever in both wild boar and domestic pigs.

A literature search in 2018 did not reveal any studies on the risk of introducing African Swine Fever to pig herds using whole corn or wheat for feed production. Heat treatment to 56 °C for 70 min or 60 °C for 20 min can mitigate the risk, but this was not a realistic option [6].

Due to the lack of experimental data, historical trade patterns of wheat and barley for feed production from areas with African Swine Fever were investigated. Large volumes of grain for feed production had been exported from Russia, Ukraine, and Eastern Europe to many parts of the EU without causing any outbreaks. Transport times from these areas of Europe to Denmark were at least 14–18 days [18].

Furthermore, outbreak data from the EU database showed that most outbreaks in pig herds in Europe occurred from June to September [19]. If newly harvested grain contributed to outbreaks, it would have been for a short period after harvest. Based on these epidemiological findings and the transport time to Denmark, it was concluded that the risk of introducing African Swine Fever through the importation of wheat or corn was limited.

In 2024, the EFSA published a study examining different feed matrices. Live viruses could only be isolated for a short period from wheat, barley, and grass, supporting the epidemiological findings [20].

In 2018 and 2019, African Swine Fever was spreading in several East Asian countries. Denmark imports organic soya and several oils from these countries. As Dee [7] showed, some of these matrices could be contaminated and contain infectious viruses for several weeks to months. Based on these results, it was decided that soya and certain fats and oils from these regions could only be used after heat treatment. Fats and oils that are solid during transport and require heating for unloading do not need further heat treatment. These requirements were implemented in the Danish Product Standard [5].

7. Discussion and Conclusions

Mitigating the risk of introducing porcine pathogens into a country with a substantial pig production and a large export of live pigs is challenging. The Danish authorities and the Danish pig industry have worked closely together to mitigate these risks. Denmark has eliminated wild boar from Danish nature areas, built a fence at the Danish border, implemented industry rules for empty livestock trucks coming back to Denmark after unloading pigs abroad, and implemented risk-mitigating measures for the importation of feed and bedding material. Compared to the estimated cost of an outbreak of EUR 335–670 million, the cost of these mitigating measures is relatively small.

Funding: This paper received no external funding.

Institutional Review Board Statement: Not applicable.

Informed Consent Statement: Not applicable.

Data Availability Statement: Only official statistics mentioned in the reference sections was used.

Conflicts of Interest: The author works for the Danish Agriculture and Food Council.

References

1. Available online: https://webgate.ec.europa.eu/tracesnt/adis/public/notification/outbreaks-current-year-report (accessed on 7 August 2024).
2. Available online: www.statistikbanken.dk (accessed on 7 August 2024).
3. Hansen, H.O.; Denver, S.; Lassholdt, M.L.; Hemmingsen, T.V.; Christensen, T. Afrikansk Svinepest i Europa (Danish Text). Available online: https://ifro.ku.dk/aktuelt/aktuelt-2019/oekonomer-afrikansk-svinepest-i-danmark-kan-koste-milliarder/ (accessed on 7 August 2024).
4. WOAH Terrestrial Health Code. Chapter 15.1. Infection with African Swine Fever Virus. Available online: https://www.woah.org/fileadmin/Home/eng/Health_standards/tahc/2018/en_chapitre_asf.htm (accessed on 7 August 2024).
5. The Danish Produktstandard and the Danish Transportstandard. Available online: www.svineproduktion.dk/viden/paa-kontoret/love-_regler-og-standarder/transportstandard (accessed on 7 August 2024).
6. WOAH Technical Disease Card. African Swine Fever. Last Update February 2022. Available online: www.woah.org/en/disease/african-swine-fever/ (accessed on 7 August 2024).
7. Dee, S.A.; Bauermann, F.V.; Niederwerder, M.C.; Singrey, A.; Clement, T.; de Lima, M.; Long, C.; Patterson, G.; Sheahan, M.A.; Stoian, A.M.M.; et al. Survival of viral pathogens in animal feed ingredients under transboundary shipping models. *PLoS ONE* **2018**, *13*, e0194509. [CrossRef] [PubMed]
8. WOAH World Animal Health Information System. Available online: www.woah.org/en/what-we-do/animal-health-and-welfare/disease-data-collection/world-animal-health-information-system/ (accessed on 7 August 2024).

9. Sauter-Louis, C.; Conraths, F.J.; Probst, C.; Blohm, U.; Schulz, K.; Sehl, J.; Fischer, M.; Forth, J.H.; Zani, Z.; Depner, K.; et al. African Swine Fever in Wild Boar in Europe—A Review. *Viruses* **2021**, *13*, 1717. [CrossRef] [PubMed]
10. EFSA Protocol for ASF Epidemiological Analysis. EFSA Supporting Publication. 2023. Available online: www.efsa.europa.eu/publications (accessed on 7 August 2024).
11. Jordt, A.M.; Lange, M.; Kramer-Schadt, S.; Nielsen, L.H.; Nielsen, S.S.; Thulke, H.H.; Vejre, H.; Alban, L. Spatio-temporal modeling of the invasive potential of wild boar—A conflict-prone species—Using multi-source citizen science data. *Prev. Vet. Med.* **2016**, *124*, 34–44. [CrossRef] [PubMed]
12. Danish Nature Agency. Available online: www.naturstyrelsen.dk (accessed on 7 August 2024).
13. European Commission. Regulation (EU) 2016/429 Of The European Parliament and of The Council of 9 March 2016 on transmissible animal diseases. *Off. J. Eur. Union* **2016**, *59*, 1–208.
14. European Commission. Commission Delegated Regulation (EU) 2020/688 of 17 December 2019 Supplementing Regulation (EU) 2016/429 of the European Parliament and of the Council, as Regards Animal Health Requirements for Movements within the Union of Terrestrial Animals and Hatching Eggs. Available online: https://eur-lex.europa.eu/legal-content/EN/TXT/?uri=CELEX:02020R0688-20231204 (accessed on 7 August 2024).
15. BEK nr 773 af 29/08/2001. Bekendtgørelse om Rengøring og Desinfektion af Transportmidler m.v. til Klovbærende Dyr Samt Om Afhentning af Klovbærende Dyr (Danish Text). Available online: www.retsinformation.dk/eli/lta/2001/773 (accessed on 7 August 2024).
16. BEK nr 1312 af 30/11/2010. Bekendtgørelse om Rengøring og Desinfektion af Transportmidler til Klovbærende Dyr. (Danish Text). Available online: www.retsinformation.dk/eli/lta/2010/1312 (accessed on 7 August 2024).
17. European Commission. Regulation (EC) No 1069/2009 of the European Parliament and of the Council of 21 October 2009 laying down health rules as regards animal by-products and derived products not intended for human consumption and repealing Regulation (EC) No 1774/2002 (Animal by-products Regulation). *Off. J. Eur. Union* **2009**, 1–33.
18. Dahl, J.; Nielsen, B.; Tybirk, P. Baggrundsmateriale Til Risikovurdering For Afrikansk Svinepest Og Muligheder for Risikohåndtering. 2018. Available online: www.svineproduktion.dk/publikationer/kilder/notater/2018/1832 (accessed on 7 August 2024).
19. Ståhl, K.; Boklund, A.E.; Podgórski, T.; Vergne, T.; Abrahantes, J.C.; Cattaneo, E.; Papanikolaou, A.; Mur, L. Epidemiological analysis of African swine fever in the European Union during 2023. *EFSA J.* **2024**, *22*, e8809. Available online: https://efsa.onlinelibrary.wiley.com/doi/epdf/10.2903/j.efsa.2024.8809 (accessed on 7 August 2024). [PubMed]
20. Blome, S.; Schäfer, M.; Ishchenko, L.; Müller, C.; Fischer, M.; Carrau, T.; Liu, L.; Emmoth, E.; Stahl, K.; Mader, A.; et al. Survival of African swine fever virus in feed, bedding materials, and mechanical vectors and their potential role in virus transmission. *EFSA Support. Publ.* **2024**, *21*, 8776E. [CrossRef]

Disclaimer/Publisher's Note: The statements, opinions and data contained in all publications are solely those of the individual author(s) and contributor(s) and not of MDPI and/or the editor(s). MDPI and/or the editor(s) disclaim responsibility for any injury to people or property resulting from any ideas, methods, instructions or products referred to in the content.

animals

Article

Biosecurity Insights from the United States Swine Health Improvement Plan: Analyzing Data to Enhance Industry Practices

Michael Harlow [1,2], Montserrat Torremorell [3], Cristopher J. Rademacher [2], Jordan Gebhardt [4], Tyler Holck [2], Leticia C. M. Linhares [2], Rodger G. Main [2] and Giovani Trevisan [2,*]

1 College of Public Health, George Mason University, Fairfax, VA 22030, USA
2 College of Veterinary Medicine, Veterinary Diagnostic and Production Animal Medicine, Iowa State University, Ames, IA 50011, USA
3 Department of Veterinary Population Medicine (VPM), College of Veterinary Medicine, University of Minnesota, St. Paul, MN 55108, USA
4 Department of Diagnostic Medicine/Pathobiology, College of Veterinary Medicine, Kansas State University, Manhattan, KS 66506, USA
* Correspondence: trevisan@iastate.edu; Tel.: +1-(515)-294-3556

Simple Summary: This article presents a snapshot of biosecurity practices implemented across the United States swine industry. Foreign animal diseases and endemic emerging and re-emerging swine diseases are serious threats. Biosecurity practices reported during enrolment in the United States Swine Health Improvement Plan are summarized here and provide insights into areas where the US Swine industry is prepared and identified opportunities and areas of opportunity for improvement in biodefense. This voluntary collaborative animal health program effort provides a baseline of biosecurity efforts over a significant portion of the swine industry. The findings can help guide the swine industry to continue to improve biosecurity, making sure the industry stays strong and sustainable despite new challenges.

Abstract: Biosecurity practices aim to reduce the frequency of disease outbreaks in a farm, region, or country and play a pivotal role in fortifying the country's pork industry against emerging threats, particularly foreign animal diseases (FADs). This article addresses the current biosecurity landscape of the US swine industry by summarizing the biosecurity practices reported by the producers through the United States Swine Health Improvement Plan (US SHIP) enrollment surveys, and it provides a general assessment of practices implemented. US SHIP is a voluntary, collaborative effort between industry, state, and federal entities regarding health certification programs for the swine industry. With 12,195 sites surveyed across 31 states, the study provides a comprehensive snapshot of current biosecurity practices. Key findings include variability by site types that have completed Secure Pork Supply plans, variability in outdoor access and presence of perimeter fencing, and diverse farm entry protocols for visitors. The data also reflect the industry's response to the threat of FADs, exemplified by the implementation of the US SHIP in 2020. As the US SHIP program advances, these insights will guide industry stakeholders in refining biosecurity practices, fostering endemic re-emerging and FAD preparedness, and ensuring the sustainability of the swine industry in the face of evolving challenges.

Keywords: swine health; biosecurity; foreign animal disease; animal health programs; Swine Health Improvement Plan

1. Introduction

The United States of America (US) is the third largest producer and consumer of pork. The US is also the second largest pork exporter, supplying almost 30% of the world's pork products [1]. The US pork industry creates over 600,000 jobs, producing over USD 35 billion in private income and USD 50 billion in gross national product (GDP) [2]. Endemic diseases, e.g., porcine reproductive and respiratory syndrome virus (PRRSV), pose a continuous threat, with PRRSV alone causing annual economic losses greater than USD 600 million to the US swine industry [3]. The potential introduction of a foreign animal disease (FAD) also poses a constant threat to the US swine industry. The most recent introduction of a FAD to the US swine industry occurred in 2013 with porcine epidemic diarrhea virus (PEDV) and in 2014 with porcine deltacoronavirus (PDCoV) [4]. Within 8 weeks of first PEDV diagnosis, the disease was already present in 12 US states [5] and caused the death of an estimated 5 million pigs [6], and it became endemic in the US [7].

FADs like African swine fever (ASF) have never been detected in the US, while classical swine fever (CSF) was eradicated in 1978, and foot-and-mouth disease (FMD) in 1929 [8,9]. ASF is species-specific, affecting only Suidae, and it is currently spreading throughout Europe, Asia, and Africa, significantly impacting swine trade, as it steadily creeps toward the US. The epidemiological link for the European ASF outbreak was identified as contaminated fomites on transport ships. In 2021, ASF was discovered in Haiti and Dominican Republic, which are located about a 1000 km from the US continental shores [10], or about 380 km from Puerto Rico, a US territory, making ASF one of the most eminent FAD threats to the US swine industry. ASF persists in the environment for long periods of time, has multiple routes of infection, and has high morbidity and mortality [11]. FAD agents, such as the ASF virus (ASFV), can be transported on fomites, including materials such as clothing, utensils, and farm tools [12]. Wild pig populations are susceptible to ASFV, and if wild pigs become infected, their presence could contribute to the maintenance of ASF in a region. ASF has been found to retain viability in feces and urine for up to eight days and in certain feed ingredients for up to 30 days [13]. ASFV has been found in decaying wild boar carcasses for at least 15 weeks, which serve as an environmental reservoir and important vector in maintaining the disease in wild boar populations because pigs are omnivorous [14]. Once an FAD is introduced to an area, it can be maintained in reservoirs of sylvatic pigs and can be transmissible through contact between sylvatic and domestic animals, and some FADs can be transmitted through tick vectors [12]. There are three species of soft ticks, *Ornithodoros coriaceus*, *O. turicata*, and *O. puertoricienses*, present in the US that are thought to be able to transmit ASFV [15]. The presence of wild boars in some regions of the US coupled with the presence of potential ASFV tick reservoirs can pose an additional level of complexity for control and eradication in case this FAD reaches the US continental territory.

Biosecurity includes measures and behaviors to mitigate or prevent the entrance of a given pathogen/disease or infestation in a farm or region and its subsequent establishment, to exclude it, and to reduce its spread to/from and within an animal population and reduce its impact once it has been introduced [16]. Biosecurity practices are broadly defined and include everything entering the premises, such as animals, people, supplies, and the vehicles that transport animals, feed, and maintenance materials [17]. Specific biosecurity practices can include requiring showering before entry and the usage of farm-specific clothes, as well as bench entry systems; meanwhile, biocontainment is related to the measures put in place to prevent the spread of the pathogen/disease to other farms, e.g., proper quarantining of new livestock [18,19]. Showering in and out of facilities is an effective procedure for reducing the possibility of pathogen entry into and exiting from a site through personnel. Trailers and transport vehicles are a concerning source of disease spread, as they travel from multiple locations and can carry organic matter, providing ample opportunities for disease to spread [20–23]. This includes feed delivery, as those vehicles travel to multiple farms per day and have a common point source, i.e., the feed mill. The thorough cleaning of transportation equipment is a priority in a successful biosecurity plan. On-farm biosecurity practices can be used to characterize the level of biosecurity on given farms and generate a score for practices that distinguish the biosecurity levels across sites [24] and assess the relative vulnerability of swine breeding herds to the introduction of pathogens such as porcine reproductive and respiratory syndrome virus [25].

The increase in the efficiency of pork production has resulted in the consolidation of pork producers. The reduced number of operations has resulted in larger operations with an increase in the number of heads per farm, mainly located in the US Midwest states [26,27]. These large facilities greatly improve production efficiency, but the nature of this design creates unique challenges for biosecurity. These challenges can be addressed through site-specific biosecurity plans, such as the US example of the Secure Pork Supply (SPS) (https://www.securepork.org, accessed on 5 December 2023) plans, providing a framework to create site-specific biosecurity, animal movement, and business continuity plans in the face of an FAD event. As an example, understanding who enters a site, when they visited, and where they were previously is essential to tracing the possible disease spread during a disease outbreak investigation. Epidemiologists and State and Federal Animal Health Officials can use visitor logbooks to trace back human contacts to find potential links of where an outbreak could have started. With this information, they can identify and address the issues that allowed for the entrance of the pathogen into the site and prevent further disease spread.

There is growing concern about the potential introduction of FADs into a country through contaminated feed ingredients [28,29]. Ingredients used in US swine diets are sourced globally, with some of those regions being affected by FADs, including ASF and/or CSF. Pathogens such as ASFV have been shown to potentially survive for weeks in both plant- and animal-based ingredients [30]. Ensuring the provision of safe food for the animals is a big step in FAD preparation. Ingredients originating from ASF-positive countries are particularly concerning given the extended survival characteristics of ASFV [28]. An extended feed ingredient holding time, a common biosecurity practice implemented by ingredient importers and feed manufacturers, allows for the natural decay of viruses to occur [31]. Holding times will depend on the type of feed/ingredient, ambient temperature, and environmental conditions. Additionally, there are chemical feed additives, also known as feed mitigants, which are safe for animal consumption that can be added to the diet, with experimental evidence demonstrating the potential to reduce the detection and infectivity of viruses.

Having a defined, consistent, broadly adopted, and internationally recognized biosecurity plan is paramount in mitigating the impact of an introduced FAD and helps to further enhance disease preparedness for foreign and endemic diseases. A new national health certification program for the swine industry was implemented in the US in late 2020, the US Swine Health Improvement Plan (US SHIP, https://usswinehealthimprovementplan.com/, accessed on 5 April 2023). US SHIP is a voluntary, collaborative effort between industry, state, and federal entities modeled after the National Poultry Improvement Plan (NPIP, https://www.poultryimprovement.org/default.cfm, accessed on 5 April 2023). US SHIP has been specifically organized to meet the unique needs and challenges of the U.S. swine industry. US SHIP is able to use the lessons learned by the NPIP over the last 80-plus years to fast-track a plan that addresses the concerns of the swine industry to prevent the introduction of trade impacting diseases such as ASF and CSF, as well as respond to and mitigate their impact if found in the US.

US SHIP is creating an industry-led effort to protect animal health, mitigate the impact of an FAD on the economy, and have a sustainable process to re-establish trade from unaffected sites outside of control zones. US SHIP is a new initiative to support FAD preparedness, with an emphasis on the demonstration of the freedom of disease outside of control zones, creating a solid basis for FAD preparedness that the United States Department of Agriculture (USDA)'s ASF Response Plan the Red Book recognizes as a new US initiative for FAD preparedness [11]. Thus, a critical step in disease preparedness is a need to understand the current state of biosecurity practices implemented by sites participating in the program. As part of US SHIP enrollment, producers are required to provide site(s) demographic(s) information, fill out a biosecurity survey, and profess their willingness to comply with program standards. The US SHIP biosecurity survey results create a snapshot of the US swine industry's current level of biosecurity, which helps identify areas of improvement, guide future studies, and assess infrastructure building. The current study summarizes the biosecurity practices reported by the producers enrolled in the US SHIP program and provides a general assessment of practices implemented across the US swine industry.

2. Materials and Methods

To capture the current biosecurity practices being implemented in swine farms at the time of US SHIP enrollment, an online survey was created and distributed to participants using a commercial survey builder (Qualtrics XM, https://www.qualtrics.com/ (accessed on 5 January 2020), Seattle, WA, USA). The survey questions were crafted to obtain a broad understanding of the status of biosecurity practices being implemented in the US swine industry. All producers' sites enrolled with US SHIP were asked to complete a biosecurity survey for all of their sites. The end goal of the biosecurity survey was to capture and benchmark the biosecurity practices implemented across the swine industry.

The survey consisted of initial screener questions to collect site demographics and the respondent's contact information. Surveys were filled out for the production type, for each producer, (e.g., boar stud, sow farm, etc.), for each state. For example, if a producer had three breeding herds in one state and two in another state, then they would fill out two surveys in total. The initial enrollment survey questions collected site demographic information for the respective state(s) in which the site(s) were located, enabling us to capture the number of participating sites in US SHIP, categorized by site type in each state. Eight site types were listed in the survey: boar stud, breeding herd, growing pig, farrow-to-feeder/finish, small holding, non-commercial, packing plant, and live animal marketing operators. Questions include a free range of responses in percentage or Lickert-scale format (Table 1) and were sectioned by site types (Table 2). A complete set of questions is provided in the Supplementary Materials (File S1).

Table 1. US SHIP enrolment biosecurity survey questions.

Questions by Site Type		Response Format
1—Sites that have completed the Secure Pork Supply plans		Response entered in percentile form from 0 to 100.
2—Sites were animals have access to the outdoors.		
3—Sites that have perimeter fences.		
4—Sites that use the following as their primary means of dead disposal	(a) Rendering	Response entered in percentile form from 0 to 100. The total should be equal to 100.
	(b) Non-rendering	
	(c) A combination of rendering and non-rendering	
5—Sites whose requirements for people to enter the farm most closely resembles on of the following	(a) No specific requirements	Response entered in percentile form from 0 to 100. The total should be equal to 100.
	(b) Only visitors are required to change into clean or site-specific clothes and footwear without a requirement to shower in	
	(c) Everyone changes into clean or site-specific clothes and footwear, but are not required to shower in	
	(d) Everyone showers and changes into clean or site-specific clothes and footwear	
6—Do visitors sign a logbook?		Lickert Scale, with responses on a 1–5 scale: (a) All of the time; (b) Most of the time; (c) About half of the time; (d) Rarely; (e) Never; (f) Don't know.
7—During the last 12 months, how often have you used the following ingredients in rations?	Plasma	
	Meat and bone meal	
	Feed mitigants to reduce the disease transmission risk	
	Uncooked food scraps	
	Regulated (cooked) garbage	
8—During the last 12 months, how often have your feed suppliers held the imported feed ingredients to reduce disease transmission risk?		
9—Have trailers been washed and disinfected since last returning from a point of concentration when being used to pick up animals from the site? *		

* For growing sites, the survey question on trailer wash was divided into two questions: (i) The first asked if the trucks being used to transport the first group of pigs (first cut) to market were washed and disinfected. (ii) The second asked if, when the remaining pigs in the barn were moved to slaughter (run out loads), the trucks were washed and disinfected. The movement of pigs between the first cut and when the barn is emptied occurs within the time frame of a couple of weeks.

Table 2. US SHIP enrolment biosecurity survey site-type definitions.

Site Types	Definition
Boar stud	Production site with mature boars that distribute semen to other production sites
Breeding herd	Production site with breeding females and house ≥1000 breeder or feeder swine
Growing pig	Production site with ≥1000 feeder swine
Farrow-to-feeder /finish	Production site with breeding females, grow feeder swine for purposes other than breeding stock replacement for this particular farm site, and house ≥1000 breeder or feeder swine
Small holding	Production sites with ≥100 and ≤1000 breeder or feeder swine
Non-commercial	Production sites with <100 pigs

Survey responses were recorded in Qualtrics. An R markdown script was written using an Application Program Interface (API) connection that is available in the R package Qualtrics to connect survey responses to a wrangling and graphical application built using the R package tidyverse, which summarizes and presents results in a standardized graphical and table-content formats, providing basic survey results. Analytical and graphical displays of survey responses were performed in R version 4.3.0 (R Core Team (2023), Vienna, Austria, https://www.R-project.org, accessed on 5 April 2023).

3. Results

From January 2022 to 1 March 2024, surveys were completed for 12,195 individual production sites across 31 states. Growing pigs made up over 86% (10,552 of 12,195) of the responding sites; 10% (1262 of 12,195) for breeding herds; 1% (141 of 12,195) for small holding; 1% (121 of 12,195) for farrow-to-finish; and the remaining for non-commercial, boar stud, and packing plants (Figure 1). The three states with the highest number of US SHIP-enrolled sites were Iowa, with 36% (4425 of 12,195); North Carolina, with 15% (1872 of 12,195); and Minnesota, with 13% (1637 of 12,195). The remaining 36% (4261 of 12,195) of the sites were in the other 28 states (Figure 2).

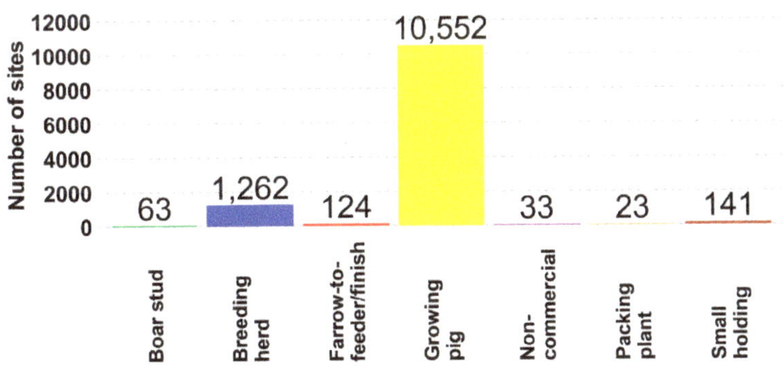

Figure 1. Number of sites by type of production site.

From participant sites, 91.5% (1154 of 1262) of responding breeding herds, 87% (9149 of 10,552) of growing pigs, 87% (54 of 63) of boar studs, 36% (12 of 33) of non-commercial sites reported having a completed SPS plan in place (Figure 3). SPS plans provide an industry-acceptable base for creating credible, provable disease-free zones in the instances of an FAD introduction. These zones could be leveraged through international trade agreements to support the continuity of business.

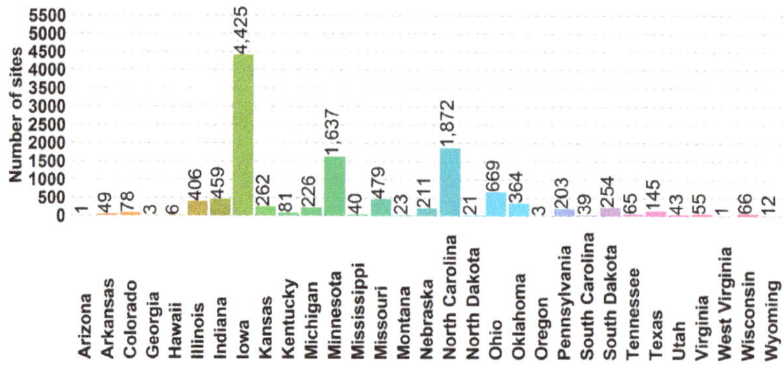

Figure 2. Number of sites that completed the US SHIP survey at enrollment by state.

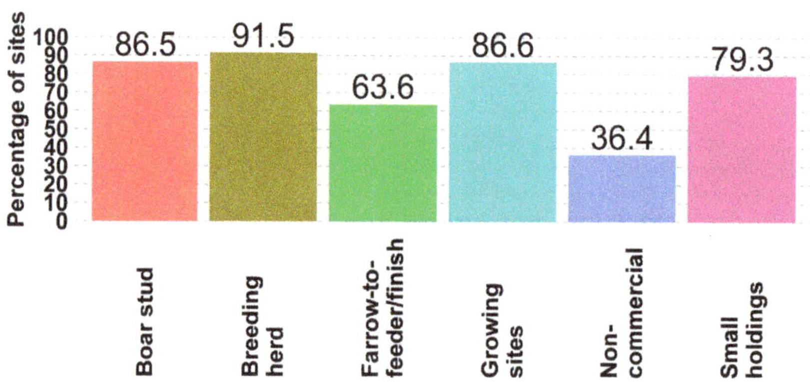

Figure 3. Percentage of sites by site type that have completed the Secure Pork Supply (SPS) plans.

While 65% (22 of 33) of non-commercial sites allow animals to access the outdoors, 64% (21 of 33) of non-commercial sites also have perimeter fencing. While only 3.2% (2 of 63) of boar stud sites allow outdoor access, 58.7% (37 of 63) of the sites have perimeter fencing (Figures 4 and 5). In general, perimeter fences are not widely adopted in the US swine industry.

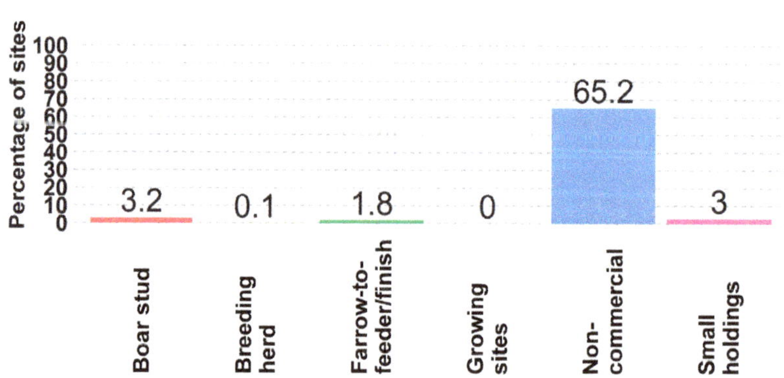

Figure 4. Percent of sites by site type where animals have access to the outdoors.

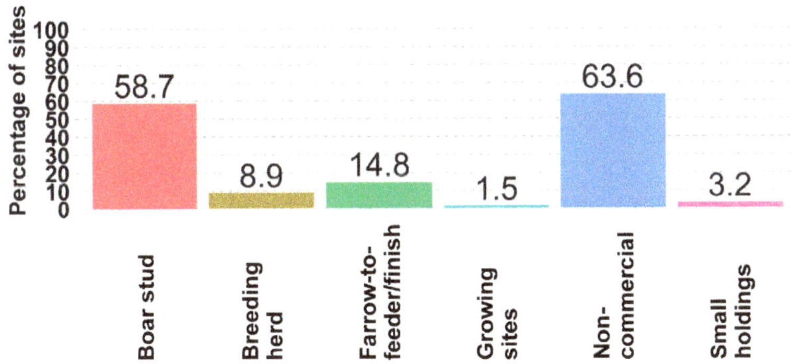

Figure 5. Percentage of sites, by site type, where sites have perimeter fences.

For tracking visitors on site, 42% (5051 of 12,172) of responding sites stated that they require visitors to sign a logbook all the time, another 46% (5550 of 12,172) reported doing so most of the time, and less than 1% (97 of 12,172) of all sites stated that they never require visitors to sign in. A breakdown by site type reveals that 94% (1186 of 1262) of the breeding herds reported that they require visitors to sign in via log-entry books all the time. Farrow-to-feeder had a widely dispersed response, with non-commercial sites reporting that 55% (18 of 33) of them do not have any entry log requirement (Figure 6).

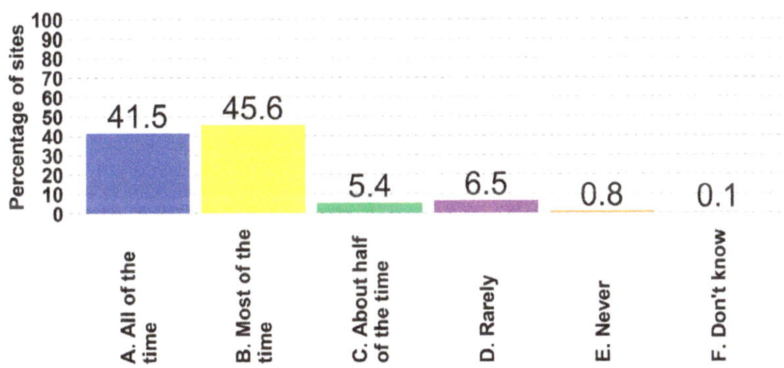

Figure 6. Percentage of responses by procedure frequency where visitors sign a log-in book prior to entering a site(s).

Of all responding sites, 51% (6219 of 12,195) require everyone to shower in and change foot and outerwear, 30% (3719 of 12,195) require everyone to change foot and outwear but not shower, 18% (2195 of 12,195) require only visitors to change clothes and footwear, and 0.5% (61 of 12,195) stated they have no entry requirements (Figure 7). Breeding herds had the largest percentage of sites requiring everyone to shower and change his/her outerwear 95% (1192 of 1262).

There is variability in the swine industry in regard to the usage of feed mitigants, where 4% (487 of 12,172) of all sites stated they always use feed mitigants, while 62% (7486 of 12,172) said they never did (Figure 8). Nineteen percent of boar studs (12 of 63) and approximately seventeen percent of breeding herds (209 of 1262) reported the usage of feed mitigants all the time. The variability continued in regard to holding times, with 59% (7157 of 12,172) of the respondents having reported holding imported feed ingredients specifically to reduce the transmission risk, and 15% (1765 of 12,172) said they never did. (Figure 9). Within this collection of participant survey data, there was no information gathered regarding specific conditions of ingredient holding, including holding time or temperature, nor was there further information about the use of feed mitigants, including specific products or inclusion levels.

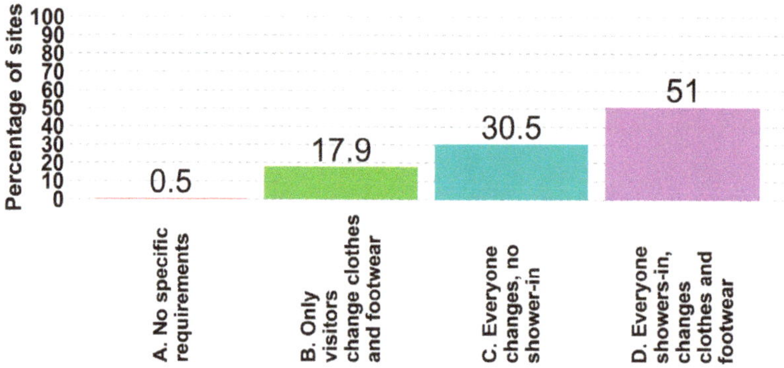

Figure 7. Percentage of sites that require a given procedure to enter a farm.

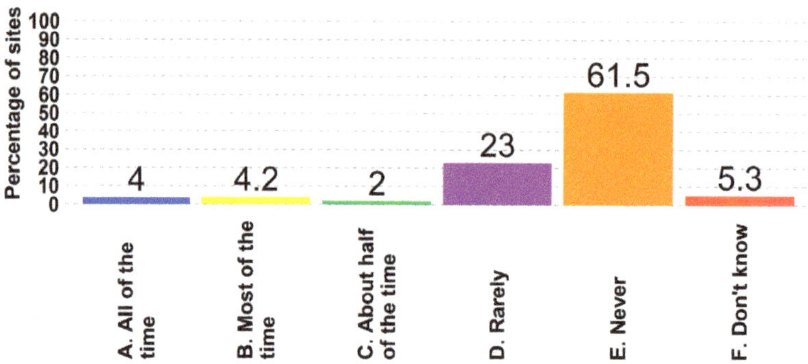

Figure 8. Frequency of feed mitigants used in feed rations to reduce disease transmission risk.

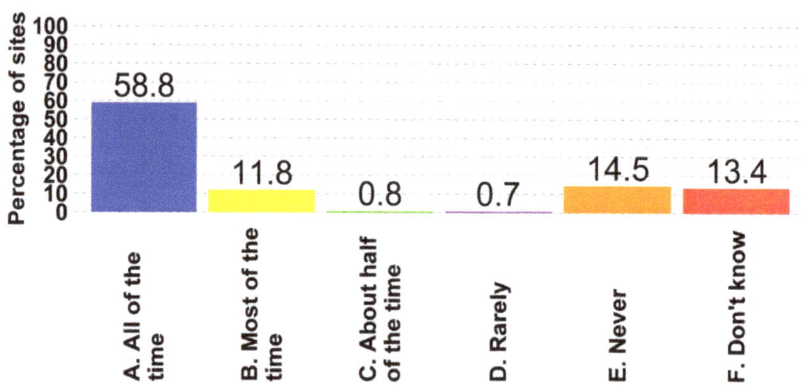

Figure 9. Frequency that feed supplier(s) have held imported feed ingredients to reduce disease transmission risk.

Related to transport biosecurity, 98.4% (62 of 63) of boar studs, 99.7% of breeding herds (1258 of 1262), and 91.7% (9676 of 10,552) of growing pigs' top-grade loads reported that the trailers were washed before returning to the point of concentration (e.g., slaughter plant and cull sow buying station). Top-graded pigs are the first loads removed from the site, with the pigs that first reach marketing criteria and usually take place from one to a few weeks before the site is emptied, i.e., loading of run-out loads. Run-out loads and non-commercial sites have a wider dispersion of responses (Figure 10).

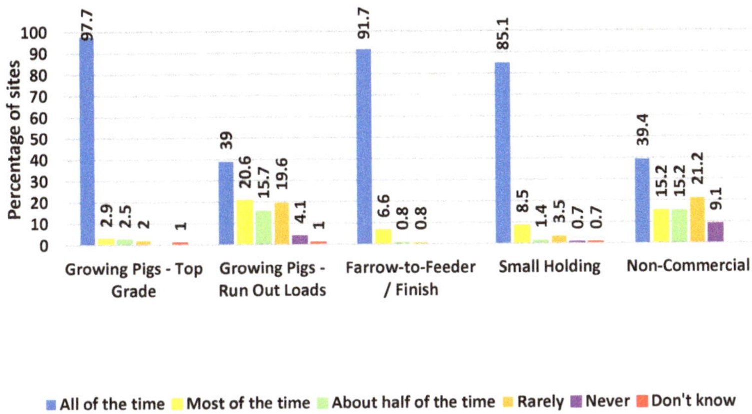

Figure 10. Frequency of transport trailers reported to have been washed before returning to the point of concentration broken down by type of production site.

4. Discussion

This study presents a synopsis of the US industry's implementation of the current practices of biosecurity that address major risks for disease introduction. Overall, there is limited quantitative information on biosecurity practices being implemented across swine production sites in the US. Having an understanding of biosecurity practices is helpful to design better disease control programs, such as the US SHIP, to ultimately prevent the introduction and spread of diseases. As part of developing a voluntary producer-led government-endorsed disease certification program in the US, named US SHIP, a set of questions were required for the participants to enroll their sites in the program. Biosecurity survey responses provided baseline information on certain biosecurity practices with the goal of identifying and quantifying areas for biosecurity improvement to prevent the introduction of FADs and the spread of endemic diseases.

In this study, a biosecurity practice survey questionnaire was distributed to pork producers through the US SHIP enrollment program, starting in January of 2022 and going through 1 March 2024, and it has been filled out for 12,195 pork production sites over multiple farm types, across 31 states. The captured survey responses reflect the diversity in biosecurity practices of the US swine industry and provide a snapshot of biosecurity practices in different production types, presenting an opportunity to assess and enhance its biosecurity measures over time to mitigate the threats posed by FADs. Commercial breeding herds and growing pig sites were the majority of the sites responding to the US SHIP biosecurity survey.

Production owners and site supervisors can use the findings of this survey to better prepare themselves to battle the FAD threat and work within the US SHIP to develop practices to address site-specific concerns and industry-wide shortfalls. An example of an action item generated by such information is the US SHIP biosecurity working group evaluating what mitigation strategies should be implemented in indoor and outdoor pigs to help decrease the risk posed by feral pigs and tick vectors in FAD transmission [32–34]. Findings, advancements, and novel approaches can be shared with producer delegates at US SHIP House of Delegates Meetings.

Having measures in place before any FAD introduction could help decrease the risks of foreign disease introduction and reduce the chance of endemic diseases spreading across farms. Every pig in the US is important in matters of FAD defense, and every producer has different challenges but shares the same responsibility. The results of this survey can better help industries and industry regulators' direct education, effort, and funds to strengthen biosecurity of the industry.

Diseases, especially FADs, are one of the most urgent concerns facing animal production. Countries such as Brazil [35], the Czech Republic [36], and Belgium [37] have been able to eradicate ASF from their borders after an introduction by using strategies and techniques described by the World Organization of Animal Health (WOAH). Efficient biosecurity and biocontainment practices take advantage of strategies and techniques developed through experience and research and proven to be effective in real-word scenarios for reducing the frequency of disease occurrence and spread across farms, regions, countries, and continents.

A multitiered approach is necessary to ensure the continued ability of pork producers to operate in the face of a FAD. The US Federal Government uses international trade agreements, customs, and border inspections to combat the introduction of FADs. Individual sites must also be prepared to prevent and mitigate the introduction of an FAD. This study identified potential areas of improvement by farm type. An example is the difference between incorporation of biosecurity best practices, such as reducing animal outdoor access and controlling visitor entry, where differences between commercial and non-commercial sites are evident. Across all sections of production, there is an opportunity to improve biosecurity practices, such as showering, the tracking of people's movements, and truck washing and disinfection, to minimize the risk of disease spread across sites. Even though breeding herds and growing sites are well-positioned to have SPS plans in place, 8.5% of breeding herds and 13.4% of growing sites are still without a biosecurity plan in place that leaves another practical opportunity for improvement.

With 18% of sites lacking entry requirements and 30% requiring only change in clothes, there exists an avoidable risk that can be effectively managed though the implementation of more stringent entry protocols. Entry protocols and sign-in books are practices requiring discipline to follow procedures and could be implemented irrespective of the current infrastructure present on the farm. Continuous training of farm personnel in following biosecurity and biocontainment practices is necessary to keep the discussions and reminders on top of the head of farm employees. Other biosecurity and biocontainment practices, such as promoting the utilization of feed mitigants (currently not employed by 61.5% of respondents) and adhering to proper holding times for imported feed ingredients (practiced by only 58.8% of respondents) are more costly but could significantly reduce the potential risks for FADs through the entry of contaminated feed. The ability of viral particles to maintain infectivity in feed has had real world impacts and is cited as the cause for the introduction of Senecavirus A into a previously naïve country [38]. Specifically for holding imported ingredients, there remains significant variation in how this practice is applied at the industry level, which warrants further investigation to generate a standardized baseline program for the US swine industry to generate a meaningful reduction in FAD introduction and dissemination risk. Another very costly procedure is the implementation of cleaning and disinfection procedures for hauling trucks. Extreme cold weather temperatures during winter pose a challenge in implementing these practices, requiring expensive infrastructure facilities to perform the procedure and treat the effluents.

In general, this study found that breeding herd sites routinely apply more biosecurity practices than other types of sites. Non-commercial sites were less likely to adopt any one specific biosecurity element discussed. This is relevant, as the swine industry is not stove-piped into vacuums, and each industry sector impacts the other and the overall security of the US swine industry [39,40]. Biosecurity improvements in one site or industry type can contribute collectively contain the introduction, establishment, and further spread of threats that affect the overall health of the swine industry.

One limitation of this descriptive study is that all results are self-reported by US SHIP participants. While the producers report their standards, the actual workday practices have not been evaluated. Together, this information provides a baseline understanding of the biosecurity practices in the US swine industry within the participant pool of US

SHIP-enrolled sites. The reporting of current stages of biosecurity practices implemented across US SHIP participant farms generates the basic foundation that can be used in future assessments for comparisons of the advancements made across participant sites.

The US SHIP program is a collaboration of swine producers and allied industry, state, and federal partners aimed at safeguarding the US swine industry, and it helps support business continuity in the event of an FAD introduction into the US. This baseline biosecurity through continued producer enrollment and engagement and the evaluation of current practices, as performed in this report, will serve as a means to generate meaningful improvement in biosecurity practices and enhance FAD preparedness and continue to decrease the detrimental impacts of endemic pathogens. The current reporting of biosecurity practices allows for information sharing and the future benchmarking of practices implemented and progress towards improving biosecurity and biocontainment measures to protect the US swine production.

5. Conclusions

In conclusion, this initial cross-sectional investigation sheds light on the biosecurity practices within the United States' swine industry, providing insights derived from the data collected through the US Swine Health Improvement Plan (US SHIP) enrollment program. The findings underscore the industry's diverse approaches to biosecurity. The establishment of US SHIP as a voluntary initiative demonstrates a collaborative effort among industry stakeholders, state, and federal entities to enhance disease preparedness. Multiple opportunities to improve the current level of biosecurity throughout all facets of the swine industry were identified and here reported. These findings offer a foundation for ongoing assessments, fostering continuous improvement in biosecurity measures and ultimately fortifying the industry's ability to combat the introduction and spread of diseases, both foreign and endemic. As the swine sector evolves, the knowledge gained from this study is pivotal for shaping targeted strategies to mitigate risks, promote sustainable practices, and contribute to the resilience of the US swine industry against emerging threats.

Supplementary Materials: The following supporting information can be downloaded at: https://www.mdpi.com/article/10.3390/ani14071134/s1, File S1: US SHIP Biosecurity Enrollment Survey.

Author Contributions: Conceptualization, R.G.M. and G.T.; methodology, R.G.M. and G.T.; software, G.T. and L.C.M.L.; validation, G.T. and L.C.M.L.; formal analysis, M.H., G.T., L.C.M.L., T.H., M.T., J.G. and C.J.R.; investigation, M.H., G.T., R.G.M., J.G., M.T., C.J.R., L.C.M.L. and T.H.; resources, R.G.M.; data curation, G.T.; writing—original draft preparation, M.H.; writing—review and editing, M.H., G.T., J.G., M.T., C.J.R., R.G.M. and T.H.; visualization, G.T. and M.H.; supervision, R.G.M.; project administration, R.G.M. and T.H.; funding acquisition, R.G.M. All authors have read and agreed to the published version of the manuscript.

Funding: This project was supported by the USDA–Animal & Plant Health Inspection Service (APHIS) Cooperative Agreement no. AP20VSSP0000C004 to Dr. Rodger Main.

Institutional Review Board Statement: The ethical approval was not required for the current study as the biosecurity practices data has only been collected from an online survey.

Informed Consent Statement: Informed consent was obtained from all subjects involved in the study.

Data Availability Statement: All the relevant collected data are provided in this manuscript. Individual enrolment survey responses are confidential and cannot be shared.

Conflicts of Interest: The authors declare no conflicts of interest.

References

1. USDA ERS. USDA ERS—Hogs & Pork. 30 June 2022. Available online: https://www.ers.usda.gov/topics/animal-products/hogs-pork/ (accessed on 2 October 2023).
2. Cook, H.; Schulz, L. The United States Pork Industry 2021: Current Structure and Economic Importance. July 2022. Available online: https://nppc.org/wp-content/uploads/2022/07/2021-NPPC-Economic-Contribution-Report-FINAL.pdf (accessed on 30 July 2022).

3. Holtkamp, D.J.; Kliebenstein, J.B.; Zimmerman, J.J.; Neumann, E.; Rotto, H.; Yoder, T.K.; Haley, C. Assessment of the economic impact of porcine reproductive and respiratory syndrome virus on United States pork producers. *J. Swine Health Prod.* **2013**, *21*, 72–84.
4. Stevenson, G.W.; Hoang, H.; Schwartz, K.J.; Burrough, E.R.; Sun, D.; Madson, D.; Yoon, K.J. Emergence of Porcine epidemic diarrhea virus in the United States: Clinical signs, lesions, and viral genomic sequences. *J. Vet. Diagn. Investig.* **2013**, *25*, 649–654. [CrossRef] [PubMed]
5. Marthaler, D.; Jiang, Y.; Collins, J.; Rossow, K. Complete Genome Sequence of Strain SDCV/US/Illinois121/2014, a Porcine Deltacoronavirus from the United States. *Genome Announc.* **2014**, *2*, 10-1128.
6. Schulz, L.; Tonsor, G.T. Assessment of the economic impacts of porcine epidemic diarrhea virus in the United States. *J. Anim. Sci.* **2015**, *93*, 5111–5118. [CrossRef] [PubMed]
7. Trevisan, G.; Linhares, L.; Schwartz, K.; Burrough, E.; Magalhaes, E.; Crim, B.; Dubey, P.; Main, R.; Gauger, P.C.; Thurn, M.; et al. Data standardization implementation and applications within and among diagnostic laboratories: Integrating and monitoring enteric coronaviruses. *J. Vet. Diagn. Investig.* **2021**, *33*, 457–468. [CrossRef] [PubMed]
8. USDA APHIS | Classical Swine Fever. (n.d.). Available online: https://www.aphis.usda.gov/aphis/ourfocus/animalhealth/animal-disease-information/swine-disease-information/classic-swine-fever/classic-swine-fever#:~:text=Although%20classical%20swine%20fever%20was,from%20the%20U.S%20in%201978 (accessed on 5 January 2024).
9. Rodriguez-Torres, J.G. International Approach to Eradication and Surveillance for Foot-and-Mouth Disease in the Americas. *Ann. N. Y. Acad. Sci.* **2000**, *916*, 194–198. [CrossRef] [PubMed]
10. African Swine Fever Confirmed in Haiti. American Veterinary Medical Association. 1 November 2021. Available online: https://www.avma.org/javma-news/2021-11-01/uropa-swine-fever-confirmed-haiti (accessed on 5 January 2024).
11. USDA. ASF Response Plan: The Red Book. April 2020. Available online: https://www.aphis.usda.gov/animal_health/emergency_management/downloads/asf-responseplan.pdf (accessed on 5 January 2024).
12. Sánchez-Vizcaíno, J.M.; Laddomada, A.; Arias, M.L. African Swine Fever Virus. In *Diseases of Swine*; Wiley: Hoboken, NJ, USA, 2019; pp. 443–452. [CrossRef]
13. Blome, S.; Franzke, K.; Beer, M. African swine fever—A review of current knowledge. *Virus Res.* **2020**, *287*, 198099. [CrossRef] [PubMed]
14. O'Neill, X.; White, A.; Ruiz-Fons, F.; Gortázar, C. Modelling the transmission and persistence of African swine fever in wild boar in contrasting European scenarios. *Sci. Rep.* **2020**, *10*, 5895. [CrossRef] [PubMed]
15. Golnar, A.J.; Martin, E.; Wormington, J.D.; Kading, R.C.; Teel, P.D.; Hamer, S.A.; Hamer, G.L. Reviewing the Potential Vectors and Hosts of African Swine Fever Virus Transmission in the United States. *Vector Borne Zoonotic Dis.* **2019**, *19*, 512–524. [CrossRef]
16. World Organisation for Animal Health. Terrestrial Code Online Access—WOAH—World Organisation for Animal Health. WOAH—World Organisation for Animal Health. 18 July 2023. Available online: https://www.woah.org/en/what-we-do/standards/codes-and-manuals/terrestrial-code-online-access/ (accessed on 5 January 2024).
17. Food and Agriculture Organization of the United Nations; World Organisation for Animal Health; World Bank; Madec, F.; Hurnik, D.; Porphyre, V.; Cardinale, E. Good practices for biosecurity in the pig sector. In *FAO Animal Production and Health*; (No. 169); Domenech, J., Lubroth, J., Eds.; FAO: Rome, Italy, 2010; Available online: https://www.fao.org/3/i1435e/i1435e00.pdf (accessed on 21 February 2024).
18. Galvis, J.A.; Machado, G. The role of vehicle movement in swine disease dissemination: Novel method accounting for pathogen stability and vehicle cleaning effectiveness uncertainties. *Prev. Vet. Med.* **2024**, *226*, 106168. [CrossRef]
19. Anderson, A.V.; Fitzgerald, C.; Baker, K.; Stika, R.; Linhares, D.; Holtkamp, D.J. Comparison of shower-in and shower-in plus bench entry protocols for prevention of environmental contamination due to personnel entry in a commercial swine facility. *J. Swine Health Prod.* **2018**, *26*, 192–199. [CrossRef] [PubMed]
20. Otake, S.; Dee, S.A.; Rossow, K.D.; Deen, J.; Joo, H.S.; Molitor, T.W.; Pijoan, C. Transmission of porcine reproductive and respiratory syndrome virus by fomites (boots and coveralls). *J. Swine Health Prod.* **2002**, *10*, 59–65.
21. Galvis, J.A.; Corzo, C.; Prada, J.M.; Machado, G. Modeling between-farm transmission dynamics of porcine epidemic diarrhea virus: Characterizing the dominant transmission routes. *Prev. Vet. Med.* **2022**, *208*, 105759. [CrossRef] [PubMed]
22. Lowe, J.; Gauger, P.C.; Harmon, K.M.; Zhang, J.; Connor, J.F.; Yeske, P.; Loula, T.; Levis, I.; Dufresne, L.; Main, R. Role of transportation in spread of porcine epidemic diarrhea virus infection, United States. *Emerg. Infect. Dis.* **2014**, *20*, 872–874. [CrossRef]
23. Ruston, C.R.; Linhares, D.; Blay, E.; Nickel, M.; Skoland, K.; Kittrell, H.; Brown, J.; Karriker, L.; Breuer, M.; McKeen, L.; et al. Evaluation of a staged loadout procedure for market swine to prevent transfer of pathogen contaminated particles from livestock trailers to the barn. *J. Swine Health Prod.* **2021**, *29*, 234–243. [CrossRef]
24. Silva, G.S.; Leotti, V.B.; De Jezus Castro, S.M.; Medeiros, R.; Silva, A.P.; Linhares, D.; Corbellini, L.G. Assessment of biosecurity practices and development of a scoring system in swine farms using item response theory. *Prev. Vet. Med.* **2019**, *167*, 128–136. [CrossRef] [PubMed]
25. Silva, G.S.; Corbellini, L.G.; Linhares, D.; Baker, K.L.; Holtkamp, D. Development and validation of a scoring system to assess the relative vulnerability of swine breeding herds to the introduction of PRRS virus. *Prev. Vet. Med.* **2018**, *160*, 116–122. [CrossRef] [PubMed]

26. USDA ERS—Sector at a Glance. USDA ERS—Sector at a Glance. 5 August 2021. Available online: https://www.ers.usda.gov/topics/animal-products/hogs-pork/sector-at-a-glance/ (accessed on 30 July 2022).
27. Davis, C.; Dimitri, C.; Nehring, R.; Collins, L.; Haley, M.; Ha, K.; Gillespie, J.U.S. *Hog Production: Rising Output and Changing Trends in Productivity Growth*; USDA: Washington, DC, USA, 2022; p. 308.
28. Dee, S.; Bauermann, F.V.; Niederwerder, M.C.; Singrey, A.; Clement, T.; De Lima, M.; Long, C.; Patterson, G.; Sheahan, M.A.; Stoian, A.M.M.; et al. Survival of viral pathogens in animal feed ingredients under transboundary shipping models. *PLoS ONE* **2018**, *13*, e0194509. [CrossRef]
29. Dee, S.; Neill, C.; Singrey, A.; Clement, T.; Cochrane, R.A.; Jones, C.K.; Patterson, G.; Spronk, G.; Christopher-Hennings, J.; Nelson, E. Modeling the transboundary risk of feed ingredients contaminated with porcine epidemic diarrhea virus. *BMC Vet. Res.* **2016**, *12*, 51. [CrossRef]
30. Stoian, A.M.M.; Zimmerman, J.; Ji, J.; Hefley, T.J.; Dee, S.; Diel, D.G.; Rowland, R.R.R.; Niederwerder, M.C. Half-Life of African Swine Fever Virus in Shipped Feed. *Emerg. Infect. Dis.* **2019**, *25*, 2261–2263. [CrossRef] [PubMed]
31. Niederwerder, M.C. Risk and Mitigation of African Swine Fever Virus in Feed. *Animals* **2021**, *11*, 792. [CrossRef] [PubMed]
32. Medrano, M.; US SHIP Feral Swine—Summary of Literature Review. US Swine Health Improvement Plan House of Delegates Meeting (US SHIP HOD). 2023, pp. 88–97. Available online: https://usswinehealthimprovementplan.com/wp-content/uploads/US-SHIP-Booklet-2023-Final_web.pdf (accessed on 21 February 2024).
33. US Swine Health Improvement Plan, Producers Congress of Industry Stakeholders & Subject Matter Experts, USDA Veterinary Services, & US SHIP Traceability Working Group. 2022. Resolutions [Report]. Available online: https://usswinehealthimprovementplan.com/wp-content/uploads/US-SHIP-Resolutions-2022.pdf (accessed on 21 February 2024).
34. US Swine Health Improvement Plan, US Swine Health Improvement Plan, USDA Veterinary Services, & US SHIP Traceability Working Group. 2023. Resolutions. Available online: https://usswinehealthimprovementplan.com/wp-content/uploads/US-SHIP-Program-Resolutions-Full-2023.pdf (accessed on 21 February 2024).
35. Moura, J.A.; McManus, C.M.; Bernal, F.E.M.; de Melo, C.B. An analysis of the 1978 African swine fever outbreak in Brazil and its eradication. *Rev. Sci. Tech.* **2010**, *29*, 549–563. [CrossRef] [PubMed]
36. Charvátová, P.; Wallo, R.; Satran, P. Lessons learned from successful eradication of ASF in the Czech Republic. *Bulletin* **2020**, *2020*, 1–5. [CrossRef]
37. African Swine Fever: Eradication in Belgium Confirmed. European Commission News Room. 20 November 2020. Available online: https://ec.europa.eu/newsroom/sante/items/694493#:~:text=All%20the%20African%20swine%20fever,infected%20with%20ASF%20was%20discovered (accessed on 8 January 2024).
38. Dee, S.; Havas, K.A.; Spronk, G. Detection of Senecavirus A in pigs from a historically negative national swine herd and associated with feed imports from endemically infected countries. *Transbound. Emerg. Dis.* **2022**, *69*, 3147–3149. [CrossRef] [PubMed]
39. Porphyre, T.; Boden, L.; Correia-Gomes, C.; Auty, H.; Gunn, G.J.; Woolhouse, M. How commercial and non-commercial swine producers move pigs in Scotland: A detailed descriptive analysis. *BMC Vet. Res.* **2014**, *10*, 140. [CrossRef]
40. Correia-Gomes, C.; Henry, M.K.; Auty, H.; Gunn, G.J. Exploring the role of small-scale livestock keepers for national biosecurity—The pig case. *Prev. Vet. Med.* **2017**, *145*, 7–15. [CrossRef]

Disclaimer/Publisher's Note: The statements, opinions and data contained in all publications are solely those of the individual author(s) and contributor(s) and not of MDPI and/or the editor(s). MDPI and/or the editor(s) disclaim responsibility for any injury to people or property resulting from any ideas, methods, instructions or products referred to in the content.

animals

MDPI

Communication

Further Evidence That Science-Based Biosecurity Provides Sustainable Prevention of Porcine Reproductive and Respiratory Syndrome Virus Infection and Improved Productivity in Swine Breeding Herds

Scott Dee [1,*], Lisa Brands [2], Roy Edler [1], Adam Schelkopf [2], Joel Nerem [2], Gordon Spronk [2], Mariana Kikuti [3] and Cesar A. Corzo [3]

1 Pipestone Research, Pipestone, MN 56164, USA; roy.edler@pipestone.com
2 Pipestone Veterinary Services, Pipestone, MN 56164, USA; lisa.brands@pipestone.com (L.B.); adam.schelkopf@pipestone.com (A.S.); joel.nerem@pipestone.com (J.N.); gordon.spronk@pipestone.com (G.S.)
3 College of Veterinary Medicine, University of Minnesota, St. Paul, MN 55108, USA; mkikuti@umn.edu (M.K.); corzo@umn.edu (C.A.C.)
* Correspondence: scottdee7255@gmail.com

Simple Summary: Porcine reproductive and respiratory syndrome virus (PRRSV) infection in sows results in reduced animal welfare, poor breeding herd performance, and economic loss to farmers. The best way to reduce the effects of PRRSV is to prevent its entry into swine breeding herds through improvements in farm biosecurity. To accomplish this goal, the concept of Next Generation Biosecurity (NGB) was developed. A recent publication demonstrated that the practice of a comprehensive, science-based program of biosecurity (NGB COMPLETE) which incorporated science-based biosecurity protocols targeting direct and indirect routes of PRRSV transmission, significantly reduced PRRSV incidence risk in breeding herds compared to a partial program (NGB INCOMPLETE). This new communication follows up on this earlier paper and brings new information by reporting significant differences in key performance indicators between NGB COMPLETE breeding herds versus NGB INCOMPLETE herds over that original 2-year period across the same swine production system. It also supplements the earlier paper with data from a third consecutive year of reduced PRRSV incidence risk, resulting in a PRRSV incidence risk of 8.0% across all breeding herds for the entire three years. In closing, this is further evidence demonstrating that NGB, while not perfect, brings value to farmers through sustainable prevention of PRRS and improved productivity.

Abstract: Porcine reproductive and respiratory syndrome virus (PRRSV) is a globally significant pathogen of pigs. Preventing the entry of PRRSV into swine breeding herds enhances animal health and welfare. A recently published retrospective cohort study reported significant differences in PRRSV incidence risk between breeding herds that practiced Next Generation Biosecurity (NGB) COMPLETE, versus herds that practiced a partial approach (NGB INCOMPLETE) over a 2-year period. This follow-up communication builds on this previous publication and brings new information regarding statistical differences in key performance indicators (KPIs) from 43 NGB COMPLETE herds and 19 NGB INCOMPLETE herds during disease years 1 and 2. Statistically significant differences included higher total born/farrow and pigs weaned/female along with a reduced pre-weaning mortality and wean to 1st service interval, as well as a 0.91 increase in the number of pigs weaned/mated female/year. In addition, this communication reports that PRRSV incidence risk throughout disease years 1–3 was 8.0%, and that the association of NGB status (COMPLETE vs. INCOMPLETE) and disease burden for the cumulative 3-year period was statistically significant ($p < 0.0001$). These findings support previously published data that NGB, while not perfect, provides sustainable prevention of PRRSV, and may help improve herd productivity.

Keywords: pigs; PRRS; virus; biosecurity; incidence; risk; key; performance; indicators; breeding herds; prevention

1. Introduction

Porcine reproductive and respiratory syndrome (PRRS), a globally significant disease of pigs, is caused by the porcine reproductive and respiratory syndrome virus (PRRSV) [1]. Infection of breeding female swine with PRRSV results in reduced animal welfare, poor reproductive performance, and significant economic loss to farmers [2]. To control PRRS, PRRSV entry into breeding herds must be prevented through the application of a comprehensive, science-based plan of biosecurity known as "Next Generation Biosecurity" (NGB) [3]. As previously described, NGB prevents viral entry through the application of specific routes of PRRSV transmission, including direct routes (infected pigs and contaminated semen) and indirect routes (mechanical/fomite-based routes), aerosols, and feed through the application of science-based biosecurity protocols [4–8]. For review, the protocols of biosecurity designed to prevent viral entry via these routes have been applied, tested, and published [3]. Specifically, NGB was applied to breeding herds managed by Pipestone Veterinary Services within a large-scale commercial swine production operation known as the Pipestone System, the third largest swine production company in the United States [3]. Historically, from July 2009 to June 2021, annual PRRSV incidence risk across Pipestone-managed breeding herds ranged from 15% to 55% (Table 1).

Table 1. Annual Pipestone PRRSV Incidence Risk 1 July 2009 to 30 June 2021.

Disease Year	% PRRSV Incidence Risk
2009–2010	55%
2010–2011	40%
2011–2012	45%
2012–2013	35%
2013–2014	15%
2014–2015	16%
2015–2016	32%
2016–2017	15%
2017–2018	23%
2018–2019	17%
2019–2020	18%
2020–2021	32%

To drive change, NGB was initially applied for a 2-year test period (1 July 2021–30 June 2023), and PRRSV incidence risk was analyzed using a retrospective cohort design across 321,013 sows from 69 managed breeding herds (disease year 1) and 381,404 sows from 76 managed breeding herds (disease year 2) [3]. Over this time, the application of NGB reduced PRRSV incidence risk to 8.6% in disease year 1 and 9.2% in disease year 2 [3]. The analysis also identified two cohorts: one that implemented protocols for all phases of NGB (NGB COMPLETE, 56 herds) and those that implemented all phases of NGB biosecurity except air filtration (NGB INCOMPLETE, 20 herds). Over the two disease years, PRRSV incidence risk in NGB COMPLETE herds was 8.9% versus 40.0% in NGB INCOMPLETE herds, and the association between NGB status and PRRSV incidence risk for the 2-year period was statistically significant (p = 0.006), favoring the NGB COMPLETE approach [3]. This paper provided the initial evidence that improvements in biosecurity resulted in a sustainable reduction of PRRSV incidence risk in breeding herds in a large commercial swine production system. However, limitations of this report included no information on differences in key performance indicators (KPIs) during the 2-year assessment period between the two cohorts, or if a reduced level of PRRSV incidence risk could be sustained for a third consecutive year (disease year 3). Therefore, the purpose of this follow-up communication was to report PRRSV incidence risk for a disease year 3, to analyze the association between biosecurity status (NGB COMPLETE vs. NGB INCOMPLETE) and disease burden (PRRSV incidence risk) over the cumulative three-year period, and to statistically assess key performance indicators between NGB COMPLETE breeding herds and NGB INCOMPLETE herds in disease years 1 and 2.

2. Materials and Methods

2.1. Ethical Review

Following the Pipestone Research IACUC review, it was determined that an ethical review of the farms reported for the pending assessments was not needed, as the original study was data-based, and this new paper was a follow-up in support. In addition, Pipestone Veterinary Services has permission to use site data for research and publication as part of the management contracts.

2.2. Descriptive Data from Participating Herds in Disease Year 3

Table 2 describes data from breeding herds that participated in disease year 3. Included are data from the total number of participating herds, as well as the number of herds in each cohort, NGB COMPLETE and NGB INCOMPLETE.

Table 2. Descriptive statistics of breeding herd inventories in disease year 3 across the total number of participating herds in the company and the number of herds in the two cohorts, NGB COMPLETE and NGB INCOMPLETE.

	TOTAL	NGB COMPLETE	NGB INCOMPLETE
# Herds	75	58	17
# Sows	384,207	318,788	65,419
Mean herd size	5123	5590	3473
Median herd size	5516	5708	3183
Maximum herd size	12,064	12,064	6249
Minimum herd size	1362	1397	1362
95% CI	4622–5624	5020–6160	2719–4226
SD	2100	2149	1466

2.3. Herd Selection and Statistical Analysis of KPIs across Cohorts

For assessment of KPIs, only herds that provided two complete years of data during disease year 1 (1 July 2021 to 30 June 2022) and disease year 2 (1 July 2022 to 30 June 2023) were selected. This dataset was controlled through the use of a standardized nutrition program (Pipestone Nutrition, Pipestone, MN, USA), and a farm management company practicing consistent employee training programs and animal handling protocols (Pipestone Management, Pipestone, MN, USA). This approach ensured that all farm workers were employees of the company, had been trained using consistent protocols, and underwent regular auditing and performance reviews, maximizing both stability in the labor force and compliance with biosecurity protocols. In addition, one record-keeping system (Porcitec, Agritecsoft.com, Barcelona, Spain) was used, and animal health programs and products were provided by one source (Pipestone Veterinary Services, Pipestone, MN, USA). To maximize sample size, all genetic sources (n = 3) employed within the company during the two years were included. This decision was justified using a generalized linear model which indicated no interaction between genetic source and treatment (NGB COMPLETE and NGB INCOMPLETE) across any of the KPIs (Edler R., Pipestone Research, Pipestone, MN, USA, personal communication, 31 July 2024). Following final selection, breeding herds were defined as either NGB COMPLETE, involving protocols to mitigate the defined designated direct and indirect routes, or NGB INCOMPLETE, protocols to mitigate direct and indirect routes in the absence of air filtration, as described [3]. Across qualifying herds, KPIs from the Porcitec performance analysis were analyzed. The KPIs evaluated included the following: farrowing rate (%); total number of pigs born/farrow; number of pigs weaned/female; % pre-weaning mortality; number of pigs weaned/mated female/year; % repeat services; average parity of sows farrowed; % multiple matings; wean to first service interval (days); % sows bred by 5 days; conception rate (%); farrowing rate (%); age at first service (days); entry to first service (days); number of live-born pigs/farrowing; % stillbirths; % mummies; % sow mortality; replacement rate (%); culling rate (%); and litters/mated female/year. Differences between cohorts were tested for significance by *t*-test.

2.4. Calculation of PRRSV Incidence Risk during Disease Year 3 and over Disease Years 1–3

As in disease years 1 and 2, PRRSV incidence risk in disease year 3 (1 July 2023 to 30 June 2024) was based on the number of new viral entries to a fully managed breeding herd divided by the number of herds at risk in each disease year, according to calculations previously published by the University of Minnesota Dr. Bob Morrison's Swine Health Monitoring Project (MSHMP) [3,9]. As published, a new introduction was defined either as the entry of PRRSV to a historically naïve breeding herd, or the entry of a PRRSV variant to a historically infected herd with the new variant being $\geq$2% heterologous to the existing variant(s) based on nucleic acid sequencing of open reading frame 5 [3,10]. In addition, the animal inventories of participating breeding herds in disease year 3 were analyzed using descriptive statistics (mean, median, maximum, minimum) and the differences in proportions of new viral entries/number of participating herds in disease year 3 were calculated and compared to years 1 and 2 using Chi square. The association of disease burden (PRRSV incidence risk) and biosecurity level (NGB COMPLETE vs. NGB INCOMPLETE) over the cumulative three-year period was tested by Chi square. Finally, the PRRSV incidence risk over the cumulative three-year period was calculated using a weighted average to account for differences in the number of participating breeding herds each year.

2.5. Re-Assessment of Neighboring Swine Herd Density during Disease Year 3

As published in disease years 1 and 2 [3], calculation of neighboring swine density in disease year 3, defined as 1 July 2023 to 30 June 2024, was repeated by identifying swine herds within an 8.3 km radius of each breeding herd using geographical software (Google Earth, version 7.3.2.5776 and Google Map Developers, Chicago, IL, USA (mapdevelopers.com)). This was important to be certain that area density did not influence the results. The distance of 8.4 km was selected based on published data regarding the ability of PRRSV to be transported via aerosols out to and beyond this distance [7]. In conjunction with mapping, site inspections by approved and trained Pipestone personnel were conducted around each participating breeding herd to validate whether the sites did or did not house pigs, and whether each site was owned and/or managed by Pipestone. This practice had been in place across the Pipestone System for more than 13 years. The evaluation of area density around these herds had been calculated twice a year, including regular updates and site inspections during the third year of the study. The difference in the mean number of neighboring swine herds within an 8.3 km radius of NGB COMPLETE herds and NGB INCOMPLETE herds was analyzed by the Mann–Whitney *U* test (Statistics Kingdom, Melbourne, Australia) (https://www.statskingdom.com/170median_mann_whitney.html, 27 July 2023), with the level of significance set at $\leq$0.05. Graphs were constructed using Microsoft Excel (Microsoft

Corporation, Redmund City, CA, USA) as published, in accordance with the format used by MNSHMP [9].

3. Results

3.1. Statistical Analysis of KPIs during Disease Years 1 and 2 across Cohorts

Data from the KPIs between the breeding herd cohorts during disease years 1 and 2 are summarized in Table 3. A total of 62 herds qualified for the comparison based on the previously described criteria, and 43 herds (208,918 sows) were categorized as NGB COMPLETE and 19 herds (64,119 sows) as NGB INCOMPLETE. Table 4 summarizes the statistical differences in the means of the 20 KPIs between the two cohorts as determined by *t*- test, with significant differences highlighted in red. Significant differences favoring NGB COMPLETE herds included an increase in total born per farrow ($p = 0.047$), and pigs weaned per farrow ($p = 0.021$), a decrease in pre-weaning mortality rate ($p = 0.013$), and a reduced wean to 1st service interval ($p = 0.007$). While not statistically significant ($p = 0.15$), there was a 0.91 increase in the number of pigs weaned/sow/year in NGB COMPLETE herds versus INCOMPLETE herds. There were also no statistically significant differences in % sow mortality ($p = 0.18$), % stillborn ($p = 0.77$) and % mummies ($p = 0.17$) between cohorts, parameters historically associated with PRRSV infection.

Table 3. Descriptive statistics of the inventories from the 62 breeding herds included in the analysis of the key performance indicators from disease years 1 and 2. Values are expressed as number of sows.

Biosecurity Level	# Herds	#Sows	Mean	Median	Min	Max
NGB complete	43	208,918	4858	5507	930	9907
NGB incomplete	19	64,119	3375	3101	1497	5993

Table 4. Statistical analysis of the KPIs across NGB COMPLETE and NGB INCOMPLETE herds.

KPI	NGB COMPLETE	NGB INCOMPLETE	*p*-Value	Difference
Number of farms	43	19	-	+24
Farrowing Rate	88.57	88.79	0.80	−0.22
Total Born Per Farrow	16.11	15.81	0.047	0.30
Pigs Weaned Per Female	12.20	11.77	0.021	0.43
Pre-weaning Mortality Rate	15.85	17.63	0.013	−1.78
Weaned/Mated Female/Year	28.64	27.73	0.15	0.91
% Sow Mortality	11.48	10.71	0.18	0.77
% Repeat Services	3.49	4.18	0.31	0.69
Average Parity Farrowed	3.52	3.55	0.77	0.03
% Multiple Matings	90.17	89.72	0.77	0.46
Wean to 1st Service Interval (days)	7.13	8.21	0.007	−1.09
% Bred by 5 Days	83.28	80.70	0.09	2.58
Conception Rate (%)	93.68	93.46	0.64	0.22
Age at First Service (days)	212.75	217.72	0.08	−4.97
Entry to 1st Service Interval (days)	9.90	11.30	0.56	−1.40
Liveborn Per Farrow	14.51	14.28	0.13	0.23
% Stillborn	6.51	6.60	0.77	−0.09
% Mummies	3.43	3.07	0.17	0.36
Litters/Mated Female/Year	2.34	2.35	0.70	−0.09
Replacement Rate (%)	59.36	55.40	0.28	3.96
Culling Rate (%)	46.66	42.66	0.11	4.00

3.2. PRRSV Incidence Risk Calculations during Disease Year 3 and over the Cumulative 3-Year Period

During disease year 3 (1 July 2023 to 30 June 2024), a total of 75 breeding herds, consisting of 384,207 sows, were involved in the project. The mean sow inventory was 5123 (min = 1362, max = 12,064) and the median herd size was 5516. The 95% CI from the mean ranged from 4622 to 5624 with a standard deviation of 2100 (Table 2). Across these herds, 11 new PRRSV introductions were recorded for a PRRSV incidence risk of 14.6%. The degree of heterology at ORF 5 across these viruses ranged from 10 to 12%. Of these 11 infected herds, six were categorized as NGB COMPLETE and five as NGB INCOMPLETE. The proportion of positive herds in disease year 3 (11 new viral entries/75 herds, 14.6%) was not significantly different ($p = 0.77$) as compared to disease year 1 (6/69 herds, 8.6% PRRSV incidence risk) or disease year 2 (7/76 herds, 9.2% PRRSV incidence risk) (Table 5). In disease year 3, 58 of the 75 herds (77.3%) were categorized as NGB COMPLETE while 17 (22.7%) were categorized as NGB INCOMPLETE. Across the cumulative 3-year period, the association of PRRSV incidence risk and level of biosecurity favored NGB COMPLETE herds ($p < 0.0001$) and the PRRSV incidence risk over the cumulative 3-year period across all herds was 8.0%. Figure 1 depicts the change in PRRSV incidence risk in the Pipestone System from disease year 2009–2010 through 2023–2024, denoting the system-wide application of NGB from July 2021 to June 2024.

Table 5. Comparison of the proportion of PRRSV positive breeding herds recorded during disease year 3 as compared to disease years 1 and 2.

	Disease Year 1	Disease Year 2	Disease Year 3
Proportion positive (# new infections/#herds)	8.69% (6/69) [a]	9.23% (7/76) [a]	14.6% (11/75) [a]
# infected herds NGB COMPLETE/total # herds NGB COMPLETE	6.25% (3/48)	3.5% (2/56)	10.3% (6/58)
# herds NGB INCOMPLETE/total # herds NGB INCOMPLETE	14.2% (3/21)	25% (5/20)	29.4% (5/17)

[a]: Differences in the proportions of PRRSV positive herds across the 3 disease years were found to be not significant by Chi square at $p = 0.77$.

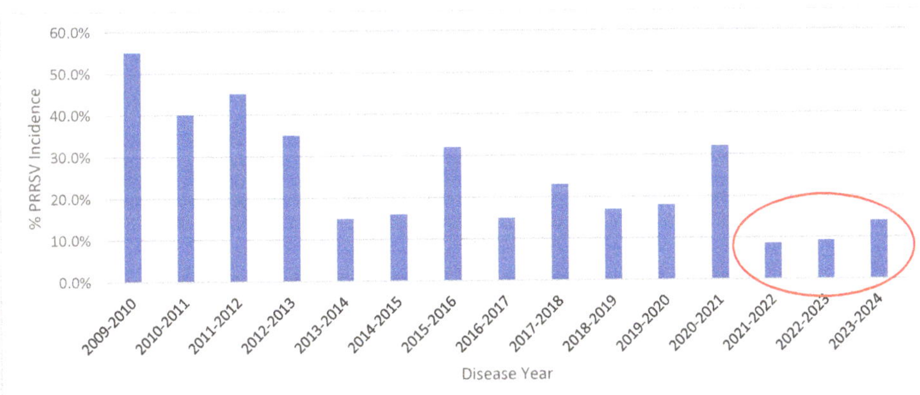

Figure 1. Trend in mean PRRSV incidence risk from disease year 2009–2010 through 2023–2024. Note reduction in PRRSV incidence risk beginning in disease year 2021–2022 through 2023–2024 (circled in red) following the application of Next Generation Biosecurity in July 2021.

3.3. *Re-Assessment of Neighboring Swine Herd Density during Disease Year 3*

During disease year 3, it was confirmed that no changes had occurred in area herd density as compared to data previously published [3]. Therefore, the median number of neighboring swine herds within an 8.3 km radius of NGB COMPLETE and NGB INCOMPLETE breeding herds were once again determined to be 2.0 and 2.0, respectively.

4. Discussion

Throughout global agriculture, herds and flocks are constantly dealing with the relentless pressure of viral diseases, be it from transboundary agents such as the African swine fever virus and highly pathogenic avian influenza virus, or from domestic pathogens such as PRRSV. In the absence of vaccines capable of inducing sterilizing immunity, the key to controlling these pathogens is to prevent introduction to susceptible populations, which results in improved health, welfare, and productivity [11,12]. Recently, in the US, there has been a significant effort towards reducing the impact of PRRSV through the application of Next Generation Biosecurity, and the initial published results are encouraging [3]. In support of this publication, this follow-up communication reports new information consisting of PRRSV incidence risk data from disease year 3, a cumulative incidence risk over disease years 1–3, the association of biosecurity level and PRRSV incidence over the cumulative 3-year period, and describes statistically significant differences in KPIs across the two cohorts during disease years 1 and 2.

As mentioned, KPIs derived from Porcitec performance analyses were used to test differences in breeding herd productivity between NGB COMPLETE and NGB INCOMPLETE cohorts. For this assessment, we selected 62 herds from a well-controlled and finely managed system of pig production. Results of this analysis indicated a statistically significant advantage to NGB COMPLETE herds regarding total born/female, pigs weaned/female, pre-weaning mortality, and wean to 1st service interval. As we attempted to control variability, it was good to see that several critical KPIs that could have biased the outcomes were similar between cohorts, including average parity farrowed (NBG COMPLETE = 3.52 vs. NGB INCOMPLETE = 3.55) and litters/mated female/year (NGB COMPLETE = 2.34 vs. NGB INCOMPLETE = 2.35). It was also good to see that the percent sow mortality was statistically insignificant ($p = 0.18$) between the cohorts, which suggests a high degree of animal care and welfare as it pertains to managing gestating and lactating sows. Possibly most impactful, despite its lack of statistical significance, was the increase of 0.91 pigs weaned/mated female/year, which resulted in 190,115 additional weaned pigs across the 208,918 sows in the 43 NGB COMPLETE herds, clearly a "biologically significant" result. These are interesting outcomes, and one conclusion that could be drawn was that the improvements in productivity observed in the NGB COMPLETE cohort may have been due to reduced PRRSV incidence risk secondary to improvements in biosecurity, supported by the significant association ($p < 0.0001$) between PRRSV incidence in NGB COMPLETE herds vs. NGB INCOMPLETE herds over the 3-year study period. However, it must be disclosed that a major limitation of using the Porcitec program was that the performance analysis did not include an abortion rate and clustered the number of weak born piglets in the pre-weaning mortality metric; therefore, these parameters, known to be characteristic of PRRSV infections in sows, could not be analyzed.

As with all studies, there were strengths and limitations that need to be reported and discussed. One strength of this communication is the generation of additional data across more sows, more herds, and more time, thereby building on the outcomes from disease years 1 and 2. Another strength is the transparency of the reporting. Specifically, while the proportion of positive herds across the three years was not statistically different across the 3-year period, it is important to disclose that the number of new viral introductions in year 3 was numerically greater than the two previous years. Upon investigation, it was observed that of the 11 new infections in disease year 3, four occurred in NGB INCOMPLETE herds which had also been infected in year 2, which is likely due to the lack of an air filtration system. This is a very important point to raise, as the presence or absence of an air filtration system was technically the only difference between the NGB COMPLETE and INCOMPLETE cohorts. Therefore, one conclusion that could be drawn was that the presence of an air filtration system played a role in reducing PRRSV incidence risk as previously reported by Havas and others [12]. In addition, two other herds, one categorized as NGB COMPLETE and another as NGB INCOMPLETE, became infected during herd repopulation, a very chaotic event when the frequency of personnel entry, and pig transport into a farm, are higher than normal, all which could have enhanced viral entry via an aerosol and/or fomite-based route. Finally, another NGB COMPLETE herd became infected shortly after a tornado caused significant damage to the roof of the farrowing facility, resulting in extensive air entry and possible viral entry.

We must also acknowledge that a challenge to the practice of biosecurity is the "human factor", and NGB is no exception to this rule [13]. The practice of NGB, while very effective, places employees under tremendous pressure and, while it was not quantified, it is possible that "biosecurity fatigue" may have occurred in disease year 3. Alternatively, the high degree of success that occurred in disease years 1 and 2 may have promoted an attitude of complacency, and people may have reduced focus. While we acknowledge the inability to completely control the "human factor", the fact that all farm personnel were employees of Pipestone and had participated in standardized training and auditing programs controlled this variable to the best of our knowledge and ability.

From the viral perspective, another limitation of NGB could be that certain routes of PRRSV transmission have not yet been identified and therefore, all the necessary protocols may not be in place. In addition, recently evolved variants such as PRRSV 144 L1C appear to be more infectious and contagious than historic variants, which could challenge the efficacy of the biosecurity protocols that were used [14]. As the 144 L1C variant was identified in several of the outbreaks over the three years, viral adaptation could have played a role. Finally, as this study only involved one pig production company, we must acknowledge that NGB may not be practical in other companies.

Clearly, the results from this paper clearly indicate that the NGB approach has not yet been perfected and additional work is needed. However, despite the acknowledged limitations, the ability to report a PRRSV incidence risk of 8.0% over a 3-year period across a pig production system of this size is not only a novel observation, but also low enough that the occasional new infection no longer compromises overall system operations. For the record, it needs to be reported that across all participating herds there was no application of bio-containment strategies. Therefore, the next steps include calculating the cost: the benefit of NGB, evaluating its ability to manage the wean-to-market aspect of biosecurity, and determining its ability to reduce incidence risk of other swine viral diseases. Along these lines, it is worth mentioning that only one of the 75 breeding herds (1.3%) became infected with the porcine epidemic diarrhea virus in disease year 3 (Brands, L., Pipestone Veterinary Services, Pipestone, MN, USA, Personal communication, 1 July 2024). With this in mind, it needs to be acknowledged that the improvements in KPIs noted in the study may have been due in part to the control of other pathogens, not only to improvements in PRRSV incidence risk. This is a very exciting hypothesis and future studies should investigate the effect of NGB on bacterial pathogens, such as *Streptococcus suis* and viral pathogens such as influenza virus A of swine. In addition, should the concept of NGB be applied across species groups, i.e., poultry or dairy cattle, it may help prevent the spread of significant pathogens that plague both livestock sectors, such as the highly pathogenic avian influenza virus.

In closing, this communication supports the previously published conclusion that Next Generation Biosecurity, while not perfect, helps to improve both animal health and productivity through the sustained reduction of PRRSV incidence risk at the level of the breeding herd [3]. This is further evidence that PRRSV can be successfully controlled through improvements in biosecurity. Therefore, the authors hope that this new way of thinking will benefit global animal agriculture, as practitioners and scientists collectively strive to manage the challenge of viral diseases of herds and flocks.

5. Conclusions

This communication provides further evidence that improved biosecurity can successfully reduce the impact of PRRS, one of the most economically significant diseases of global agriculture, as well as improve breeding herd productivity. The information from this paper, in combination with its predecessor [3], provides hope and guidance for farmers and veterinarians regarding the management of PRRSV, as well as other domestic and transboundary diseases across herds and flocks throughout the global livestock industry.

Author Contributions: Conceptualization, S.D.; methodology, S.D., L.B. and R.E.; validation, C.A.C., M.K., J.N., A.S. and G.S.; formal analysis, R.E.; writing—original draft preparation, S.D.; writing—review and editing, R.E., L.B., C.A.C., M.K., J.N., A.S. and G.S.; visualization, S.D.; supervision, S.D. and A.S.; project administration, S.D. All authors have read and agreed to the published version of the manuscript.

Funding: This research received no external funding.

Institutional Review Board Statement: Not applicable.

Informed Consent Statement: Not applicable.

Data Availability Statement: All data are included in the paper.

Acknowledgments: The authors acknowledge the late Bob Morrison, a great scientist and friend, for his vision that led to the creation of the MSHMP, a Swine Health Information Center (SHIC) funded project.

Conflicts of Interest: Scott Dee and Roy Edler are employees of Pipestone Research; Lisa Brands, Adam Schelkopf, Joel Nerem, and Gordon Spronk are employees of Pipestone Veterinary Services; the authors declare no further conflicts of interest.

References

1. Wensvoort, G.; Terpstra, C.; Pol, J.M.; ter Laak, E.A.; Bloemraad, M.; de Kluyver, E.P.; Kragten, C.; van Buiten, L.; den Besten, A.; Wagenaar, F.; et al. Mystery swine disease in The Netherlands: The isolation of Lelystad virus. *Vet. Q.* **1991**, *13*, 121–130. [CrossRef] [PubMed]
2. Meulenberg, J.J. PRRSV, the virus. *Vet. Res.* **2000**, *31*, 11–21. [CrossRef] [PubMed]
3. Dee, S.; Brands, L.; Nerem, J.; Schelkopf, A.; Spronk, G.; Kikuti, M.; Corzo, C.; Havas, K. Improvements in swine herd biosecurity reduce the incidence risk of porcine reproductive and respiratory syndrome virus in breeding herds in the Midwestern United States. *J. Am. Vet. Med. Assoc.* **2024**, *262*, 520–525. [CrossRef] [PubMed]
4. Wills, R.W.; Zimmerman, J.J.; Yoon, K.J.; Swenson, S.L.; McGinley, M.J.; Hill, H.T.; Platt, K.B.; Christopher-Hennings, J.; Nelson, E.A. Porcine reproductive and respiratory syndrome virus: A persistent infection. *Vet. Microbiol.* **1997**, *55*, 231–240. [CrossRef] [PubMed]
5. Christopher-Hennings, J.; Nelson, E.A.; Hines, R.J.; Nelson, J.K.; Swenson, S.L.; Zimmerman, J.J.; Chase, C.C.L.; Yaeger, M.J.; Benfield, D.A. Persistence of porcine reproductive and respiratory syndrome virus in serum and semen of adult boars. *J. Vet. Diagn. Investig.* **1995**, *7*, 456–464. [CrossRef] [PubMed]
6. Dee, S.A.; Deen, J.; Rossow, K.D.; Mahlum, C.; Otake, S.; Joo, H.S.; Pijoan, C. Mechanical transmission of porcine reproductive and respiratory syndrome virus throughout a coordinated sequence of events during cold weather. *Can. J. Vet. Res.* **2002**, *66*, 232–239. [PubMed]
7. Dee, S.A.; Otake, S.; Oliviera, S.; Deen, J. Evidence of long-distance airborne spread of porcine reproductive and respiratory syndrome virus and *Mycoplasma hopnuemoniae*. *Vet. Res.* **2009**, *40*, 39. [CrossRef] [PubMed]
8. Dee, S.; Clement, T.; Nelson, E. Transmission of porcine reproductive and respiratory syndrome virus in domestic pigs via oral ingestion of feed material. *J. Am. Vet. Med. Assoc.* **2024**, *262*, 1–4. [CrossRef] [PubMed]
9. Vilalta, C.; Arruda, A.G.; Tousignant, S.J.P.; Valdes-Donoso, P.; Muellner, P.; Muellner, U.; Alkhamis, M.A.; Morrison, R.B.; Perez, A.M. A Review of Quantitative Tools Used to Assess the Epidemiology of Porcine Reproductive and Respiratory Syndrome in U.S. Swine Farms Using Dr. Morrison's Swine Health Monitoring Program Data. *Front. Vet. Sci.* **2017**, *4*, 94. [CrossRef] [PubMed]
10. Paploski, I.A.D.; Corzo, C.; Rovira, A.; Murtaugh, M.P.; Sanhueza, J.M.; Vialta, C.; Schroeder, D.C.; VanderWaal, K. Temporal dynamics of co-circulating lineages of porcine reproductive and respiratory syndrome virus. *Front. Microbiol.* **2019**, *10*, 2486. [CrossRef] [PubMed]
11. Dee, S.A.; Joo, H.S. Recurrent reproductive failure associated with porcine reproductive and respiratory syndrome in a swine herd. *J. Am. Vet. Med. Assoc.* **1994**, *205*, 1017–1018. [CrossRef] [PubMed]
12. Havas, K.A.; Brands, L.; Cochrane, R.; Spronk, G.D.; Nerem, J.; Dee, S.A. An assessment of enhanced biosecurity interventions and their impact on porcine reproductive and respiratory syndrome virus outbreaks within a managed group of farrow-to-wean farms. *Front. Vet. Sci.* **2023**, *9*, 952383. [CrossRef] [PubMed]
13. Smith, J.M.; Saegermann, C.; Vaillancourt, J.P. Editorial: Promoting compliance with biosecurity in animal agriculture. *Front. Vet. Sci.* **2023**, *10*, 1215433. [CrossRef] [PubMed]
14. Melini, C.M.; Kikuti, M.; Torremorell, M.; VanderWaal, K.; Rossow, S.; Corzo, C.A. Evaluation of the infectiousness level of three wildtype PRRSV variants. In Proceedings of the 2023 NAPRRS/NC229 International Conference on Swine Disease, Chicago, IL, USA, 30 November–2 December 2023; p. 64.

Disclaimer/Publisher's Note: The statements, opinions and data contained in all publications are solely those of the individual author(s) and contributor(s) and not of MDPI and/or the editor(s). MDPI and/or the editor(s) disclaim responsibility for any injury to people or property resulting from any ideas, methods, instructions or products referred to in the content.

MDPI AG
Grosspeteranlage 5
4052 Basel
Switzerland
Tel.: +41 61 683 77 34

Animals Editorial Office
E-mail: animals@mdpi.com
www.mdpi.com/journal/animals

Disclaimer/Publisher's Note: The title and front matter of this reprint are at the discretion of the Guest Editors. The publisher is not responsible for their content or any associated concerns. The statements, opinions and data contained in all individual articles are solely those of the individual Editors and contributors and not of MDPI. MDPI disclaims responsibility for any injury to people or property resulting from any ideas, methods, instructions or products referred to in the content.

www.ingramcontent.com/pod-product-compliance
Lightning Source LLC
Chambersburg PA
CBHW040813250926
51850CB00007BA/1154
* 9 7 8 3 7 2 5 8 5 8 1 5 6 *